HISTOCHEMISTRY AND CELL BIOLOGY OF AUTONOMIC NEURONS, SIF CELLS, AND PARANEURONS

Advances in Biochemical Psychopharmacology
Volume 25

Advances in Biochemical Psychopharmacology

Series Editors

Erminio Costa, M.D.
Chief, Laboratory of Preclinical Pharmacology
National Institute of Mental Health
Washington, D.C.

Paul Greengard, Ph.D.
Professor of Pharmacology
Yale University School of Medicine
New Haven, Connecticut

Histochemistry and Cell Biology of Autonomic Neurons, SIF Cells, and Paraneurons

Advances in Biochemical
Psychopharmacology
Volume 25

Editors

Olavi Eränkö, M.D., Ph.D., D.Sc.
Seppo Soinila, M.L.
Heikki Päivärinta, M.Sc., M.L.

Department of Anatomy
University of Helsinki
Siltavuorenpenger
Helsinki, Finland

Raven Press ∎ New York

Raven Press, 1140 Avenue of the Americas, New York, New York 10036

Made in the United States of America

Great care has been taken to maintain the accuracy of the information contained in the volume. However, Raven Press cannot be held responsible for errors or for any consequences arising from the use of the information contained herein.

Library of Congress Cataloging in Publication Data
Main entry under title:

Histochemistry and cell biology of autonomic neurons,
    SIF cells, and paraneurons.

    Includes bibliographies and index.
    1. Ganglia, Autonomic. 2. Nonchromaffin paraganglia.
3. Neurons. 4. Histochemistry. I. Eränkö, Olavi.
II. Soinila, Seppo. III. Päivärinta, Heikki.
QP368.8.H57   599.01′88   79-5506
ISBN 0-89004-495-3

Preface

This volume presents an up-to-date account of the histochemistry and cell biology of autonomic neurons, SIF cells, and paraneurons. The functions of these cells and the factors regulating the development of primitive sympathetic cells into principal neurons, interneurons, endocrine cells, or chemoreceptor cells are discussed. Such cells have been examined in sympathetic ganglia and paraganglia, adrenal medulla, carotid body and other glomi, and peripheral autonomically innervated organs such as the gut, uterus, and iris, as well as the brain.

The contributors are scientists actively involved in the recent rapid developments in the study of the structure, function, and histochemistry of neurons and paraneurons. These studies have been made possible by using sophisticated methods of organ and cell culture, tissue transplantation, isolation of growth factors, immunohistochemistry, electron microscopy, and neurophysiology.

It has been known for a long time that sympathetic neurons and adrenal medullary chromaffin cells are related and have a common origin in the neural crest. Chromaffin cells are also known to be present in sympathetic ganglia, notably in the abdominal region. Recently, increasing attention has been given to small cells in sympathetic ganglia which exhibit an intense formaldehyde-induced fluorescence as a result of the high catecholamine concentration in the cytoplasm. These cells are commonly called small, intensely fluorescent cells, or SIF cells, and they are often nonchromaffin. Another name for these cells is small granule-containing cells, or SGC cells, so-named because of the presence of granular, amine-storing vesicles in their cytoplasm (observed by electron microscopy). The fine structural features of these cells are intermediate between sympathetic neurons and adrenal medullary chromaffin cells; they have been thought to be interneurons, endocrine cells, or chemoreceptor cells. It is now evident that SIF cells vary in structure and function.

The first section of the volume deals with the phenotypic plasticity and development of sympathetic cells. It is shown that the nerve growth factor can transform immature cells that would normally become adrenal medullary cells into sympathetic neurons. On the other hand, glucocorticoids are shown to guide prospective sympathetic neurons into becoming SIF cells, rather than principal neurons. Denervation of sympathetic ganglia is also shown to affect the normal responses of developing neurons and SIF cells. Phenotypic plasticity of developing and adult sympathetic cells has been demonstrated in experiments showing that adult adrenal medullary cells transplanted into the eye can functionally innervate transplanted target organs. Experiments with cell and organ cultures described in other chapters give further evidence of the remarkable phenotypic plasticity of sympathetic cells.

The fine structure of SIF cells and their functions are examined in the second section, in which SIF cells in the ganglia of several species are discussed. The contributions to this section are somewhat controversial, but they clearly show that there are wide variations in the fine structural features of these cells, all of which are relevant to cell function. Some SIF cells seem to be interneurons, receiving synapses afferent to them and forming efferent synapses to principal neurons. Evidence is presented for, and against, such cells being responsible for the slow inhibitory postsynaptic potentials evoked by preganglionic volleys. On the other hand, structurally reciprocal but functionally sensory synapses have been reported on similar SIF cells, which have therefore been thought to serve chemoreceptor function. Still other SIF cells are believed to be endocrine, exclusively or combined with an interneuron function.

The third section is devoted to the histochemistry and experimental cytology of adrenal chromaffin cells and glomus cells; the effects of a variety of experimental stimuli and drugs have been described. It includes observations with a new method for immunohistochemical demonstration of norepinephrine and epinephrine in the adrenal medulla and others on the histochemically demonstrable adenine nucleotides in the carotid body. The profound influence of drugs on the pituitary adrenal axis on chromaffin cells is substantiated in one chapter; the effects of perinatal injections of anti-nerve-growth-factor, guanethidine, and 6-hydroxydopamine are examined in another.

The fourth section deals with granular vesicles (amine-storing organelles contained in the cytoplasm of sympathetic neurons, SIF cells, and chromaffin cells) and with the experimental cell biology of sympathetic neurons and paraneurons. Biochemical, histochemical, and fine structural observations are described on the formation, storage, and release of catecholamines and indoleamines under a variety of conditions, and evidence is given of the existence of functionally different compartments in these vesicles. Other chapters describe histochemical and ultrastructural studies on sympathetic ganglia and peripheral nerves under various experimental conditions and during pregnancy. The paraneuron concept is presented and is illustrated by secretory processes of a cell-secreting cholecystokinin–pancreozymin.

The final section is concerned with the immunohistochemistry of opioid and other peptide transmitters in sympathetic ganglia, adrenal medulla, intestinal neurons, and the central nervous system. By using the recently developed immunohistochemical methods for the localization of transmitter peptides (e.g., beta-endorphin, methionine-enkephalin, and leucine-enkephalin), it is shown that some nerve fibers containing these peptides are in close contact with amine-storing nerve cells of sympathetic ganglia, and that some sympathetic and adrenal medullary cells contain enkephalin in the cytoplasm, in addition to catecholamines. The morphological and pharmacological significance of opioid mechanisms regulating catecholamine-containing systems in the brain are also examined and discussed.

This volume is mainly concerned with basic neuronal and neuroendocrine mechanisms, and with the structure and function of different types of sympathetic cells and the effect of drugs on them. It is thus of primary interest to basic neuroscientists working in anatomical, physiological, pharmacological, or biological laboratories. However, it is also expected to be of interest to clinicians in the fields of neurology, psychiatry, endocrinology, and neuropsychopharmacology, because knowledge of the basic mechanisms will lead to understanding normal and diseased functions of the nervous system as well as the mode of action of neuroleptic and psychotropic drugs.

Olavi Eränkö
Seppo Soinila
Heikki Päivärinta

Acknowledgments

The editors are greatly indebted to Mrs. Paula Vepsäläinen, who, with never-failing enthusiasm and great skill, has taken care of the typing involved in the preparation of this volume. This volume has been supported by grants from the Sigrid Jusélius Foundation and the Finnish Cultural Fund, both of which are gratefully acknowledged.

Contents

Granular Vesicles and Experimental Cell Biology of Sympathetic Neurons and Paraneurons

Contributors

Klaus Addicks
Department of Anatomy
University of Köln
D-5000 Köln, Federal Republic of Germany

Liisa Ahtee
Division of Pharmacology
Department of Pharmacy
University of Helsinki
Kirkkokatu 20
SF-00170 Helsinki 17, Finland

Karl-Erik Åkerman
Department of Medical Chemistry
University of Helsinki
Helsinki, Finland

Hannu Alho
Department of Biomedical Sciences
University of Tampere
33101 Tampere 10, Finland

Per Alm
Department of Histology
University of Lund
Biskopsgatan 5
S-223 62 Lund, Sweden

Luigi Aloe
Laboratory of Cell Biology, CNR
Via Romagnosi 18A
00196 Rome, Italy

L. M. J. Attila
Division of Pharmacology
Department of Pharmacy
University of Helsinki
Kirkkokatu 20
SF-00170 Helsinki 17, Finland

Amapola Autillo-Touati
Groupe de Neurocytobiologie
Laboratoire de Biologie Cellulaire et Histologie
*   (Secteur Sud)*
Faculté de Médecine
13385 Marseille, France

Celia Barraza
Department of Anatomy
University of New Mexico
Albuquerque, New Mexico 87131

Anders Björklund
Department of Histology
University of Lund
Biskopsgatan 5
S-223 62 Lund, Sweden

Asa C. Black, Jr.
Department of Anatomy
College of Medicine
University of Iowa
Iowa City, Iowa 52242

H. Blaschko
University Department of Pharmacology
South Parks Road
Oxford OX1 3QT, England

Peter Böck
Department of Anatomy
Technical University of Munich
D-8000 Munich 40, Federal Republic of Germany

Carol Bradley
Department of Anatomy
University of New Mexico
Albuquerque, New Mexico 87131

F. Bustami
Department of Human Morphology
University of Nottingham
Nottingham, England

Tanemichi Chiba
Department of Anatomy
Saga Medical School
Nabeshima
Saga 840-01, Japan

M. Costa
Department of Human Physiology
School of Medicine
Flinders University
Bedford Park
S.A. 5042 Australia

R. E. Coupland
Department of Human Morphology
University of Nottingham
Nottingham, England

M. Da Prada
Pharmaceutical Research Department
F. Hoffmann-La Roche & Co. Ltd.
CH-4002 Basel, Switzerland

William Dail
Department of Anatomy
University of New Mexico
Albuquerque, New Mexico 87131

Ted Ebendal
Department of Zoology
Uppsala University
Uppsala, Sweden

Lars-G. Elfvin
Department of Anatomy
Karolinska Institutet
10401 Stockholm 60, Sweden

Liisa Eränkö
Department of Anatomy
University of Helsinki
Siltavuorenpenger 20
Helsinki, Finland, FIN 00170

Olavi Eränkö
Department of Anatomy
University of Helsinki
Siltavuorenpenger 20
Helsinki, Finland FIN 00170

Tsuneo Fujita
Department of Anatomy
Niigata University School of Medicine
Asahimachi
Niigata, 951 Japan

J. B. Furness
Centre for Neuroscience
Departments of Human Morphology, Human
*   Physiology, and Medicine*
School of Medicine
Flinders University
Bedford Park
S.A. 5042 Australia

Norbert Gerold
Department of Anatomy
University of Heidelberg
Im Neuenheimer Feld 307
D-6900 Heidelberg, Federal Republic of Germany

Nedzad Gluhbegovic
Department of Anatomy
College of Medicine
University of Iowa
Iowa City, Iowa 52242

Lloyd A. Green
Department of Pharmacology
New York University Medical Center
New York, New York 10016

Mats Grönblad
Department of Anatomy
University of Helsinki
Siltavuorenpenger 20
00170 Helsinki 17, Finland

Pauli Helén
Department of Biomedical Sciences
University of Tampere
33101 Tampere 10, Finland

Ian A. Hendry
Department of Pharmacology
John Curtin School of Medical Research
Australian National University
Canberra City ACT 2601, Australia

Antti Hervonen
Department of Biomedical Sciences
University of Tampere
33101 Tampere 10, Finland

Christine Heym
Department of Anatomy
University of Heidelberg
D-6900 Heidelberg
Im Neuenheimer Feld 307
Federal Republic of Germany

Caryl E. Hill
Department of Pharmacology
John Curtin School of Medical Research
Australian National University
Canberra City ACT 2601, Australia

Barry Hoffer
Department of Pharmacology
University of Colorado
Denver, Colorado 80220

Tomas Hökfelt
Department of Histology
Karolinska Institutet
S-10401 Stockholm, Sweden

P. R. C. Howe
Centre for Neuroscience
Departments of Human Morphology, Human
*   Physiology, and Medicine*

School of Medicine
Flinders University
Bedford Park
S.A. 5042 Australia

Jean Y. Jew
Department of Anatomy
College of Medicine
University of Iowa
Iowa City, Iowa 52242

Tong H. Joh
Laboratory of Neurobiology
Department of Neurology
Cornell Medical College
New York, New York 10021

H. W. M. Joosten
Department of Anatomy and Embryology
University of Nijmegen
Geert Grooteplein
Noord 21
6500 HB Nijmegen, The Netherlands

Lasse Kanerva
Department of Anatomy
University of Helsinki
Siltavuorenpenger 20
00170 Helsinki 17, Finland

Tomio Kanno
Department of Physiology
Faculty of Veterinary Medicine
Hokkaido University
Sapporo 060, Japan

Susanne Khoudary
Department of Anatomy
University of New Mexico
Albuquerque, New Mexico 87131

Shigery Kobayashi
Department of Anatomy
Niigata University School of Medicine
Niigata 951, Japan

Hisatake Kondo
Department of Anatomy
College of Medicine
University of Illinois
Chicago, Illinois 60612

Rainer König
Department of Anatomy
University of Heidelberg
Im Neuenheimer Feld 307
D-6900 Heidelberg, Federal Republic of Germany

J. D. Lever
Department of Anatomy
University College
Cardiff, CF1 1XL, Wales

Rita Levi-Montalcini
Laboratory of Cell Biology, CNR
Via Romagnosi 18A
Rome, Italy

Benjamin Libet
Department of Physiology
School of Medicine
University of California
San Francisco, California 94143

Ilona Linnoila
Endocrinology Group
Laboratory of Pulmonary Function and Toxicology
National Institute of Environmental Health
    Sciences
Research Triangle Park, North Carolina 27709

K-S. Lu
Department of Anatomy
College of Medicine
National Taiwan University
Taipei, Taiwan

Jan M. Lundberg
Department of Histology
Karolinska Institutet
S-10401 Stockholm 60, Sweden

Joe A. Mascorro
Department of Anatomy
Tulane University School of Medicine
New Orleans, Louisiana 70112

Margaret R. Matthews
Department of Human Anatomy
University of Oxford
Oxford OX1 3QX, England

Ian S. McLennan
Department of Pharmacology
John Curtin School of Medical Research
Australian National University
Canberra City ACT 2601, Australia

Richard J. Miller
Department of Pharmacological and Physiological
    Sciences
University of Chicago
Chicago, Illinois 60637

Heather M. Murray
Department of Anatomy
University of New Mexico
Albuquerque, New Mexico 87131

Lars Olson
Department of Histology
Karolinska Institutet
10401 Stockholm 60, Sweden

Christer Owman
Department of Histology
University of Lund
Biskopsgatan 5
S-223 62 Lund, Sweden

Heikki Päivärinta
Department of Anatomy
University of Helsinki
Siltavuorenpenger 20
00170 Helsinki 17, Finland

S. A. Palmer
Department of Anatomy
University College
Cardiff, CF1 1XL, Wales

Raymond E. Papka
Department of Anatomy
University of Kentucky Medical School
Lexington, Kentucky 40536

Matti Partanen
Department of Biomedical Sciences
University of Tampere
33520 Tampere 52, Finland

Amanda Pellegrino de Iraldi
Instituto de Biologia Celular
Facultad de Medicina
Universidad Nacional de Buenos Aires
(1121) Buenos Aires, Argentina

Markku Pelto-Huikko
Department of Biomedical Sciences
University of Tampere
33520 Tampere 52, Finland

Botond Penke
Department of Medical Chemistry
University of Szeged
Dom Tér 8
6720 Szeged, Hungary

Virginia M. Pickel
Cornell University Medical College
Department of Neurology and Laboratory of
    Neurobiology
New York, New York 10021

J. S. Ploem
Department of Histochemistry and Cytochemistry
Medical Faculty
Leiden University
Wassenaarseweg 72
2333 Al Leiden, The Netherlands

Donald J. Reis
Laboratory of Neurobiology
Department of Neurology
Cornell Medical College
New York, New York 10021

J. G. Richards
Pharmaceutical Research Department
F. Hoffmann-La Roche & Co. Ltd.
CH-4002 Basel, Switzerland

Bibiana Rieffert
Department of Anatomy and Cell Biology
Philipps-Universität
D-3550 Marburg, Federal Republic of Germany

Atsushi Saito
Department of Physiology
Faculty of Veterinary Medicine
Hokkaido University
Sapporo 060, Japan

Dean Sandquist
Department of Anatomy
College of Medicine
University of Iowa
Iowa City, Iowa 52242

Robert M. Santer
Department of Anatomy
University College
Cardiff, Wales

J. Schipper
Department of Pharmacology
Medical Faculty
Free University
Van der Boechorststraat 7
1081 BT Amsterdam, The Netherlands

Hanns-Jörg Schröder
Department of Anatomy
University of Heidelberg
Im Neuenheimer Feld 307
D-6900 Heidelberg, Federal Republic of Germany

Marianne Schultzberg
Department of Histology
Karolinska Institutet
S-10401 Stockholm 60, Sweden

Åke Seiger
Department of Histology
Karolinska Institutet
10401 Stockholm 60, Sweden

Raumond Seite
Groupe de Neurocytolbiologie
Laboratoire de Biologie Cellulaire et Histologie
    (Secteur Sud)

Faculté de Médecine
13385 Marseille, France

Yuriko Serizawa
Department of Anatomy
Niigata University School of Medicine
Niigata 951, Japan

Peter A. Smith
Laboratory of Preclinical Studies
National Institute of Alcohol Abuse and
  Alcoholism
Rockville, Maryland 20852

Seppo Soinila
Department of Anatomy
University of Helsinki
Siltavuprenpenger 20
00170 Helsinki 17, Finland

Harry Steinbusch
Department of Anatomy and Embryology
University of Nijmegen
Geert Grooteplein
Noord 21
6500 HB Nijmegen, The Netherlands

Teuvo Takala
Department of Biomedical Sciences
University of Tampere
SF-33101 Tampere 10, Finland

Gunnar Thorbert
Department of Histology
University of Lund
Biskopsgatan 5
S-223 62 Lund, Sweden

F. J. H. Tilders
Department of Pharmacology
Medical Faculty
Free University
Van der Boechorststraat 7
1081 BT Amsterdam, The Netherlands

Arthur S. Tischler
Department of Pathology
Tufts University School of Medicine
Boston, Massachusetts 02111

Klaus Unsicker
Department of Anatomy and Cell Biology
Philipps-Universität
D-3550 Marburg, Federal Republic of Germany

Janos Varga
Department of Medical Chemistry
University of Szeged
Dom Tér 8
6720 Szeged, Hungary

Albert Verhofstad
Department of Anatomy and Embryology
University of Nijmegen
Geert Grooteplein
Noord 21
6500 HB Nijmegen, The Netherlands

Hiroshi Watanabe
Department of Anatomy
Tohoku University
School of Medicine
Sendai 980, Japan

Christa Wedel
Department of Anatomy
University of Heidelberg
Im Neuenheimer Feld 307
D-6900 Heidelberg, Federal Republic of Germany

Forrest Weight
Laboratory of Preclinical Studies
National Institute of Alcohol Abuse and
  Alcoholism
Rockville, Maryland 20852

James R. West
Department of Anatomy
College of Medicine
University of Iowa
Iowa City, Iowa 52242

Terence H. Williams
Department of Anatomy
College of Medicine
University of Iowa
Iowa City, Iowa 52242

Robert D. Yates
Department of Anatomy
Tulane University School of Medicine
New Orleans, Louisiana 70112

Hidetoshi Yonezawa
Department of Physiology
Faculty of Veterinary Medicine
Hokkaido University
Sapporo 060, Japan

Ryogo Yui
Department of Anatomy
Niigata University School of Medicine
Niigata, 951 Japan

Wolfgang Ziegler
Department of Anatomy and Cell Biology
Philipps-Universität
D-3550 Marburg, Federal Republic of Germany

Phenotypic Plasticity and Development of Sympathetic Cells

*Histochemistry and Cell Biology of
Autonomic Neurons, SIF Cells, and
Paraneurons,*
edited by O. Eränkö et al.
Raven Press, New York © 1980

Tropic, Trophic, and Transforming Effects of Nerve Growth Factor

Rita Levi-Montalcini and Luigi Aloe

Laboratorio di Biologia Cellulare CNR, Rome, Italy

> Experiments are the only means of knowledge at our disposal. The rest is poetry, imagination.
>
> *Max Planck*

It is the rule rather than the exception in experimental sciences that the results seldom if ever measure up to the expectation. It is likewise a widespread habit to ignore those findings that do not fit in preconceived schemes. The history of the nerve growth factor (NGF) from its rather remote beginnings to the present day exemplifies such constant refusal to conform to predictions based on previous findings and on firmly established dogmas in developmental neurobiology. In the case of NGF, however, the discrepancies between anticipation and results followed a reverse trend from that mentioned above: The findings almost invariably went beyond the expectation, uncovering new facets of the phenomenon under investigation and forcing us to extend these studies and at the same time to focus our attention on new features of this remarkable stimulus–response system.

Although it is not the purpose of this presentation to review the main steps of this never ending investigation, we shall, in a way of introduction to the report of recent developments, briefly mention the most striking deviations of the earlier findings from those that had been anticipated.

It was the excessive production and atypi-cal distribution of sympathetic nerve fibers into the viscera and inside the vascular bed of avian embryos bearing grafts of mouse sarcomas 180 or 37 (27) that called for a revision of the hypothesis that implantation and vigorous growth of these tumors elicits a growth response from sensory (10) and sympathetic (39) ganglia innervating these tumors which is similar even if more impressive than that elicited by implantation of additional limb buds borrowed from other embryos. The finding that extraembryonic transplants of these two tumors elicited the same effects as intraembryonic transplants (27,40) strongly supported the hypothesis that these mouse sarcomas produce a growth effect on sensory and sympathetic embryonic ganglia by releasing a humoral agent which reaches the nerve cells through vascular channels rather than through, or concomitance with, the conventional generally accepted mediation through nerve fibers.

The tissue culture hanging-drop technique provided an ideal system to explore the morphological effects called forth by (at that time) still unknown tumoral factor (28,41) and to obtain preliminary information on its chemical nature, tentatively identified in a nucleoprotein moiety. The same *in vitro* technique made it possible to assay other tissues and humoral organic agents as potential sources of the nerve growth tumoral factor which ever since 1954 became known as the nerve growth factor (18). It was this technique which revealed the existence of two much more potent NGF sources than mouse sarcomas in snake venom (17) and in mouse submaxillary salivary glands (29). These discoveries not only invalidated the hypothesis that the nerve growth-promoting effect elicited by the two mouse sarcomas could reside in a specific property of neoplastic cells (possibly a virus particle) but made available much larger quantities of NGF, which afforded the possibility of much more extensive and rigorous studies of its biochemical properties and of the growth response elicited in its target cells. In 1956 NGF was identified by Cohen in a protein molecule isolated from snake venom (15), and in 1958 an almost identical molecule endowed with the same biological properties was extracted and purified from mouse salivary glands (16,29). With the elucidation in 1971 of the primary amino acid sequence of the 2.5S salivary NGF achieved by Angeletti and Bradshaw (6), this protein molecule became the first growth specific factor to be characterized at the chemical level.

Studies during this last decade centered on the exploration of the binding properties of NGF to its membrane receptors (8,23,42), the evidence of its retrograde axonal transport (22) from target tissues, and the molecular basis of its mechanism of action (12,13). Here we first consider some of the recent developments which brought to our attention new features of the NGF trophic effects and uncovered two other NGF properties: (a) its neurotropic effects on developing sympathetic fibers, and (b) its transforming effects of neoplastic or immature chromaffin normal cells in sympathetic nerve cells.

RECENT STUDIES ON NGF TROPHIC AND PROTECTIVE EFFECTS ON IMMATURE SYMPATHETIC NERVE CELLS

Previous studies (9,30,36,37,45) gave unequivocal evidence for the magnitude of the nerve growth-promoting effect elicited by the NGF purified from snake venom and mouse salivary glands, and at the same time proved that NGF displays a much more essential role in the life of target cells than was first conceived. It was in fact shown that a specific antiserum to NGF injected in neonatal rodents and other neonatal mammals produces massive destruction of sympathetic para- and prevertebral ganglia (38). This dramatic effect, which deprives the animal of the sympathetic function for its entire life and which became known as immunosympathectomy (35), gave evidence for the essential role played by NGF on the sympathetic cells during their early phases of growth and differentiation. *In vitro* experiments provided additional evidence in favor of a NGF unique role. Sensory and sympathetic nerve cells dissociated from ganglia of 8-day chick embryos do not survive when cultured in minimal essential media deprived of serum. They survive instead indefinitely and build a dense fibrillar network which covers the entire surface of the culture dish when minute quantities (10 ng/ml culture) of NGF are added daily to these otherwise inadequate media (34). The much more marked vulnerability of immature sympathetic nerve cells to the effects of a specific antiserum to NGF than that of differentiated noradrenergic neurons suggested that these studies be extended to the effects of pharmacological compounds which in one or another way interfere with the transmission of the nerve impulse from the end terminals of sympathetic cells to their target tissues in fully developed organisms. The first of these compounds to be tested was a dopamine derivative known as 6-hydroxy-dopamine (6-OHDA), which blocks the transmission of the nerve impulse in sympathetic nerve fibers for a 6- to 8-week period (46). The cause of this effect is the accumulation of this compound in the adrenergic nerve end terminals where it selectively accumulates, producing degenerative reversible lesions of these endings (47). Studies performed in our laboratory in 1969 showed that 6-OHDA injections in neonatal rodents result in the massive death of para- and prevertebral sympathetic ganglia comparable to that produced by a specific antiserum to NGF. This effect became known as chemical sympathectomy (4). Subsequent studies provided strong evidence in favor of the hypothesis that nerve cell death is due to the block of retrograde axonal transport of the NGF produced by peripheral tissues. This process, which is essential for growth and differentiation of the immature nerve cells, is impeded by the accumulation of 6-OHDA in the growing adrenergic endings. During the following years it was discovered that two other compounds, guanethidine and vinblastine, also produce, although through entirely different mechanisms (5,20,44), the destruction of para- and prevertebral sympathetic ganglia when injected in neonatal rodents. These findings raised the question of whether an exogenous supply of NGF would counteract the degenerative effects produced by these three agents (6-OHDA, guanethidine, and vinblastine) in immature sympathetic nerve cells. As reported in previous articles (25,33), the results fully confirmed our expectation of an NGF protective effect on these cells. Of particular interest and worth some comment was the finding that, in the case of a combined 6-OHDA and NGF treatment, the NGF not only prevented the death of the immature sympathetic nerve cells but produced an extraordinary volume increase of sympathetic para- and prevertebral sympathetic ganglia, which over a 3-week period of daily treatment reached a volume 30 times larger than that of the same ganglia in control littermates and about three times larger than that of ganglia of littermates injected with NGF alone (3). Structural, ultrastructural, fluorescence, and biochemical studies provided an explanation of the cause of this paradoxical growth effect. Even in the presence of NGF, 6-OHDA accumulates in the noradrenergic nerve endings and blocks the

transmission of the nerve impulse and the retrograde axonal transport of NGF from the endogenous peripheral NGF sources. The simultaneous supply of NGF far in excess of that which normally reaches the cells from peripheral tissues produces an increase in size and number of nerve cells comparable to that elicited by injections of the same NGF doses in intact neurons. In addition, and at variance with the latter effects, the chemically axotomized axons produce an enormous amount of neurofibrillar material which is routed in collaterals sprouting from the proximal segment of the chemically transected axons. A similar effect obtains in surgically axotomized and NGF-treated ganglia (2). The formation of this extraordinary number of collaterals, which force the cells apart and form a dense fibrillar capsule around the ganglia, is due to the lack of negative feedback control mechanisms which under normal conditions prevent extra production of nerve fibers, as shown in ingenious and rigorous experiments performed by Diamond and co-workers (19). Neither guanethidine nor vinblastine interfere with the two-way traffic at the terminal endings of axons with their target tissues; the combined treatment of NGF with either one or the other of these two compounds results therefore in an effect comparable to, but not exceeding, that elicited by NGF alone (25,44).

NGF NEUROTROPIC EFFECT

One of the most debated and controversial problems in developmental neurobiology is defining the role of endogenous and exogenous factors in guiding nerve fibers toward their target organs. Three basic mechanisms have been proposed over the years to explain the formation of specific neuronal circuitry: (a) an elaborate predetermined program encoded genetically in each neuron that unfolds according to rigid and unmodifiable rules; (b) a random process of trial and error in which growing nerve fibers that make the right connections are consolidated and those that fail are reabsorbed; and (c) a general program of circuit formation that is brought to completion by an interplay between genetic and extrinsic factors. The first mechanism can be discounted since it would require very large amounts of genetic information—much more than could be encoded in the entire complement of DNA in the cell nucleus of each neuron. The second mechanism is also most unlikely because such a random process of circuitry formation would be extremely time-consuming and wasteful of energy and resources. The third mechanism therefore seems the most probable: The circuitry between neuronal cell population in the central nervous system and between motor, sensory, and autonomic neurons and their peripheral effectors or receptors is established through a combination of genetic and extrinsic factors. The existence of such extrinsic factors was first suggested by Cajal, who visualized them as chemical signals issued from peripheral tissues that would direct the growing nerve fibers toward their matching target cells (11). At that time and during subsequent years, however, the methodology for detecting these hypothetical chemical factors in the developing embryos was not yet available, and the neurotropism theory (the name suggested by Cajal to designate this effect) was dismissed in favor of an even more putative role played by mechanical "contact guidance factors," or adhesion of the fibers to a solid substrate in the growth of nerve fibers and formation of correct end connections. The discovery of NGF, however, offered the opportunity of reconsidering the concept of neurotropism under far more favorable conditions than had been previously available. Here we only briefly mention experiments performed in our laboratory and described in detail in previous publications (32,43) which gave unequivocable evidence in favor of an NGF neurotropic role on growing sympathetic nerve fibers in neonatal rodents. NGF was injected with the aid of a microcapillary glass pipette into the brain of neonatal mice and rats in the proximity of the two nuclei cerulei. The experiments were aimed at exploring whether intracerebral monoaminergic nerve cells, and in particular those of the locus ceruleus, would undergo a size increase in the presence of NGF. The results did not confirm this hypothesis but brought to light a different and unforeseeable effect. Extensive studies performed with the fluorescence microscope on the serially sectioned cerebrospinal axis processed according to the Falck–Hillarp technique showed in fact that sympathetic nerve fibers emerging from the greatly enlarged paravertebral sympathetic ganglia gained access inside the spinal cord and the brainstem through the dorsal roots of spinal nerves and the sensory-motor roots of the lowest cephalic nerves, and settled, respectively, in the dorsal funiculi of the spinal cord and ventrolateral cordons of the brainstem. Collaterals emerging from these ectopic fiber bundles entered the white and gray matter in close association with blood vessels. Discontinuation of the NGF intracerebral injections resulted in the fading and final disappearance of these parasitic systems. The hypothesis that the entrance of sympathetic nerve fibers into the central nervous system of intracerebrally NGF-injected rodents is produced by an NGF diffusion gradient from the site of injection into the floor of the fourth ventricle to the spinal cord realized through its transport along sensory and motor roots to the adjacent ganglia is supported by the marked volume increase of these ganglia and the demonstration of a high NGF level in the brain and spinal cord of the injected pups 24 to 30 hr after the last injection. More recent ingenious *in vitro* experiments performed in two other laboratories provided strong additional support in favor of a clear-cut neurotropic NGF effect on sympathetic sensory and nerve fibers growing out of dissociated nerve cells of both types (14,26).

NGF IN VITRO TRANSFORMING EFFECT OF PHEOCHROMOBLASTS IN SYMPATHETIC NERVE CELLS

Greene and Tischler discovered that cells of the PC 12 line isolated from a rat pheochromocytoma tumor respond to NGF by acquiring properties characteristic of sympathetic neurons upon addition of NGF to the culture medium. This in turn brought to light a new feature of the biological activity of this protein molecule (21). Subsequent studies by Unsicker et al. (48) provided evidence for a similar *in vitro* effect elicited by NGF on explants of rat's adrenal medulla and on the chromaffin cells dissociated from this gland. These studies proved the property of NGF to divert neoplastic and normal chromaffin cells cultured *in vitro* in sympathetic nerve cells. On the basis of these findings, Unsicker and co-workers submitted the hypothesis that these *in vitro* effects were at least in part attributed to the fact that chromaffin and sympathetic cells, which have their origin in a common precursor, under normal conditions undergo transformation in glandular cells under the influence of glucocorticoid hormones released from the adjacent cortical cells of the adrenal gland. Although this is a plausible hypothesis, experiments performed *in vivo* in our laboratory showed that NGF injections in intact rat fetuses and chick embryos also elicit this effect in an even more massive and impressive way than on dissociated neoplastic or normal chromaffin cells *in vitro*.

TRANSFORMATION OF IMMATURE CHROMAFFIN CELLS IN SYMPATHETIC NERVE CELLS IN RAT PUPS AFTER PRE- AND POSTNATAL NGF INJECTIONS

As reported in a previous article (1), NGF injections in 16 to 17-day-old rat fetuses administered through the intact tubes, and resumed immediately after birth on a daily basis from the first day to days 10 to 20, result in massive transformation of the immature chromaffin cells in sympathetic nerve cells. In the experiments briefly considered here and reported more extensively elsewhere, we increased the prenatal treatment to two NGF injections rather than one, given at 15.5 and 17 days; we also extended the studies of the effects of this pre- and postnatal treatment to the chromaffin cells of the abdominal paraganglia and the carotid bodies. The effects of this more intensive NGF prenatal treatment on the adrenal medulla is illustrated in Fig. 1a,b, which compares the two largest cross sections of the adrenal gland in a control and an experimental 2-week-old pup. The experimental pup received two intrauterine injections between the end of the 15th and 17th gestational days and then daily postnatal injections of NGF (10 μg/g body weight) until the day of sacrifice. It is also evident from comparison of the two microphotos that the intensive NGF treatment resulted not only in massive transformation of chromaffin tissue into sympathetic nerve cells but also in the production of an enormous number of nerve fibers, which encapsulated the transformed nerve cells and enlarged the medullary area at the expense of the surrounding cortical zones. The latter are infiltrated with large nerve fiber bundles, which further reduces the space available for cortical cells of the three lateral, medial, and internal zones. Figure 1c,d compares the two largest sections of the adrenal glands of a control and an experimental rat of the same age as those portrayed in Fig. 1a,b. The glands were processed according to the Falck–Hillarp technique, and the sections were photographed with the fluorescence microscope. Note the enormous enlargement of the fluorescent medullary area in Fig. 1d, as compared to Fig. 1c and the presence of intensely fluorescent fiber bundles in the cortical zones of the experimental but not the control gland. The fluorescence of the control adrenal medulla was intensely yellow, whereas

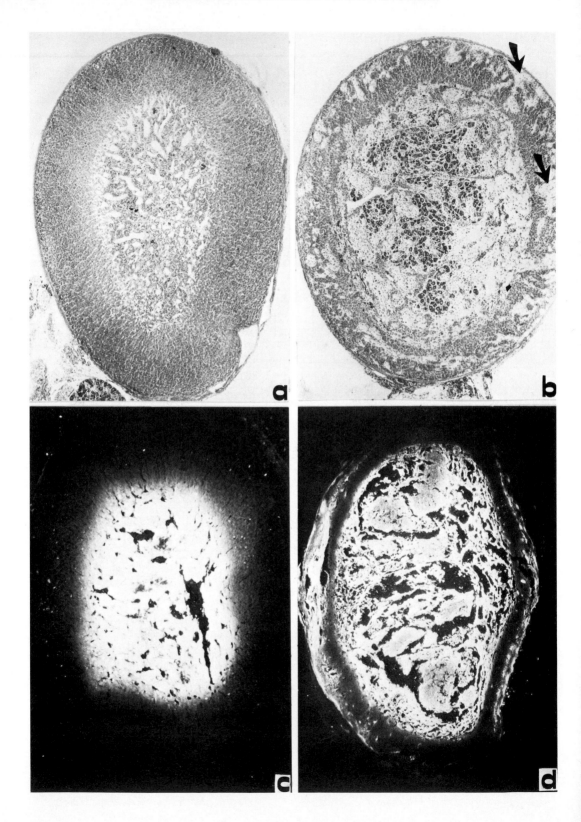

the experimental section exhibited the typical green color of noradrenaline-containing cells and fiber bundles. Figure 2 compares at low and high magnification the carotid bodies and surrounding tissues in a control and an experimental rat pup of the same series as those illustrated in Fig. 1. The small spindle-shaped glomus cells of the control body intermingled with interstitial cells of similar size and appearance are, in the experimental organ, transformed into a homogenous population of much larger cells indistinguishable from sympathetic neurons of the adjacent superior cervical ganglion apparent in Fig. 1a,b. The marked volume increase of the experimental carotid body, obvious from a comparison of Fig. 2a and b, is due to the increased size of individual cells, but foremost to the production of a large number of nerve fibers which join the postganglionic nerves of the adjacent superior cervical ganglion.

Studies still in progress on the abdominal paraganglia showed the identical pattern of transformation of glandular cells into nerve cells as depicted in the adrenal medulla and the carotid bodies.

TRANSFORMATION OF IMMATURE CHROMAFFIN CELLS IN NERVE CELLS IN CHICK EMBRYOS

The technique devised by Auerbach and co-workers (7) to cultivate chick embryos in Petri dishes from the 3rd to the 19th day of incubation provided an invaluable tool to inspect the effects of intraembryonic NGF injections from the 4th to the 17th day of embryonic development. These studies could not be performed *in ovo* because of the movements of the embryo, its envelopment in the chorioallantoic membranes, and the hemorrhage that follows to the blind insertion of the injection needle into the embryos. Under excellent conditions of visibility it was possible to inject microgram quantities of NGF directly into the body of the embryos with a microcapillary needle. We report on the preliminary results of these studies.

Figure 3 illustrates two transverse sections of 5-day-old chick embryos processed according to the Falck technique and photographed in the fluorescence microscope. On the fourth day the control embryo received an injection of physiological solution and the experimental embryo received 5 μg NGF. The marked increase in fluorescent cells migrating from the neural crest is apparent in the experimental embryo as compared to the control. Of particular interest is the formation of two intensely fluorescent cell aggregates in close proximity to the medial aspect of the primordium of the adrenal gland in the experimental embryo, whereas these cells are entirely absent in the control.

Older control and experimental embryos sacrificed daily between the 7th and the 18th days were fixed according to the Cajal De Catro silver technique and sectioned serially. Figure 4a,b illustrates two transverse sections of 11-day-old control and experimental embryos at the level of the adrenal medulla and mesonephros, and Fig. 4c,d shows sections at the same level in two control and experimental embryos sacrificed on the 18th day of incubation. Before commenting on the results, it is worth noting that the adrenal gland in the avian species differs markedly from that of mammals. Chromaffin cells in long thin strands or islands are intermingled with cords of cortical cells. The two cell types are

---

◄
FIG. 1. a and **b**: Transverse sections of adrenal glands of 3-week-old rat pups. **a**: Control. **b**: Pup injected during fetal and postnatal life with NGF as explained in the text. Arrows in **b** point to fiber bundles produced by the transformed sympathetic cells in the adrenal medulla, infiltrating the three zones of the adrenal cortical gland. Further explanation in text. Toluidine blue. ×50. **c** and **d**: Histofluorescence preparations of transverse sections of control (**c**) and NGF-injected (**d**) pups of the same age as the specimen in **a** and **b**. Further explanation in text. ×50.

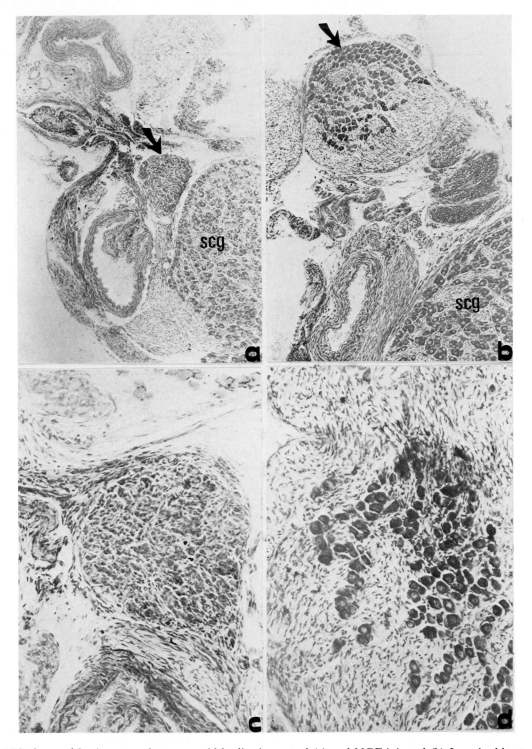

FIG. 2. a and **b**: Arrows point to carotid bodies in control (**a**) and NGF-injected (**b**) 3-week-old rat pups. (scg) Superior cervical ganglion. Further explanation in text. Toluidine blue. ×60. **c** and **d**: Carotid bodies of control (**c**) and NGF-injected (**d**) pups shown at higher magnification than in **a** and **b**. ×200.

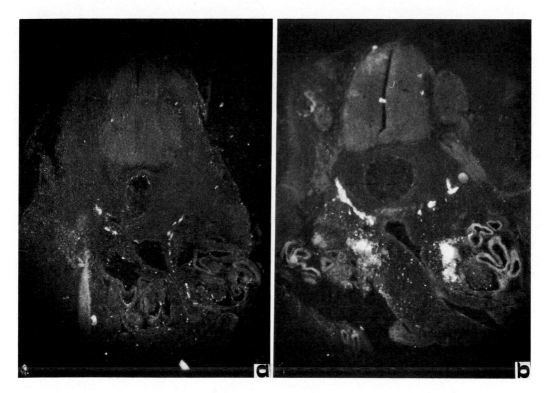

FIG. 3. Histofluorescence preparations of cross sections of 5-day-old chick embryos: (**a**) control and (**b**) embryo injected with NGF as explained in text. Notice marked increased number of fluorescent cells in **b** and the formation of two fluorescent cell aggregates in proximity to the medial aspect of the primordium of the unstained adrenal gland. Further explanation in text. ×36.

distinguishable from each other by the different shades of color in routine histological preparations and silver-stained sections. When the specific chromaffin staining technique is used, the chromaffin cells exhibit a yellow-green color on the light background of the unstained cortical cells.

The still not fully differentiated adrenal glands in 11-day embryos differ markedly in the control and experimental specimens. In the former (Fig. 4a), the glands appear as two large ovoidal bodies in close apposition to the adjacent mesonephros. In the experimental embryo (Fig. 4b), the two glands at this low magnification appear to consist of two components: a lateral irregularly shaped body (marked with a continuous solid line) and a medial intensely stained body (marked with a white dotted line). Inspection at

higher magnification showed that the latter darker areas consist of dense aggregates of nerve cells similar in all respects to the cells of para- and prevertebral sympathetic ganglia, apparent in the same microphoto. Figure 4c,d depicts the adrenal glands in a control and an experimental 18-day embryo. The dotted line encircles the intensely darkly stained agglomerates of sympathetic nerve cells which surround, as large irregularly shaped cell aggregates, the greatly reduced adrenal gland. Note that, besides this marked volume reduction, the experimental gland consists of a uniform and apparently homogenous cell population, whereas the larger control gland consists of dark and light colored cords and cell islands. The former are cortical cells, and the latter (also examined in chromaffin-stained sections) are glandular

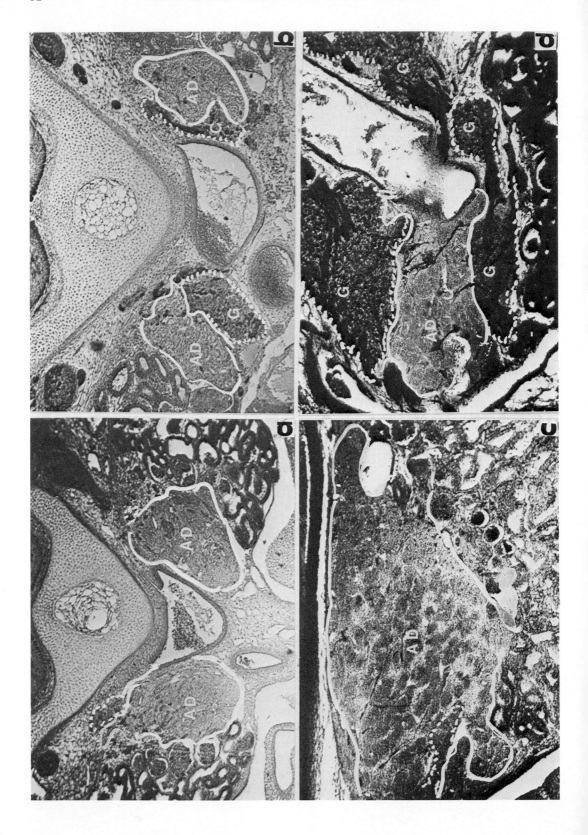

medullary cells. This cell population is absent in the experimental gland. In the control gland a diminutive dark cell aggregate consists of sympathetic nerve cells. In the experimental specimen fiber bundles assembled in large fascicles direct their course toward adjacent blood vessels, which are wrapped by these newly formed nerve fibers and invade the adjacent mesonephros in large number.

CONCLUDING REMARKS

From the beginning of these studies we realized that the effects elicited by a humoral factor released by two mouse sarcomas did not fit in any of the preconceived schemes of stimulus-response systems described by previous investigators who had watched the growth response of spinal ganglia elicited in amphibian larvae and avian embryos after implantation of additional organs or other tissues. We could not anticipate at that time, however, to what extent this newly discovered phenomenon departed from other known effects, nor could we foresee the magnitude and plurality of the responses elicited by this molecule which is synthesized in large amount in some glands and released in minute quantities by a large variety of vertebrate cell lines. Perhaps the most challenging aspect of this never-ending investigation is to have shed light not on one but on an ever-increasing list of events which materialize during the early phases of growth and differentiation of sympathetic nerve cells and cells stemming from the same neural crest precursors. Studies now in progress on the effects of a monospecific antiserum to NGF prospect a new and even broader field of action of NGF than that outlined in this chapter.

As stated by Hunt, "In Science . . . novelty emerges with difficulty, manifested by resistance, against a background provided by expectation" (24). In the case of NGF–target cell interaction, 30 years of investigations have seen this barrier broken down and, again to use an expression of Hunt: "The initially anomalous has become the anticipated." It is in this spirit of expectation of other novelties that we conclude this report of the recent developments in the "NGF uncharted route" (31).

REFERENCES

1. Aloe, L., and Levi-Montalcini, R. (1979): Nerve growth factor in-vivo induced transformation of immature chromaffin cells in sympathetic neurons: Effect of antiserum to the nerve growth factor. *Proc. Natl. Acad. Sci. USA,* 76:1246–1250.
2. Aloe, L., and Levi-Montalcini, R. (1979): Nerve growth factor induced overgrowth of axotomized superior cervical ganglia in neonatal rats: Similarity and differences with NGF effects in chemically axotomized sympathetic ganglia. *Arch. Ital. Biol. (in press).*
3. Aloe, L., Mugnaini, E., and Levi-Montalcini, R. (1975): Light and electron microscope studies on the excessive growth of sympathetic ganglia in rats injected daily from birth with 6-OHDA and NGF. *Arch. Ital. Biol.,* 113:326–353.
4. Angeletti, P.U., and Levi-Montalcini, R. (1970): Sympathetic nerve cell destruction in newborn mammals by 6-hydroxydopamine. *Proc. Natl. Acad. Sci. USA,* 65:114–121.
5. Angeletti, P.U., and Levi-Montalcini, R. (1972): Growth inhibition of sympathetic cells by some adrenergic blocking agent. *Proc. Natl. Acad. Sci. USA,* 69:86–88.
6. Angeletti, R.H., and Bradshaw, R.A. (1971): Nerve growth factor from mouse submaxillary gland: Amino acid sequence. *Proc. Natl. Acad. Sci. USA,* 68:2417–2420.
7. Auerbach, R., Kubai, L., Knighton, D., and Falkman, J. (1974): A simple procedure for long-term cultivation of chick embryos. *Dev. Biol.,* 41:391–394.

◄

FIG. 4. **a** and **b**: Transverse sections of control (**a**) and NGF-injected (**b**) 11-day-old chick embryos at the level of the adrenal glands. The contour of the adrenal gland (AD) in the control and the experimental case is shown with the solid white line. The dotted line in **b** encircles large sympathetic cell aggregates (G) lacking in the control case. Further explanation in text. Silver technique. ×80. **c** and **d**: Adrenal glands of control (**c**) and NGF-injected (**d**) chick embryos sacrificed at 18 days. The contour of the adrenal gland (AD) is shown in the control and the experimental case with a solid white line. The dotted line encircles the large sympathetic nerve cell (G) aggregates in **d** and the diminutive sympathetic nerve cell cluster in **c**. Silver technique. ×60.

8. Banerjee, S.P., Snyder, S.H., Cuatrecasas, P., and Greene, L.A. (1973): Binding of nerve growth factor receptors in sympathetic ganglia. *Proc. Natl. Acad. Sci. USA,* 70:2519-2523.

9. Boyd, F.L., Bradshaw, A.R., Frazier, A.W., Hogue-Angeletti, A.R., Jeng, I., Pulliam, M.W., and Szutowicz, A. (1975): Nerve growth factor. In: *Minireviews of Neurosciences: Life Sciences,* p. 29. Pergamon Press, New York.

10. Bucker, E.D. (1948): Implantation of tumors in the hind limb field of the embryonic chick and developmental response of the lumbosacral nervous system. *Anat. Rec.,* 102:369-390.

11. Cajal y Ramon (1913): *Degeneration and Regeneration of the Nervous System,* translated by R. May. Reprinted 1968 by MIT Press, Cambridge, Mass. (original Spanish edition, 1913).

12. Calissano, P., Levi, A., Alemà, S., Chen, J.S., and Levi-Montalcini, R. (1975): Studies on the interaction of the nerve growth factor with tubulin and actin. In: *26 Colloquium Mosbach, Molecular Basis of Motility,* edited by L. Heilmeyer, J.C. Rüegg, and Th. Wieland. Springer Verlag, New York.

13. Calissano, P., Monaco, G., Castellani, L., Mercanti, D., and Levi, A. (1978): Nerve growth factor potentiates actomyosin adenosine triphosphatase. *Proc. Natl. Acad. Sci. USA,* 75:2210-2214.

14. Campenot, R. (1977): Local control of neurite development by nerve growth factor. *Proc. Natl. Acad. Sci. USA,* 74:4516-4519.

15. Cohen, S. (1958): A nerve growth promoting protein. In: *Chemical Basis of Development,* edited by W.D. McElroy and S. Glass, pp. 665-667. Johns Hopkins Press, Baltimore.

16. Cohen, S. (1960): Purification of a nerve growth promoting protein from the mouse salivary gland and its neurotoxic antiserum. *Proc. Natl. Acad. Sci. USA,* 46:302-311.

17. Cohen, S., and Levi-Montalcini, R. (1956): A nerve growth stimulating factor isolated from snake venom. *Proc. Natl. Acad. Sci. USA,* 42:571-574.

18. Cohen, S., Levi-Montalcini, R., and Hamburger, V. (1954): A nerve growth stimulating factor isolated from sarcoma 37 and 180. *Proc. Natl. Acad. Sci. USA,* 40:1014-1018.

19. Diamond, J., Cooper, E., Turner, C., and MacIntyre, L. (1976): Trophic regulation of nerve sprouting. *Science,* 193:371-377.

20. Eränkö, O., and Eränkö, L. (1971): Histochemical evidence of chemical sympathectomy of guanethidine in newborn rats. *Histochem. J.,* 3:451-456.

21. Greene, L.A., and Tischler, A.S. (1976): Establishment of a noradrenergic clonal line of rat adrenal pheochromocytoma cells which respond to nerve growth factor. *Proc. Natl. Acad. Sci. USA,* 73:2424-2428.

22. Hendry, I.A., Stoeckel, K., Thoenen, H., and Iversen, L.L. (1974): Retrograde axonal transport of the nerve growth factor. *Brain Res.,* 68:103-121.

23. Herrup, K., and Shooter, E.M. (1973): Properties of nerve growth factor receptors of avian dorsal root ganglia. *Proc. Natl. Acad. Sci. USA,* 70:3884-3888.

24. Hunt, T.S. (1975): *The Structure of Scientific Revolution.* International Encyclopedia of University Sciences. University of Chicago Press, Chicago, Illinois. Second edition.

25. Johnson, E.M., and Aloe, L. (1974): Suppression of the in vitro and in vivo cytotoxic effects of guanethidine in sympathetic neurons by nerve growth factor. *Brain Res.,* 81:519-538.

26. Letourneau, P.C. (1978): Chemotactic response of nerve fiber elongation to nerve growth factor. *Dev. Biol.,* 66:183-196.

27. Levi-Montalcini, R. (1952): Effects of mouse tumor transplantation on the nervous system. *Ann. NY Acad. Sci.,* 55:330-343.

28. Levi-Montalcini, R. (1953): In vivo and in vitro experiments on the effect of mouse sarcoma 180 and 137 on the sensory and sympathetic system of the chick embryo. In: *Proceedings XIV International Congress of Zoology,* Copenhagen, p. 309.

29. Levi-Montalcini, R. (1958): Chemical stimulation of nerve growth. In: *Chemical Basis of Development,* edited by W.D. McElroy and S. Glass, pp. 646-664. Johns Hopkins Press, Baltimore.

30. Levi-Montalcini, R. (1966): The nerve growth factor: Its mode of action on sensory and sympathetic nerve cells. *Harvey Lect.,* 60:217-259.

31. Levi-Montalcini, R. (1975): An uncharted route. In: *Neuroscience: Paths of Discovery,* pp. 245-255. MIT Press, Cambridge, Mass.

32. Levi-Montalcini, R. (1976): The nerve growth factor: Its role in growth, differentiation and function of the sympathetic adrenergic neuron. *Prog. Brain Res.,* 45:235-258.

33. Levi-Montalcini, R., Aloe L., Mugnaini, E., Oesch, F., and Thoenen, H. (1975): Nerve growth factor induced volume increase and enhanced tyrosine hydroxylase synthesis in the chemically axotomized sympathetic ganglia of newborn rats. *Proc. Natl. Acad. Sci. USA,* 72:595-599.

34. Levi-Montalcini, R., and Angeletti, P.U. (1963): Essential role of the nerve growth factor in the survival and maintenance of dissociated sensory and sympathetic nerve cells in vitro. *Dev. Biol.,* 7:653-659.

35. Levi-Montalcini, R., and Angeletti, P.U. (1966): Immunosympathectomy. *Pharmacol. Rev.,* 18:619-628.

36. Levi-Montalcini, R., and Angeletti, P.U. (1968): Nerve growth factor. *Physiol. Rev.,* 48:534-565.

37. Levi-Montalcini, R., Angeletti, R.H., and Angeletti, P.U. (1972): The nerve growth factor. *Struct. Function Nerv. Tissue,* 5:1-38.

38. Levi-Montalcini, R., and Booker, B. (1960): Destruction of the sympathetic ganglia in mammals by an antiserum to the nerve growth promoting factor. *Proc. Natl. Acad. Sci. USA,* 42:384-391.

39. Levi-Montalcini, R., and Hamburger, V. (1951): Selective growth-stimulating effects of the mouse sarcoma on the sensory and sympathetic nervous system of the chick embryo. *J. Exp. Zool.,* 118:321-362.

40. Levi-Montalcini, R., and Hamburger, V. (1953): A diffusible agent of mouse sarcoma producing hyperplasia of sympathetic ganglia and hyperneurotization of chick embryo. *J. Exp. Zool.,* 123:233-288.

41. Levi-Montalcini, R., Meyer, H., and Hamburger, V. (1954): In vitro experiments on the effects of mouse sarcoma 180 and 137 on the spinal and sympathetic ganglia of the chick embryo. *Cancer Res.,* 14:49-57.

42. Levi-Montalcini, R., Revoltella, R., and Calissano, P. (1974): Microtubule proteins in the nerve growth factor mediated response. *Recent Prog. Horm. Res.,* 38:635-669.

43. Menesini Chen, M.G., Chen, J.S., and Levi-Montalcini, R. (1978): Sympathetic nerve fibers ingrowth in the central nervous system of neonatal rodents upon intracerebral NGF injections. *Arch. Ital. Biol.,* 116:53-84.

44. Menesini Chen, M.G., Chen, J.S., Calissano, P., and Levi-Montalcini, R. (1977): Nerve growth factor prevents vinblastine destructive effects on sympathetic ganglia in newborn mice. *Proc. Natl. Acad. Sci. USA,* 74:5559-5563.

45. Mobley, W.C., Server, A.C., Ishii, D. N., Riopelle, R.J., and Shooter, E.M. (1977): Nerve growth factor. *N. Engl. J. Med.,* 297:1149-1158.

46. Porter, C.C., Totaro, J.S., and Stone, C.A. (1963): Effect of 6-hydroxydopamine and some other compounds on the concentration of norepinephrine in the hearts of mice. *J. Pharmacol. Exp. Ther.,* 140:308-316.

47. Tranzer, J.P., and Thoenen, H. (1968): An electron microscopic study of selective acute degeneration of sympathetic nerve terminals after administration of 6-hydroxydopamine. *Experientia,* 24:155-156.

48. Unsicker, K., Krisch, B., Otten, U., and Thoenen, H. (1978): Nerve growth factor induced fiber outgrowth from isolated adrenal chromaffin cells: Impairment by glucocorticoids. *Proc. Natl. Acad. Sci. USA,* 75:3498-3502.

Histochemistry and Cell Biology of Autonomic Neurons, SIF Cells, and Paraneurons,
edited by O. Eränkö et al.
Raven Press, New York © 1980.

Induction of SIF Cells by Hydrocortisone or Human Cord Serum in Sympathetic Ganglia and Their Subsequent Fate In Vivo and In Vitro

Olavi Eränkö and Liisa Eränkö

Department of Anatomy, University of Helsinki, Helsinki, Finland, FIN 00170

Sympathetic ganglia contain, in addition to sympathetic nerve cell bodies or principal neurons, small cells with a very bright catecholamine fluorescence (12). These cells are now commonly called small intensely fluorescent (SIF) cells, and they can be found in the sympathetic ganglia of several species (3,6, 14).

Hydrocortisone (HC) injections cause a dramatic increase in the number of SIF cells in the superior cervical ganglion of a newborn rat (7). Organotypic culture of trypsinized newborn rat sympathetic chain ganglia in a medium containing HC likewise causes the appearance of large numbers of very intensely fluorescent SIF cells in the ganglion (9). The increase in the number of SIF cells is reversible *in vivo:* If newborn rats are injected with 0.2 mg HC daily for 7 days, which causes an approximately 10-fold increase in the number of SIF cells, and are thereafter allowed to recuperate for another 7 days without injections, the number of SIF cells returns close to normal (4). The present study was undertaken to examine how far the increase in the number of SIF cells induced by HC *in vivo* and that *in vitro* are reversible when the ganglia are subsequently cultured *in vitro*. The SIF cell-inducing effect *in vitro* of human cord serum and human placental serum (8) is similarly examined.

MATERIALS AND METHODS

Whole superior cervical ganglia of newborn rats, descendants of the Sprague-Dawley strain, were cultured in air–fluid interface cultures. The ganglia were placed on a Millipore filter strip, which served as a wick for the culture medium, on a stainless steel profile in a Petri dish.

The ganglia were dissected aseptically; connective tissue and fat were carefully removed with watchmaker's forceps; and the ganglia were placed without trypsinization on a Millipore filter in a Petri dish. Modified Medium 199 was used, supplemented with 20% fetal calf serum, insulin (50 IU/liter), penicillin sodium G (100,000 IU/liter), and glucose (5 g/liter), with or without hydrocortisone sodium succinate (10 mg/liter) (9). In some culture experiments, fetal calf serum was replaced by the same volume of human cord (placental) serum, human maternal (parturient) serum obtained at delivery, or human adult male serum. The culture dishes were kept at 37°C in an incubator continuously supplied with a mixture of 5% carbon dioxide in air, bubbled through water. At least 20 cultures of any kind were examined.

For the demonstration of catecholamine fluorescence, the ganglia were frozen in liquid propane, freeze-dried, and exposed first for 30 min at 50°C, then for 1 hr at 80°C to formaldehyde vapor from paraformaldehyde powder equilibrated with 60% relative humidity (5). After the exposure, the tissues were embedded in paraffin wax and cut at 6 μm. The fluorescence was examined and photographed with a Leitz Ortholux microscope fitted with an epi-illuminator (16). Some cultures were examined electron microscopically after fixation in a mixture of 3.5% formaldehyde and 0.5% glutaraldehyde in Krebs–Ringer–glucose solution.

RESULTS

Effect of Hydrocortisone *In Vitro*

Figure 1 shows a section of a whole non-trypsinized superior cervical ganglion of a newborn rat after culture for 7 days in the control medium. A few single SIF cells and some small clusters of SIF cells can be seen. In a complete series of sections, only a few SIF cells were found in any of more than 100 cultured ganglia.

Figure 2 illustrates a ganglion cultured in the presence of HC (10 mg/liter) for 7 days in the culture medium. It shows a large number of SIF cells scattered individually and in clusters in the section. The SIF cells usually migrated during culture and formed large clusters in a half-moon-shaped area of the rounded ganglion. Hence some sections of the cultured ganglion contained only a few SIF cells, and serial sections were necessary to obtain a reliable idea of the presence of SIF cells in the ganglion. There were a large number of SIF cells in the more than 100 ganglia cultured with HC, whereas sections from the corresponding control ganglia did not show many of these cells. Thus HC had an all-or-none effect which was highly reproducible and readily obvious without cell

counts. The following description of other cultures also refers to such clear changes only, and eventual minor changes have been overlooked.

Effect of Vinblastine on the Response to Hydrocortisone

If vinblastine sulfate at 0.05 mg/liter was added to the control (Fig. 1, inset) or the HC-containing culture medium (Fig. 2, inset), a drastic diminution in the size of the ganglia resulted. However, there were still many more SIF cells in the ganglia cultured with HC than those in ganglia cultured in the control medium. Vinblastine sulfate at 0.01 mg/liter had little effect on either HC-containing or control cultures. Notably, it did not prevent the marked SIF cell increase due to HC, which was similar to that seen in cultures without vinblastine.

Effect of Human Cord or Maternal Serum

If fetal calf serum used in the control culture medium was replaced by an equal volume of either human cord (placental) serum or human maternal serum, a great increase was regularly observed in the number of SIF cells, essentially similar to that obtained with HC. Addition of HC to the medium prepared with either maternal or cord serum did not further increase the number of SIF cells. When human male serum was used instead of the fetal calf serum, no large increase in the number of SIF cells was observed.

Granular vesicles in cultures with HC and those with cord serum or maternal serum (Fig. 4) had diameters of 80 to 280 nm, whereas the range was 40 to 140 nm for SIF cells in control cultures with fetal calf serum (Fig. 3).

Consecutive Cultures in Various Media

To study the *in vitro* effect of conditioning on the SIF cell response of ganglion cultures and the subsequent fate of SIF cells induced

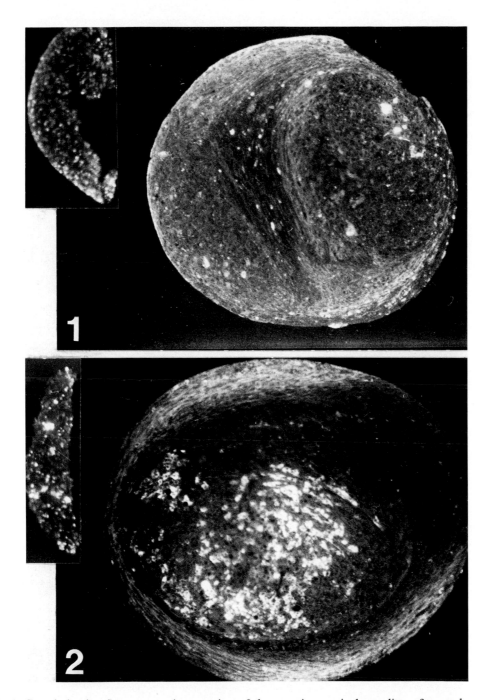

FIG. 1. Catecholamine fluorescence in a section of the superior cervical ganglion of a newborn rat cultured for 7 days in the control medium. Note three small SIF cell clusters in the right upper part of the section. **Inset**: Similarly made photomicrograph of a ganglion cultured for 7 days in a medium containing vinblastine 0.05 mg/liter. Note the small size of the cultured ganglia. No SIF cells are seen. ×115.

FIG. 2. Fluorescence photomicrograph of a ganglion cultured for 7 days in a medium containing hydrocortisone 10 mg/liter. There are numerous SIF cells in the center of the rounded ganglion. **Inset**: A ganglion cultured for 7 days in a medium containing hydrocortisone 10 mg/liter and vinblastine 0.05 mg/liter. Note the numerous small SIF cell clusters. ×115.

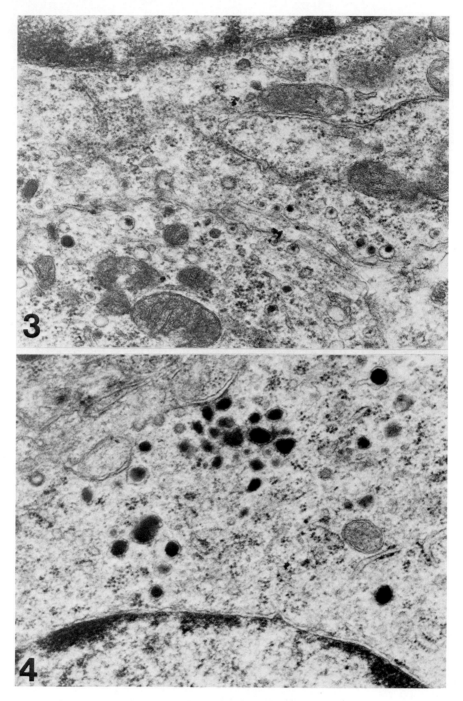

FIG. 3. Electron micrograph of two adjacent SIF cells in a 7-day control culture of a superior cervical ganglion. The diameter of the granular vesicles in SIF cells is about 100 nm. Fixation with the formaldehyde and glutaraldehyde mixture and osmium tetroxide. ×35,000.

FIG. 4. Electron micrograph of an SIF cell from a similarly fixed ganglion after culture for 7 days in the presence of hydrocortisone 10 mg/liter. The granular vesicles are much larger and more electron-dense than those in Fig. 3. Note the clustering of vesicles near the cell membrane, coated vesicles, and coated pits suggestive of exo- and endocytosis. ×35,000.

by HC, newborn rat ganglia were first cultured for 7 days either in the control medium or an HC-containing medium and then again for 7 days in the control medium or an HC-containing medium.

The results of the 2-week culture experiments with and without hydrocortisone are summarized in Fig. 5, and two are documented in Figs. 6 and 7. Not surprisingly, the number of SIF cells did not increase in the ganglia first cultured in the control medium for 7 days and then in the same medium for 7 days. Ganglia cultured for two successive 7-day periods in the HC-containing medium showed, as expected, a large increase in the number of SIF cells. On the other hand, few SIF cells were seen in the ganglia first cultured in the control medium for 7 days and subsequently in the HC-containing medium for 7 days (Fig. 6), whereas a large number of SIF cells was regularly seen in ganglia cultured first in the HC-containing medium and then in the control medium (Fig. 7).

The results of similar culture experiments with two 7-day periods but using human cord serum instead of the fetal calf serum can be expressed in a simple way: Media containing umbilical serum, with or without added HC, had the same effect as the medium containing HC and fetal calf serum. Thus very large numbers of SIF cells were present in ganglia cultured first with cord serum alone and then with any kind of culture medium, alone or with HC; few SIF cells were seen in ganglia cultured first with fetal calf serum alone and then in any SIF-cell-inducing medium, including that containing HC and cord serum. Many SIF cells were also seen in ganglia cultured first in a medium containing umbilical serum and subsequently in a medium containing fetal calf serum, but few SIF cells were seen if this order was reversed.

Fate *In Vitro* of SIF Cells Induced by Hydrocortisone *In Vivo*

SIF cells induced by HC *in vitro* also survived when subsequently cultured in HC-free

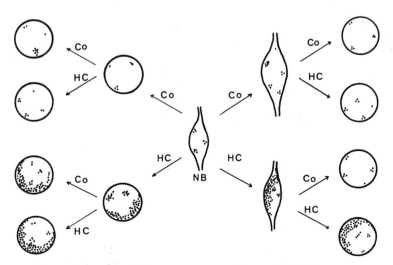

FIG. 5. Summary of the main results of the two-stage experiments. SIF cells are represented by black dots in the elongated ganglia taken from living rats and in the rounded ganglia after culture. The starting point in the center is the newborn rat superior cervical ganglion (NB). To the left of it are ganglia cultured for 7 days without (Co) or with (HC) hydrocortisone, and on the extreme left are similar ganglia further cultured for 7 days without or with HC. At the right of center are ganglia left *in vivo* in uninjected rats for 7 days (Co) or rats injected since birth with HC daily for 7 days; on the extreme right are ganglia thereafter removed and cultured for 7 days without or with HC.

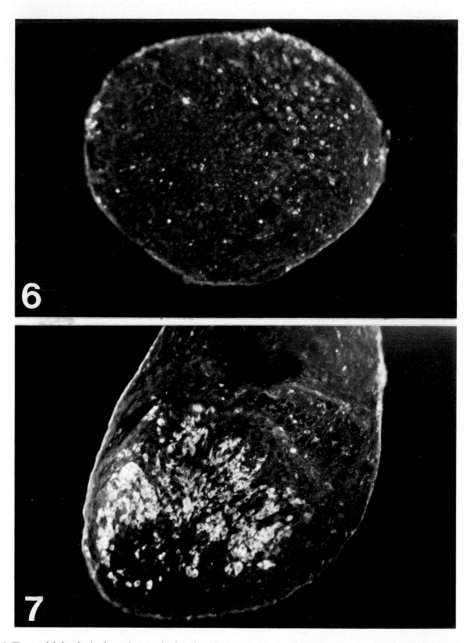

FIG. 6. Formaldehyde-induced catecholamine fluorescence in the superior cervical ganglion of a newborn rat cultured first for 7 days in the control medium and subsequently for 7 days in a medium containing hydrocortisone 10 mg/liter. Only a few SIF cells are visible in the section. ×115.
FIG. 7. Fluorescence photomicrograph of the superior cervical ganglion cultured first for 7 days in the presence of hydrocortisone 10 mg/liter and subsequently for 7 days in the control medium. Note the large number of SIF cells in the section. ×115.

control medium. Because the increase in the number in the SIF cells caused *in vivo* in a newborn rat by HC injections disappeared after the hydrocortisone injection had been discontinued, the fate of SIF cells newly formed *in vivo* was also examined after subsequent culture in various media. Ganglia of 7-day-old rats which had been injected daily with 0.2 mg HC since birth and those of uninjected 7-day-old controls were cultured for 7 days *in vitro* in the medium with fetal calf serum alone (control medium), in the medium with fetal calf serum and HC, or in the medium with human cord serum. The results are illustrated in Fig. 5 and documented in Figs. 8 to 13.

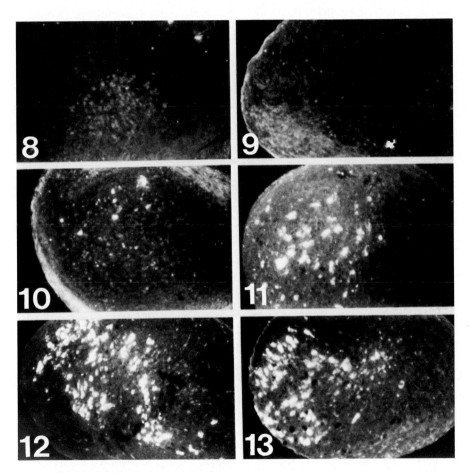

FIG. 8-13. Formaldehyde-induced fluorescence in the superior cervical ganglia of 7-day-old control or HC-injected rats after culture in various media for 7 days. ×75.
FIG. 8. Ganglion of 7-day-old control rat cultured for 7 days in the control medium. Few SIF cells are visible in the right upper part of the photograph.
FIG. 9. Ganglion of a 7-day-old control rat cultured for 7 days in the presence of HC 10 mg/liter for 7 days. Only one SIF cell cluster is visible in the section.
FIG. 10. Ganglion of a 7-day-old HC-injected rat cultured for 7 days in the control medium. A few SIF cells can be seen in the upper part of the explant.
FIG. 11. Ganglion of a 7-day-old HC-injected rat cultured for 7 days in the HC-containing medium. There are several SIF cell clusters in the explant.
FIG. 12. Ganglion of a 7-day-old HC-injected rat cultured for 7 days in a medium containing human cord serum and HC 10 mg/liter. There are numerous SIF cell clusters.
FIG. 13. Ganglion of a 7-day-old HC-injected rat cultured for 7 days in a medium containing human cord serum. There are as many SIF cells as in Fig. 12.

Only a few SIF cells were seen in the ganglia of 7-day-old control rats after culture with either HC and fetal calf serum (Fig. 8), HC and human umbilical serum (Fig. 9), or fetal calf serum only. Thus the ganglia had, during 7 days *in vivo*, lost their ability to react *in vitro* with an increase in the number of SIF cells to HC in the culture medium.

On the other hand, few SIF cells were also seen in the ganglia of HC-injected rats which had thereafter been cultured in the control medium (Fig. 10), whereas a large number of SIF cells were seen after culture of similar ganglia in a medium containing fetal calf serum and HC (Fig. 11), umbilical serum and HC (Fig. 12), or umbilical serum alone (Fig. 13).

DISCUSSION

Mechanism of Induction of SIF Cells by Hydrocortisone *In Vitro*

The observation that hydrocortisone 10 mg/liter caused a great increase in the number of SIF cells in cultures of whole nontrypsinized superior cervical ganglia confirms earlier observations reported on cultures of trypsinized (9) and nontrypsinized (8) ganglia. This HC-induced increase was not prevented by the presence of such highly toxic concentrations of vinblastine sulfate at 0.05 or 0.01 mg/liter in the culture medium; the latter concentration should also be sufficient to inhibit mitotic division but did not prevent the larger HC-induced increase in the number of SIF cells *in vitro*. These observations support the view that HC causes induction of catecholamine-synthesizing enzymes in potential precursors of SIF cells, which thus become SIF cells, rather than proliferation of already differentiated SIF cells by mitotic cell division. It seems likely that HC also exerts its action *in vivo* by causing (through enzyme induction) the differentiation of immature sympathetic cells into SIF cells, as has been proposed (7,9). For further information on phenotypic plasticity of cultured sympathetic cells, the reader is referred to chapters elsewhere in this volume (17,18).

Although SIF cells were readily induced by HC in organ cultures of whole ganglia taken from newborn rats, HC did not have this effect *in vitro* in ganglia taken from 7-day-old control rats or in ganglia of newborn rats cultured *in vitro* for 7 days without HC. These observations fit well with the previous *in vivo* observations showing that the increase caused by injections of glucocorticoids to newborn rats in the number of extraadrenal chromaffin cells (15) or SIF cells in the ganglia (2) was soon lost after birth. They suggest that the SIF-cell-increasing effect of HC depends on the presence in the ganglion of sufficiently primitive stem cells, which differentiate soon after birth into nerve cells and cannot then be converted into SIF cells with HC.

Induction of SIF Cells by Human Cord Serum or Maternal Serum

Replacing fetal calf serum in the culture medium by human cord (placental) or maternal (parturient) serum had the same SIF-cell-increasing effect on cultures of newborn rat sympathetic ganglia as addition of HC into the culture medium, whereas human male serum failed to cause such an increase in the number of SIF cells. Human cord serum has been used as an ingredient of culture medium for nervous tissues (1). We originally tried to "improve" the catecholamine content of the principal ganglion cells in culture, but an unexpected increase in the number of SIF cells led to further studies with other media. Since the number of SIF cells dramatically increased in ganglia cultured in a medium with cord serum or maternal serum but not in a medium with adult male serum (8), it is obvious that the factor responsible was a special feature of cord serum and maternal serum, rather than a general property of human serum.

There were several similarities in the responses of SIF cells in the ganglia cultured with hydrocortisone and fetal calf serum, cord serum, or maternal serum, instead of fetal calf serum, as compared with the SIF cells cultured in the control medium with fetal calf serum or adult male serum alone: (a) Addition of HC to a medium containing cord serum or maternal serum did not further enhance the increase in the number of SIF cells. (b) The magnitude and pattern of the SIF cell responses were similar. (c) The SIF cells thus induced contained immunohistochemically demonstrable phenylethanol-amine-N-methyl transferase (13). (d) The granular vesicles were larger than those in normal SIF cells. (e) The SIF cells induced by HC *in vivo* survived in ganglia cultured in these media *in vitro*.

These similarities can be taken to suggest that the site and mechanism of action of hydrocortisone and of the factor in cord and maternal serum are the same, even if the same hormone is not responsible. Indeed it is unlikely that the glucocorticoid concentration, which is at most 0.2 mg/liter in normal human individuals would ever reach levels comparable to those employed in the present *in vivo* and *in vitro* experiments, 20 mg/kg and 10 mg/liter, respectively. Further studies are necessary, taking into consideration that steroids such as estrogens or progesterone, known to increase during pregnancy, may mimic or support the effect of glucocorticoids.

Fate *In Vitro* of SIF Cells Induced *In Vivo* or *In Vitro*

It was shown in the present study that newly induced SIF cells in the ganglia of HC-injected young rats survive *in vitro* if HC is present in the culture medium but disappear when cultured for 7 days in an HC-free medium. This may not seem surprising in view of the earlier observation that the increase in the number of SIF cells caused by HC injections is reversible *in vivo* upon a further period of 7 days without injections (4). However, the increase in the number of SIF cells induced by HC *in vivo* is not reversible during a subsequent 7-day recovery period *in vivo* if the ganglia have been decentralized at birth (4). Furthermore, as was shown in the present study, SIF cells newly formed *in vitro* in cultures containing HC (or placental serum instead of fetal calf serum) do not disappear during further culture for 7 days in a HC-free control medium.

Further studies are necessary to find out the reasons for these differences in the survival *in vitro* between the SIF cells induced by HC in intact ganglia *in vivo* and those induced by HC *in vitro*. In this respect it may be relevant that the SIF cells induced *in vivo* by HC injections contain granular vesicles of about the same size (100 nm) as the ganglia of untreated controls (11), whereas the granular vesicles of the SIF cells induced by culture of newborn rat ganglia in HC-containing media are not only more numerous but also much larger (up to 200 nm) than those in the control ganglia cultured in HC-free medium (10). Although these studies were made on trypsinized cultures, our preliminary observations on nontrypsinized ganglia cultured in media containing HC (or in media containing cord or maternal serum) have shown that the SIF cells in them contain larger granular vesicles than those in the control cultures. SIF cells with larger granular vesicles can be expected to be further differentiated, and thus more stable, than those containing granular vesicles about 100 nm in diameter (e.g., SIF cells in normal superior cervical ganglion of the rat).

Further studies are planned to determine the size of granular vesicles in SIF cells induced by HC in decentralized ganglia, and to see if such SIF cells survive when cultured in HC-free control medium. Although it may be reasonable to assume that better survival *in vitro* depends partly on formation of SIF cells with larger granular vesicles, survival may also depend on other factors. It has been shown that newly formed SIF cells induced by HC *in vivo* in intact ganglia, which disappear after subsequent culture in an HC-free medium *in vitro,* survive and form large clusters when the ganglia of the HC-treated rats are transplanted into the anterior chamber of the eye of an adult rat. For the factors involved the reader is referred to the original work (Eränkö, Päivärinta, and Eränkö, *this volume*).

ACKNOWLEDGMENTS

The use of human placental serum in the tissue culture medium to improve catecholamine synthesis was proposed by Dr. Margaret Murray. We are greatly indebted to her for many stimulating discussions. The present study was supported by a grant from the Sigrid Jusélius Foundation. Samples of human sera were kindly supplied by Dr. Jarl Eklund, Finland's Red Cross Blood Transfusion Center, Helsinki. We thank Mrs. Maija Järvinen and Mr. Erkki Muona for their skilled technical assistance.

REFERENCES

1. Benitez, H.H., Masurovsky, E.B., and Murray, M.R. (1974): Interneurons of the sympathetic ganglia, in organotypic culture: A suggestion as to their function, based on three types of studies. *J. Neurocytol.,* 3:363-384.
2. Ciaranello, R.D., Jacobowitz, D., and Axelrod, J. (1973): Effect of dexamethasone in phenylethanolamine-N-methyltransferase in chromaffin tissue of the neonatal rat. *J. Neurochem.,* 20:799-805.
3. Coupland, R.E., and Fujita, T., editors (1976): *Chromaffin, Enterochromaffin and Related Cells.* Elsevier, Amsterdam.
4. Eränkö, L., and Eränkö, O. (1979): SIF cells in decentralized developing ganglia. In: *Catecholamines: Basic and Clinical Frontiers,* edited by E. Usdin, I.J. Kopin, and J. Barchas, pp. 842-844. Pergamon Press, New York.
5. Eränkö, O. (1967): The practical histochemical demonstration of catecholamines by formaldehyde-induced fluorescence. *J. R. Microsc. Soc.,* 87:259-276.
6. Eränkö, O., editor (1976): *SIF Cells. Structure and Function of the Small Intensely Fluorescent Sympathetic Cells,* Fogarty International Center Proceed-

ings No. 30. Government Printing Office, Washington, D.C.
7. Eränkö, L., and Eränkö, O. (1972): Effect of hydrocortisone on histochemically demonstrable catecholamines in the sympathetic ganglia and extra-adrenal chromaffin tissue of the rat. *Acta Physiol. Scand.,* 84:125-133.
8. Eränkö, O., and Eränkö, L. (1979): Increase in vivo and in vitro of catecholamines and catecholamine-synthesizing enzymes in SIF cells of newborn rat superior ganglia. In: *Catecholamines: Basic and Clinical Frontiers,* edited by E. Usdin, I.J. Kopin, and J. Barchas, pp. 821-823. Pergamon Press, New York.
9. Eränkö, O., Eränkö, L., Hill, C.E., and Burnstock. G. (1972): Hydrocortisone-induced increase in the number of small intensely fluorescent cells and their histochemically demonstrable catecholamine content in cultures of sympathetic ganglia of a newborn rat. *Histochem. J.,* 4:49-58.
10. Eränkö, O., Heath, J., and Eränkö, L. (1972): Effect of hydrocortisone on the ultrastructure of the small, intensely fluorescent, granule containing cells in cultures of sympathetic ganglia of newborn rat. *Z. Zellforsch.,* 134:297-310.
11. Eränkö, O., Heath, J.W., and Eränkö, L. (1973): Effect of hydrocortisone on the ultrastructure of the small, granule-containing cells in the superior cervical ganglion of the newborn rat. *Experientia,* 29:457-459.
12. Eränkö, O., and Härkönen, M. (1963): Histochemical demonstration of fluorogenic amines in the cytoplasm of sympathetic ganglion cells of the rat. *Acta Physiol. Scand.,* 58:285-286.
13. Eränkö, O., Pickel, M.V., Härkönen, M., Eränkö, L., Joh, T.H., and Reis, D.J. (1980): Effect of hydrocortisone on catecholamine content and immunohistochemically demonstrable tyrosine hydroxylase, dopamine beta-hydroxylase and phenylethanolamine-N-methyl transferase in superior cervical ganglion of developing rat. *In preparation.*
14. Kobayashi, S., and Chiba, T., editors (1977): *Paraneurons, New Concepts in Neuroendocrine Relatives.* Daiichi Printing Co., Niigata, Japan.
15. Lempinen, M. (1964): Extra-adrenal chromaffin tissue of the rat and the effect of cortical hormones on it. *Acta Physiol. Scand., [Suppl. 231]* 62.
16. Ploem, J.S. (1971): The microscopic differentiation of the colour of formaldehyde induced fluorescence. *Prog. Brain Res.,* 34:27-38.
17. Tischler, A.S., and Greene, L.A. (1980): Phenotypic plasticity of pheochromocytoma and normal adrenal medullary cells. *This volume.*
18. Unsicker, K., Rieffert, B., and Ziegler, W. (1980): Effects of cell culture conditions, nerve growth factor, dexamethasone, and cyclic AMP on adrenal chromaffin cells in vitro. *This volume.*

*Histochemistry and Cell Biology of
Autonomic Neurons, SIF Cells, and
Paraneurons,*
edited by O. Eränkö et al.
Raven Press, New York © 1980.

Comparisons of Nerve Fiber Growth from Three Major Catecholamine-Producing Cell Systems: Adrenal Medulla, Superior Cervical Ganglion, and Locus Coeruleus

Lars Olson, Åke Seiger, *Ted Ebendal, and †Barry Hoffer

*Department of Histology, Karolinska Institutet, Stockholm, Sweden; *Department of Zoology, Uppsala University, Uppsala, Sweden; and †Department of Pharmacology, University of Colorado, Denver, Colorado 80220*

Locus coeruleus (LC), the superior cervical ganglion (SCG), and the adrenal medulla (AM) are three of the most conspicuous groups of catecholamine (CA) synthesizing cell systems in the body. Only the first two are nerve cell groups, having widely distributed terminal networks in the central nervous system (CNS) and periphery, respectively. The third, AM, does not normally give rise to nerve fibers, but transplantation experiments have demonstrated that mature chromaffin cells can produce nerve fibers under certain conditions (6). Thus although different in many respects, these three cell systems share the property of being able to produce CA-containing nerve fibers.

The most important determinant of nerve fiber growth and nerve terminal patterning is the target tissue. The three major CA cell systems are normally exposed to entirely different types of target tissue. Therefore it becomes difficult to draw conclusions about any similarities or differences in their fiber growth responses from *in situ* observations.

Transplantation experiments make it possible to expose differently located CA cell types to the same targets. Systematic combination of the various CA cell groups with several types of target tissue in a controlled environment is thus one way to disclose factors regulating nerve fiber growth. It is the purpose of this chapter to review such experiments using transplantations to the anterior chamber of the eye in rats. The recent findings that adrenal chromaffin tissue can innervate brain tissue are particularly emphasized. Finally, the question of NGF effects on the developing LC is addressed using a direct comparison of LC and SCG in tissue culture.

MATERIAL AND METHODS

Methods for homologous grafting of tissue to the anterior chamber of the rat eye have been described (for a recent review of the grafting technique see ref. 15). LC and other types of CNS tissue grafts were obtained from fetal rats, and SCG and AM from adult donors (Sprague-Dawley). Intraocular grafts become attached to, and vascularized from, the anterior surface of the host iris. Vascularization and growth of the grafts can be followed postoperatively by stereomicroscopy through the cornea. At sacrifice, CA was localized by Falck-Hillarp histochemistry (2,3) of freeze-dried grafts and whole mounted irides. Fiber growth from grafts on host irides can be precisely quantified (18) provided the sympathetic adrenergic innervation of the host iris is removed by extirpation of the host SCG. Techniques for electrophysiological recordings in grafts have been described (4,8,12,21). Techniques for tissue culture of SCG and LC are found in ref. 7.

RESULTS

All immature CNS areas hitherto tested (all cortical areas, several subcortical areas, spinal cord) survive grafting and continue development *in oculo,* provided they are taken at sufficiently early stages of development. In addition, central monoamine neuron containing grafts (e.g., LC) form nerve fiber plexuses on the host iris that can be quantified. Similarly, adult SCG and AM survive grafting and produce CA nerve fibers that invade the host iris. In the following, we focus on (a) the ability of LC, SCG, and AM grafts to innervate the host iris and (b) the possibility of introducing additional peripheral and central targets which the CA grafts may also innervate. The main results of these experiments are summarized in Fig. 1.

Single Graft Experiments

An SCG graft effectively reinnervates a sympathetically denervated host iris. If placed on a sympathetically innervated host iris (Fig. 1a, second experiment), outgrowth from the SCG graft is greatly inhibited, as disclosed by acute removal of the host SCG 2 days prior to sacrifice. An immature LC graft forms a halo of nerve fibers that covers approximately one-third of the host iris regard-

less of whether sympathetic nerves are present in the host iris (Fig. 1a, third and fourth experiments). However, if the sensory innervation of the host iris is removed by lesioning the trigeminal nerve (Fig. 1a, Trig-X), LC immediately starts forming nerve fibers again and innervates the whole host iris. Finally, chromaffin AM tissue is also able to reinnervate the host iris. Some chromaffin cells migrate onto the host iris, become elongated or polygonal, and produce nerve fibers that taper off to become almost indistinguishable from normal sympathetic fibers at a distance from their cellular origin.

Additions of Peripheral Targets

When added to LC-carrying eyes, several types of peripheral tissue that are normally innervated by sympathetic nerves become effectively innervated by LC. In the first experiment of Fig. 1b, this is exemplified with an iris graft (small rectangle). Although this iris graft becomes fully innervated, the innervation of the host iris remains restricted. If peripheral tissues are grafted to normal eyes, the sympathetic nerves of the host eye reinnervate the grafts effectively.

The reinnervation of peripheral targets by sympathetic nerves from the host iris is specific in the sense that only tissues normally innervated by sympathetic nerves become reinnervated when grafted. One striking difference between LC- and SCG-derived fibers has been found: Heart tissue becomes effectively reinnervated by sympathetic nerves whereas LC-derived fibers do not penetrate the heart tissue (Fig. 1b, lower two experiments).

Additions of Central Cortical Target Tissues

When immature cortical tissue is grafted to normal eyes (Fig. 1c, upper experiments) it becomes innervated by ingrowth of sympathetic nerve fibers from the host iris. This innervation mimics the normally present LC-derived fibers in such areas. The density and distribution of ingrown sympathetic fibers remains unchanged also at long survival times (over 1 year, far right part of upper experiments). The ingrowing sympathetic fibers change their morphology to look more like the normally present central adrenergic fibers (Fig. 2a). If cortical tissue (e.g., parietal cerebral cortex or hippocampus) is added to LC-carrying eyes (Fig. 1c, middle experiment), the cortical grafts become effectively reinnervated by the LC graft. Interestingly, LC fibers continue to invade the cortical targets until an extreme degree of adrenergic hyperinnervation has been reached (Fig. 1c, far right part of middle experiment). Reinnervation of hippocampal targets by LC grafts at medium and long postoperative times are illustrated in Fig. 2c,d.

An interesting recent finding is the fact that chromaffin tissue of AM is also able to form plexuses in CNS tissue. This is illustrated in the lower experiment of Fig. 1c. In this case grafts of the hippocampal formation or cerebral cortex were allowed to develop to maturity intraocularly. AM grafts were then placed in close contact with the cortical grafts; 2.5 months later all sympathetic nerves of the host eyes were removed by extirpation of the host SCG and the animals sacrificed after 2 days. By doing this "reversed-order" type of experiment, as compared to the upper and middle experiment of Fig. 1c, three conclusions can be drawn: (a), Adrenal chromaffin tissue is able to form

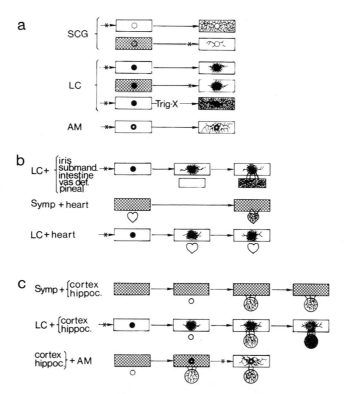

FIG. 1. Summary of the main results. In all diagrams, host irides are depicted as large rectangles, and the presence of a normal sympathetic nerve plexus on the host irides is indicated by a raster. Time flows from left to right as indicated by arrows. Removal of the sympathetic nerve plexus in the host iris is performed at various stages of the experiments and indicated by an asterisk. Open circles on host irides: grafts of the SCG. Filled circles: grafts of the developing LC. Filled circles with a star: grafts of the AM. All drawings demonstrate the amount and distribution of CA nerve fibers as seen in whole mounts of irides and freeze-dried sections of other types of grafts with Falck–Hillarp fluorescence histochemistry. For further descriptions see the text.

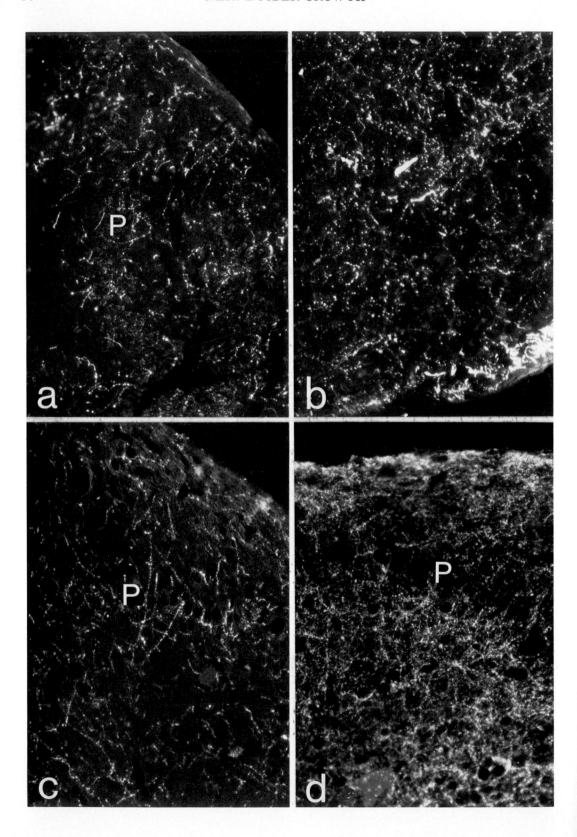

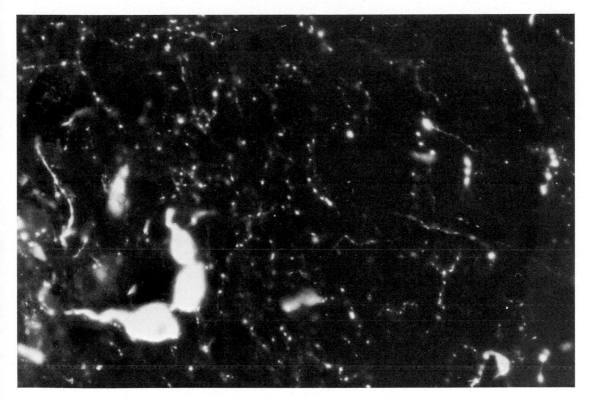

FIG. 3. Close-up view of cortical graft combined with AM graft. Area is close to the AM graft. Two strongly fluorescent chromaffin cells, one with a barely visible nucleus, form coarse processes. Thin and thick varicose fibers innervate the graft. Fluorescence microscopy.

nerve fibers on the host irides to a certain extent also in the presence of sympathetic fibers. (b), AM grafts can effectively innervate cortical CNS tissue. (c), This innervation of cortical tissue can occur as an ingrowth of AM-derived fibers into already mature CNS tissue. AM fibers in brain tissue had an interesting appearance. Although many fibers were thin with small rounded varicosities, and thereby similar to normal LC-derived fibers, some were coarse with larger varicosities and a very strong fluorescence intensity. Such fibers were more often seen in association with blood vessels (Figs. 2b and 3).

Functional Aspects

Adrenergic neurotransmission has been proved in several of the various types of reinnervation experiments described. When sympathetic (and parasympathetic) fibers from the host iris innervate a graft, activity of

◄
FIG. 2. Fluorescence histochemistry of grafts of cortical (**b**) or hippocampal (**a,c,d**) tissue that has become innervated by CA nerve fibers of different types. **a**: Ingrowth of sympathetic fibers from host iris into hippocampal target. Graft surface is at upper right. Delicate varicose fibers are present throughout the graft tissue. Postoperative time is approximately 6 months, as in **d**. **b**: Cortex cerebri graft innervated by fibers formed by AM grafts. Dense innervation of host iris seen at lower right. The cortical tissue is innervated by a plexus of fine varicose fibers. In addition, some coarse fibers are found *(middle)*. **c**: Hippocampal graft reinnervated by an LC graft. Shorter postoperative time. Density of adrenergic nerve fibers is normal and similar to that in **a**. **d**: Hippocampal graft reinnervated by LC, long postoperative time. Extreme degree of adrenergic hyperinnervation. (P) Pyramidal cell layer.

these fibers can be controlled by the light input to the host eye. Thus, for example, intraocular hearts beat with a frequency that becomes fully dependent on the light input to the host eye. Similarly, activity recorded electrophysiologically from hippocampal grafts depends on activity in ingrowing sympathetic and parasympathetic fibers, which in turn is regulated by the light input to the eye. It was recently shown that the extreme degree of hyperinnervation caused at long postoperative times by LC grafts in hippocampal targets is functionally effective. Such hippocampal grafts are so effectively inhibited by the adrenergic innervation that hypersynchronous seizure activity cannot be elicited in them unless the LC graft is inhibited (5).

Tissue Culture Experiments

To compare the sensitivity of SCG and LC to nerve growth factor (NGF), tissue culture experiments were performed as summarized in Fig. 4. Using the classical chick ganglia bioassay, it was first established that grafts of fetal rat LC did not produce any detectable NGF. Fetal rat cerebral cortex was also negative in this bioassay, whereas adult rat iris tissue elicited a strong response in chick spinal and sympathetic ganglia, which was inhibited by anti-NGF. When LC, cerebral cortex, and SCG from the same neonatal animals were cultured together in control dishes, LC and cortex explants produced rich halos of nerve fibers whereas SCG produced none. Addition of NGF to the medium caused a strong fiber response in SCG and no increase in fibers from LC or cortex. Addition of antiserum against NGF did not abolish the fiber growth response from CNS explants seen in control medium. Fluorescence microscopy performed after 4 days of culture showed a highly significant growth response of SCG to NGF, whereas there was no effect on LC.

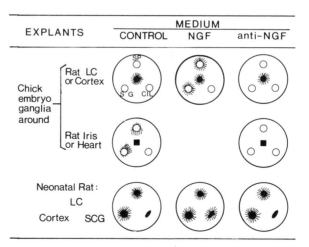

FIG. 4. Summary of tissue culture experiments. Chick and rat tissues were cultured in a collagen lattice containing Eagle's basal medium plus 1% fetal calf serum. Chick spinal, sympathetic (SG), and ciliary ganglia were combined with fetal or neonatal rat LC or cortex tissue to test for the presence of NGF-like activity in rat tissues. The chick ganglia were not stimulated by the presence of rat CNS tissues. Spinal and sympathetic ganglia could be stimulated by NGF (1 BU) also in the presence of rat CNS tissue, and this stimulation was inhibited by anti-NGF. Peripheral rat tissues (iris and heart) contained NGF-like activity, as shown by directed fiber growth from spinal and sympathetic ganglia, a response that was inhibited by anti-NGF. When neonatal rat LC, cortex, and SCG were cultured together in control medium, only LC and cortex showed outgrowth. This outgrowth was not stimulated by NGF (1 to 50 BU) and was not inhibited by anti-NGF (1:5 to 1:50; 1:500). SCG from the same donors reacted to NGF with a strong fiber growth response. (Data from ref. 7 and Ebendal, Olson, and Seiger, *unpublished.*)

DISCUSSION

Taken together, the experiments reviewed here (4–15,17–21) clearly demonstrate that grafts of the three CA-producing cell systems (LC, SCG, and AM) are able to innervate peripheral and central target tissues. In the case of LC and SCG, this innervation has been shown to be functional in several cases (5,9,10,15,21). However, important differences exist in fiber growth regulatory mechanisms between the three cell systems. Thus LC-derived adrenergic fibers do not seem to recognize the presence of SCG-derived adrenergic fibers. On the other hand, LC but not SCG seems to be able to grow to many nonadrenergic synaptic sites as suggested by the profound hyperinnervation by LC of the grafted, and thus heavily deafferentated, hippocampal formation (13). Another important difference between LC and SCG is the fact that NGF is not effective on the developing LC (7). Experimental evidence is still too meager to say if fiber growth regulation in AM is more similar to SCG than to LC.

Intracranial grafts of CA neurons can also innervate brain target tissue *in situ* (1). Recent experiments demonstrated that nigral brain grafts can reduce motor abnormalities caused by lesions of the nigrostriatal dopamine system (16). The fact that AM is able to innervate brain tissue (11) is particularly interesting in this respect since adrenal medulla represents a source of CA cells in the body which can be used for autologous grafting.

ACKNOWLEDGMENTS

Supported by the Swedish Medical and Natural Science Research Councils (04X-03185 and B2021-017), Magn. Bergvalls Stiftelse, Karolinska Institutets fonder, and USPHS NS09199.

REFERENCES

1. Björklund, A., and Stenevi, U. (1979): Regeneration of monoaminergic and cholinergic neurons in the mammalian central nervous system. *Physiol. Rev.,* 59:62–100.
2. Corrodi, H., and Jonsson, G. (1967): The formaldehyde fluorescence method for the histochemical demonstration of biogenic monoamines: A review on the methodology. *J. Histochem. Cytochem.,* 15:65–78.
3. Falck, B., Hillarp, N-Å., Thieme, G., and Torp, A. (1962): Fluorescence of catecholamines and related compounds condensed with formaldehyde. *J. Histochem. Cytochem.,* 10:348–354.
4. Hoffer, B., Seiger, Å., Freedman, R., Olson, L., and Taylor, D. (1977): Electrophysiology and cytology of hippocampal formation transplants in the anterior chamber of the eye. II. Cholinergic mechanisms. *Brain Res.,* 119:107–132.
5. Hoffer, B., Taylor, D., Freedman, R., Olson, L., and Seiger, Å. (1980): Conditions for adrenergic hyperinnervation in hippocampus. II. Electrophysiological evidence from intraocular double grafts. *Exp. Brain Res. (in press).*
6. Olson, L. (1970): Fluorescence histochemical evidence for axonal growth and secretion from transplanted adrenal medullary tissue. *Histochemie,* 22:1–7.
7. Olson, L., Ebendal, T., and Seiger, Å. (1979): NGF and anti-NGF: Evidence against effects on fiber growth in locus coeruleus from cultures of perinatal CNS tissues. *Dev. Neurosci.,* 2:160–176.
8. Olson, L., Freedman, R., Seiger, Å., and Hoffer, B.J. (1977): Electrophysiological and cytology of hippocampal formation transplants in the anterior chamber of the eye. I. Intrinsic organization. *Brain Res.,* 119:87–106.
9. Olson, L., and Malmfors, T. (1970): Growth characteristics of adrenergic nerves in the adult rat: Fluorescence histochemical and ^{3}H-noradrenaline uptake studies using tissue transplantations to the anterior chamber of the eye. *Acta Physiol. Scand. [Suppl.],* 348:1–112.
10. Olson, L., and Seiger, Å., (1976): Beating intraocular hearts: Light-controlled rate by autonomic innervation from host iris. *J. Neurobiol.,* 7:193–203.
11. Olson, L., Seiger, Å., Freedman, R., and Hoffer, B. (1980): Adrenal medullary chromaffin cells can innervate brain tissue: Fluorescence histochemical evidence from double intraocular grafts. *Brain Res. (in press).*
12. Olson, L., Seiger, Å., Hoffer, B., and Taylor, D. (1979): Isolated catecholaminergic projections from substantia nigra and locus coeruleus to caudate, hippocampus and cerebral cortex formed by intraocular sequential double brain grafting. *Exp. Brain Res.,* 35:47–67.
13. Olson, L., Seiger, Å., Taylor, D., and Hoffer, B. (1980): Conditions for adrenergic hyperinnervation in hippocampus. I. Histochemical evidence from intraocular double grafts. *Exp. Brain Res. (in press).*
14. Olson, L., Seiger, Å., and Ålund, M. (1978): How nerve fiber outgrowth from intraocular locus coeruleus grafts is controlled by the state of innervation

of the host iris. In: *Formshaping Movement in Neurogenesis,* edited by C.O. Jacobsson and T. Ebendal, pp. 245-256. Almqvist & Wiksell, Uppsala.

15. Olson, L., Seiger, Å., Ålund, M., Freedman, R., Hoffer, B., Taylor, D., and Woodward, D. (1980): Intraocular brain grafts: A method for differentiating between intrinsic and extrinsic determinants of structural and functional development in the central nervous system. In: *Neural Growth and Differentiation,* edited by E. Meisami and M.A.B. Brazier, pp. 223-235. Raven Press, New York.

16. Perlow, M.J., Freed, W.J., Hoffer, B.J., Seiger, Å., Olson, L., and Wyatt, R.J. (1979): Brain grafts reduce motor abnormalities produced by destruction of nigrostriatal dopamine system. *Science,* 204:643-647.

17. Seiger, Å., and Olson, L. (1975): Brain tissue transplanted to the anterior chamber of the eye. 3. Substitution of lacking central noradrenaline input by host iris sympathetic fibers in the isolated cerebral cortex developed in oculo. *Cell Tissue Res.,* 159:325-338.

18. Seiger, Å., and Olson, L. (1977): Quantitation of fiber growth in transplanted central monoamine neurons. *Cell Tissue Res.,* 179:285-316.

19. Seiger, Å., and Olson, L. (1977): Growth of locus coeruleus neurons in oculo independent of simultaneously present adrenergic and cholinergic nerves in the iris. *Med. Biol.,* 55:209-223.

20. Seiger, Å., and Olson, L. (1977): Reinitiation of directed nerve fiber growth in central monoamine neurons after intraocular maturation. *Exp. Brain Res.,* 29:15-44.

21. Taylor, D., Seiger, Å., Freedman, R., Olson, L., and Hoffer, B. (1978): Electrophysiological analysis of reinnervation of transplants in the anterior chamber of the eye by the autonomic ground plexus of the iris. *Proc. Natl. Acad. Sci. USA,* 75:1009-1012.

Histochemistry and Cell Biology of Autonomic Neurons, SIF Cells, and Paraneurons,
edited by O. Eränkö et al.
Raven Press, New York © 1980.

Fate of Newly Formed Hydrocortisone-Induced SIF Cells in Rat Superior Cervical Ganglion Transplanted into the Anterior Chamber of the Eye

Olavi Eränkö, Heikki Päivärinta, and Liisa Eränkö

Department of Anatomy, University of Helsinki, 00170 Helsinki 17, Finland

Hydrocortisone (HC) injections cause an approximately 10-fold increase in the number of SIF cells in the superior cervical ganglion (SCG) in newborn rats (2). *In vivo,* this increase is reversible, the number of SIF cells diminishing to approximately normal 1 week after discontinuing the HC treatment (3). *In vitro,* newly formed SIF cells react differently depending on the conditions in which they were induced. In organ culture of sympathetic ganglia of newborn rats, addition of HC to the culture medium causes an increase in the number of SIF cells (5), and large numbers of SIF cells persist for a week after the culture medium is replaced by an HC-free medium. On the other hand, newly formed SIF cells induced *in vivo* by HC injections disappear within a week when subsequently cultured *in vitro* without HC (6).

In the present study it is shown that SIF cells induced by HC injections *in vivo* survive in ganglia transplanted into the anterior chamber of the eye of adult rats.

MATERIALS AND METHODS

Newborn Sprague-Dawley rats were injected with 0.2 mg HC acetate (Lääke Oy) per animal (about 20 mg/kg body weight) daily for 7 days. One day after the last injec-tion, the rats were decapitated, the SCG was removed, the pre- and postganglionic trunks were cut near the ganglion, and the ganglion was transplanted into the right anterior chamber of the eye of an adult female rat using the technique described by Olson and Malmfors (7). Transplants of SCG from 7-day-old untreated rats served as controls. Only one ganglion of the 7-day-old rat, untreated or injected, was always transplanted, and the contralateral ganglion from the same rat was freeze-dried for the study of catecholamine fluorescence. To destroy the normal adrenergic innervation of the iris in the right eye, the right SCG was removed 1 day before transplantation.

The ganglion transplants were examined 1 week (ganglia of seven HC-injected rats and eight controls), 1 month (seven HC, six control ganglia), or 3 months (three HC, seven control ganglia) after transplantation. The host animals were killed under ether anesthesia and the eyes removed. The iris with the attached transplant was dissected and the transplant cut out of the iris. The left and right irides were stretched on a slide, dried in a desiccator containing phosphorus pentoxide for 1 hr, and exposed at 40°C for 60 min and at 60°C and 80°C for 30 min each to vapor from paraformaldehyde powder equilibrated with 80% relative humidity.

For the demonstration of catecholamine fluorescence (4), the ganglia were frozen in liquid propane cooled with liquid nitrogen and dried in a vacuum at $-40°C$ for a week. They were then exposed at $60°C$ for 30 min and at $80°C$ for 60 min to vapor from paraformaldehyde powder equilibrated with 60% relative humidity, and embedded in paraffin wax. The sections were cut at 6 μm. The sections and the iris stretches were mounted in Entellan.

The ganglia were examined with a Leitz Dialux 20 fluorescence microscope using a mercury lamp and epi-illumination with filter block D. Iris stretches were examined using transillumination with darkfield condenser and excitation filters BG 38 and BG 3 and emission filter K 470. Photographs were taken on Kodak Tri-X film using a Leitz Orthomat microscope camera.

RESULTS

Effect of Hydrocortisone *In Vivo*

In the SCG of the 7-day-old untreated control rats, at most a few individual SIF cells and small clusters were seen in a section; sympathicoblasts and young nerve cells were the dominant cell types (Fig. 4), and many sections were entirely devoid of SIF cells. In the ganglia of the 7-day-old rats injected daily with 0.2 mg HC, large numbers of SIF cells, singly and in small groups, were seen in every section (Fig. 1).

Effect of Transplantation

In the ganglia of normal 7-day-old rats transplanted into the anterior chamber of the eye of an adult rat for a week or longer, there was a marked loss of nerve cell bodies (Figs. 4 and 5). The transplant also diminished in size and became rounded during the period of transplantation.

Although few SIF cells were seen in the ganglia of normal 7-day-old rats after transplantation for 7 days (Fig. 2), 1 month (Fig. 5), or 3 months, large numbers of SIF cells were observed in all ganglia of the 7-day-old HC-treated rats 7 days (Fig. 3), 1 month (Fig. 6), or 3 months after transplantation.

Seven days after transplantation the SIF cells in the ganglia of the HC-treated rats were individual or in small clusters, as they were before the transplantation (Fig. 1), but were crowded in an half-moon-shaped area in the periphery of the transplant (Fig. 3). This appearance is suggestive of SIF cell migration toward a common area.

In older transplants of the HC-treated ganglia, the SIF cells formed large clusters of intimately connected cells. Figure 6 shows one such cluster in a 1-month transplant of a ganglion from an HC-treated rat; it is to be compared also with Fig. 5, which is an equally old transplant of a control ganglion. Such SIF cell clusters in old transplants were often larger than the largest clusters observed in normal ganglia of adult rats or in transplants of ganglia of untreated 7-day-old rats.

Although there was a considerable loss of nerve cells due to transplantation, enough nerve cells survived to innervate the adult iris, which had been denervated before transplantation (Figs. 7–14). No significant differences were observed in the innervation capacity of the ganglion transplants from HC-injected or untreated 7-day-old rats. It was not possible to determine if the SIF cells contributed to the innervation of the iris.

DISCUSSION

Survival of SIF Cells in Transplants

Since the ganglia of the HC-injected 7-day-old rats always contained large numbers of SIF cells, whereas the ganglia of the 7-day-old control rats contained but a few, it seems safe to assume that the contralateral transplanted ganglia contained, respectively, large numbers or few SIF cells. The main observation of this study is that during the transplantation period ranging from 1 week to 3 months, large numbers of SIF cells were always observed in the transplanted ganglia of the HC-injected rats whereas few SIF cells were seen in the transplanted ganglia of the control rats. In other words, the SIF cells induced *in vivo* by HC injections survived in the ganglia after transplantation in spite of a massive loss of neurons from them.

This observation is somewhat surprising because SIF cells similarly induced *in vivo* disappear *in vivo* when the rats are allowed to recover uninjected for another week (3), and *in vitro* when the ganglia taken from the 7-day-old HC-injected animals are cultured for a week in organ cultures in an HC-free cul-

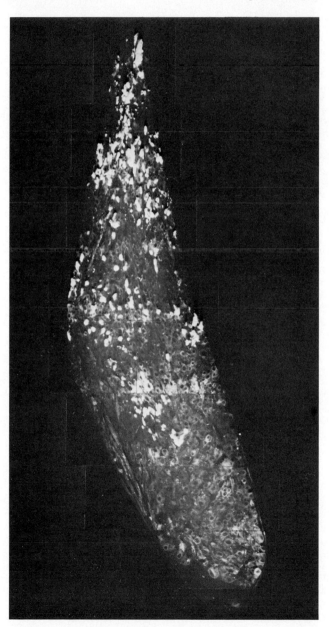

FIG. 1. Montage of fluorescence photomicrographs of the SCG of a 7-day-old rat injected since birth with 0.2 mg HC daily for 1 week. Large numbers of SIF cells are seen singly and in small clusters. ×110.

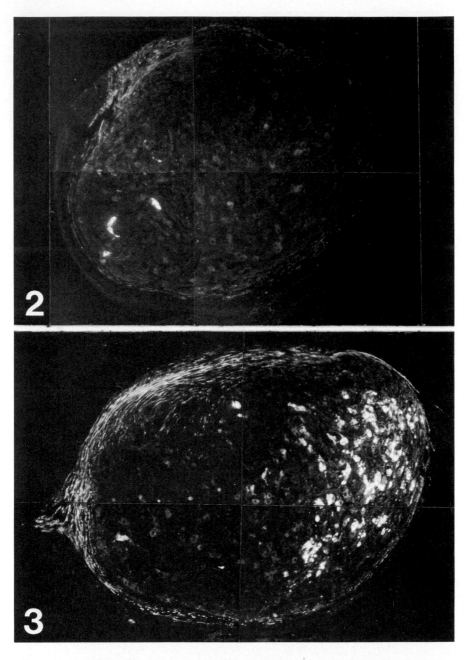

FIG. 2. Fluorescence photomicrograph of the SCG of a 7-day-old untreated rat 7 days after transplantation into the eye of an adult rat. On the left are two small clusters of SIF cells. ×110.
FIG. 3. Fluorescence photomicrograph of the SCG of a 7-day-old HC-treated rat 7 days after transplantation into the eye of an adult rat. On the right are large numbers of SIF cells, appearing singly and in small groups. ×110.

ture medium (6). Thus there is something in transplantation into the anterior chamber of the eye of an adult rat that either favors preservation of SIF cells or does not stimulate their transformation into another type of cell, e.g., nerve cells. Such a change may occur if the ganglion is left to recover *in vivo* after discontinued HC administration.

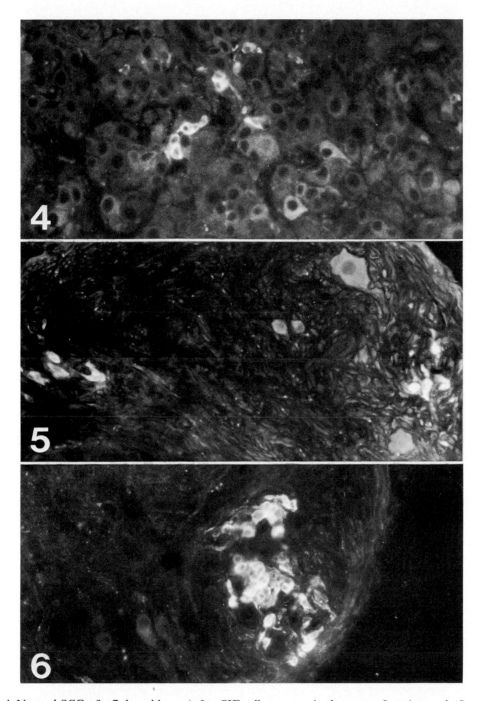

FIG. 4. Normal SCG of a 7-day-old rat. A few SIF cells are seen in the center. Less intensely fluorescent young sympathicoblasts and young nerve cells are the prevalent cell types. ×350.
FIG. 5. SCG of an untreated 7-day-old rat after transplantation for 1 month. The loss of nerve cell bodies is pronounced as compared with Fig. 4. A few SIF cells are seen on both sides of the figure. ×350.
FIG. 6. SCG of a 7-day-old HC-treated rat after transplantation for 1 month. A large cluster of SIF cells is seen. ×350.

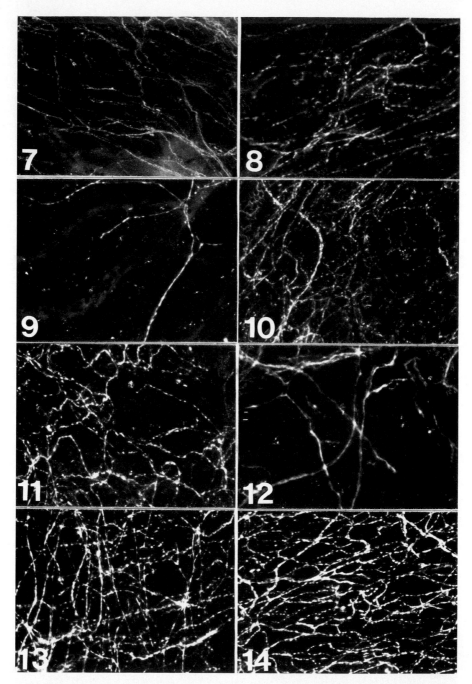

FIGS. 7-14. Reinnervation of host irises from transplanted SCG:ia of 7-day-old rats. Figures 7-11 show fibers from transplants of HC-treated rats. One week after transplantation, the iris is innervated mainly by nerve fibers of smooth appearance (Fig. 7). One month after transplantation, almost the whole iris is innervated; near the transplant the fibers form a dense network of varicose fibers (Fig. 8). Further away from the same 1-month transplant, varicose fibers grow toward the periphery (Fig. 9). Three months after transplantation, the whole iris is innervated and the nerve plexus is dense near (Fig. 10) and further away from (Fig. 11) the transplant. Figures 12 and 13 show host irises reinnervated from SCG:ia of untreated 7-day-old control rats. Nerve fibers of smooth appearance are near a 7-day transplant in Fig. 12. Dense network of varicose nerve fibers are near a 3-month transplant in Fig. 13. Figure 14 shows a normal adrenergic nerve plexus of iris from an adult rat. ×350.

If aqueous humor would contain a sufficient concentration of HC or another glucocorticoid, it would be readily understandable that the SIF cells in the ganglion transplanted into the eye survive because SIF cells induced *in vivo* also survive in ganglia cultured *in vitro* if the medium contains HC (6). We have not found any information on the HC concentration of the aqueous humor, but it is unlikely that it is of the same magnitude as that employed in HC-containing cultures, 10 mg/liter. However, this concentration is high enough to cause the appearance of newly formed SIF cells *in vitro,* and it is likely that lower concentrations would be sufficient for the maintenance of already existing SIF cells.

It is also possible that hormones or substances other than HC supporting the survival of SIF cells in transplanted ganglia may be present in the aqueous humor. It has been shown that large numbers of SIF cells are induced in the normal ganglia of newborn rats cultured in a medium containing human cord (placental) or human maternal serum instead of the normally used fetal calf serum but no additional HC (6). These sera may contain hormones or substances other than HC which induce, and in lower concentration maintain, the SIF cells. Some such substances may also be present in the aqueous humor.

The possibility must also be considered that proximity to the iris of the recipient eye and the different types of nerve fiber innervating it may somehow help the SIF cells to survive better in the eye than in culture. Further studies are obviously necessary to clarify this point.

Distribution Patterns of SIF Cells

Although newly formed SIF cells are mainly present individually and in small clusters throughout the ganglia of developing rats injected with HC (2,3), they first formed half-moon-shaped clusters and then tight conglomerates of numerous SIF cells in the ganglia transplanted into the anterior chamber of the eye. A similar tendency to SIF cell clustering has been observed under several conditions: in nontrypsinized organ cultures of newborn rat ganglia in a culture medium containing HC, human cord serum, or human maternal serum (6); in ganglia of developing rats whose sympathetic neurons had been killed by repeated injections of guanethidine (1); in decentralized ganglia of HC-injected developing rats (3); and in developing rat ganglia whose postganglionic trunks had been divided soon after birth (8). Under all these conditions, the normal development of sympathetic nerve cells is greatly disturbed, the number of nerve cell bodies being much smaller than that in the corresponding control ganglia. It seems possible that SIF cells seek contact with other SIF cells when fewer neurons are available and thus form large clusters.

ACKNOWLEDGMENTS

We are very grateful to Mrs. Maija Järvinen and Ms. Asta Irla for skillful technical assistance. This work was supported by a grant from the Sigrid Jusélius Foundation to O.E.

REFERENCES

1. Eränkö, L., and Eränkö, O. (1971): Effect of guanethidine on nerve cells and small intensely fluorescent cells in sympathetic ganglia of newborn and adult rats. *Acta Pharmacol. Toxicol. (Kbh),* 30:403–416.
2. Eränkö, L., and Eränkö, O. (1972): Effect of hydrocortisone on histochemically demonstrable catecholamines in the sympathetic ganglia and extra-adrenal chromaffin tissue of the rat. *Acta Physiol. Scand.,* 84:125–133.
3. Eränkö, L., and Eränkö, O. (1979): SIF cells in decentralized developing sympathetic ganglia. In: *Catecholamines: Basic and Clinical Frontiers,* edited by E. Usdin, I.J. Kopin, and J. Barchas, pp. 842–844. Pergamon Press, New York.
4. Eränkö, O. (1967): The practical histochemical demonstration of catecholamines by formaldehyde-induced fluorescence. *J. R. Microsc. Soc.,* 87:259–276.

5. Eränkö, O., Eränkö, L., Hill, C.E., and Burnstock, G. (1972): Hydrocortisone-induced increase in the number of small intensely fluorescent cells and their histochemically demonstrable catecholamine content in cultures of sympathetic ganglia of the newborn rat. *Histochemistry,* 4:49-58.

6. Eränkö, O., Eränkö, L. (1980): Induction of SIF cells by hydrocortisone or human cord serum in sympathetic ganglia and their subsequent fate in vivo and in vitro. *This volume.*

7. Olson, L., and Malmfors, T. (1970): Growth characteristics of adrenergic nerves in the adult rat. *Acta Physiol. Scand. [Suppl. 348],* 4-112.

8. Soinila, S., Eränkö, O. (1980): Effect of postganglionic nerve division on developing and adult SIF cells. *This volume.*

Histochemistry and Cell Biology of
Autonomic Neurons, SIF Cells, and
Paraneurons,
edited by O. Eränkö et al.
Raven Press, New York © 1980.

Effect of Postganglionic Nerve Division on Developing and Adult SIF Cells

Seppo Soinila and Olavi Eränkö

Department of Anatomy, University of Helsinki, 00170 Helsinki 17, Finland

In a previous study (4) it was demonstrated that the number of SIF cells in the superior cervical ganglion of the rat increases from about 200 cells/ggl at birth to about 700 cells/ggl in 23-day-old rats. Thereafter there was little change in the SIF cell number. Division of the postganglionic nerves did not essentially affect the increase in the number of SIF cells, although almost 90% of the total nerve cell population had disappeared by 5 days after axotomy (6,7).

Large conglomerates of SIF cells were observed in the axotomized ganglia (4). These conglomerates and their formation are further investigated in the present study in which axotomy was performed in 3-day-old and adult rats. Postganglionic nerve division is called "axotomy" in this chapter, although it is realized that the processes of the SIF cells may not have been divided by this operation.

MATERIAL AND METHODS

The main efferent branches of the left superior cervical ganglion of 3-day-old and adult rats were cut under ether anesthesia. The right ganglia were left intact and served as controls. The animals were anesthetized with ether and decapitated 5, 10, 20, 60, 80, 144, or 175 days after operation. The ganglia

were removed, frozen in liquid propane precooled with liquid nitrogen, and processed for fluorescence microscopy as described earlier (4). The sections were examined with a Leitz Dialux 20 fluorescence microscope using an HBO 50 mercury lamp with filter block D and emission filter K 470. Photographs were taken with Leitz Orthomat camera on Kodak Tri-X film.

For SIF cell counting, serial sections were cut at 6 μm and the nuclear profiles of all SIF cells counted. Correction of the error caused by section thickness was obtained by the Floderus formula (3). The statistical analysis was carried out with Student's t-test (2).

RESULTS

Axotomy in 3-Day-Old Rats

In 5-day-old rats (2 days after operation), the axotomized ganglia showed fewer nerve cell bodies than the control ganglia, especially in the distal part. The SIF cells were mostly solitary in control and axotomized ganglia, and SIF cell clusters were few. No essential change in the distribution of SIF cells was observed between operated and control ganglia.

In 8-day-old rats (5 days after operation), reduced neuron density was marked throughout the axotomized ganglia. The SIF cells were still scattered mainly as solitary cells, only few clusters being observed. Their distribution in the operated ganglia was not essentially different from that in the control ganglia.

In 13-day-old rats (10 days after operation), the SIF cells were mainly in clusters, rather than solitary cells. Visual examination did not reveal any differences in the SIF cell patterns between control and axotomized ganglia.

In 23-day-old rats (20 days after operation), cluster formation was further advanced (Figs. 1 and 2). Many SIF cell clusters were unusually large in the axotomized ganglia, especially those in the distal end of the ganglion (compare Figs. 3 and 4) and often showed irregular outlines. The SIF cell clusters were frequently seen in the immediate vicinity of blood vessels, especially in the axotomized ganglia.

In 63-day-old rats (60 days after operation), many very large, composite SIF cell clusters were seen in the axotomized ganglia (Fig. 6), distinctly different from those in the control ganglia (Fig. 5). In a complete series of sections through a SIF cell cluster in an axotomized ganglion, a total of 467 SIF cells (corrected value) were found.

In adult rats (144 and 175 days after operation), large SIF cell clusters were conspicuous in the axotomized ganglia, as compared with the corresponding control ganglia.

Axotomy in Adult Rats

In the operated ganglia (5, 10, 20, and 80 days after axotomy) the number of nerve cells and the fluorescence intensity of individual nerve cell bodies were reduced. These changes were obvious by mere visual examination.

The mean number of SIF cells and the standard error of the mean are shown in Table 1. It is evident that axotomy did not cause any significant change in the number of SIF cells, although it caused a conspicuous loss of nerve cells. Most SIF cells were in clusters in control and axotomized ganglia. They were commonly situated around blood vessels, as well as in regions of the axotomized ganglia where many nerve cells seemed to have disappeared.

No discernible size difference was observed in the SIF cell clusters between control and operated ganglia. However, the pattern of the SIF cell clusters was affected by axotomy: Five and 10 days after the operation the SIF cells formed compact clusters, as in the control ganglia, but 20 days after the operation several irregularly shaped groups of SIF cells exhibiting a variable fluorescence intensity were common, especially in the areas of reduced nerve cell density (Fig. 8). Such SIF cell patterns were never seen in the control ganglia (Fig. 7). In rats killed 80 days after axotomy, the appearance of the SIF cell clusters had returned to normal.

TABLE 1. *Effect of postganglionic nerve division in adult rats on the number of SIF cells in the superior cervical ganglion*

Time after operation (days)	Control			Nerve division			
	No.	Mean	SEM	No.	Mean	SEM	p^a
5	6	563	97	6	391	84	ns
10	6	612	92	6	641	66	ns
20	5	589	120	5	710	124	ns
80	4	417	130	5	520	137	ns

[a]The probability *(p)* that the difference of the means was due to mere chance was in all cases larger than 0.1, the difference thus being nonsignificant (ns).

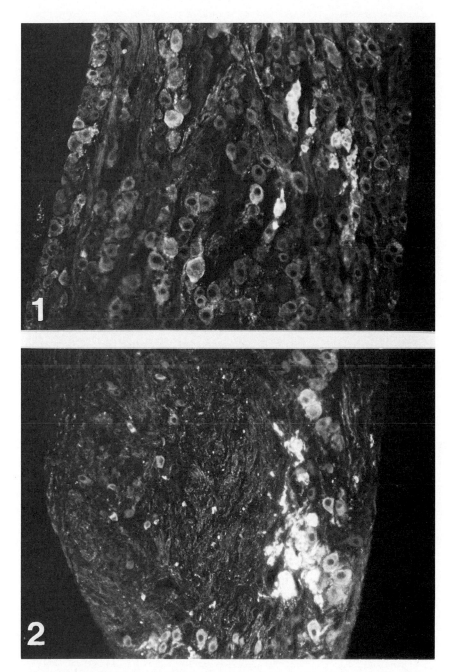

FIG. 1. Formaldehyde-induced catecholamine fluorescence in a midsection of the control ganglion of a 23-day-old rat. ×139.
FIG. 2. Similarly prepared photomicrograph of the midsection of the axotomized ganglion of the same rat as that in Fig. 1. ×139.

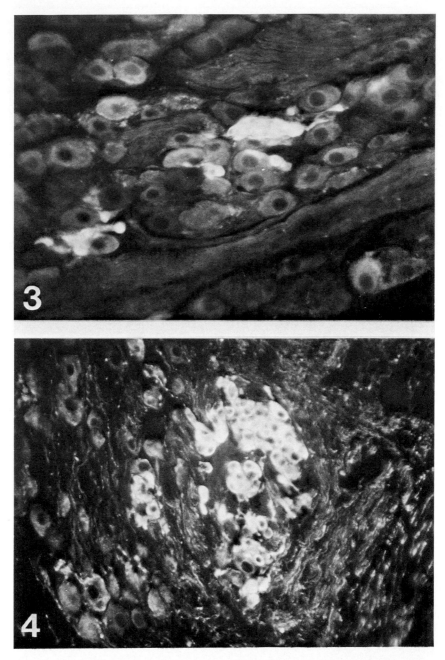

FIG. 3. A typical compact SIF cell cluster in a control ganglion of a 23-day-old rat. ×277.
FIG. 4. A large, irregularly outlined SIF cell cluster in an axotomized ganglion of a 23-day-old rat. ×277.

DISCUSSION

Axotomy in Developing Rats

Axotomy in young rats is known to cause a rapid, irreversible destruction of most neu- rons (6,7). Five days or more after division of the postganglionic nerve trunks, only a small fraction of the original number of nerve cell bodies is left in the superior cervical ganglion (4). However, postganglionic nerve division

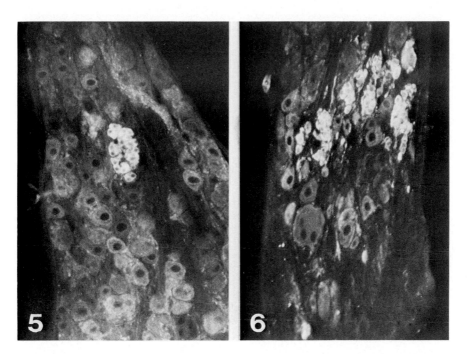

FIG. 5. Control ganglion of a 63-day-old rat showing one compact SIF cell cluster. ×146.
FIG. 6. An SIF cell conglomerate in the axotomized ganglion of a 63-day-old rat consisting of several small clusters. ×146.

does not essentially affect the normal postnatal development of SIF cells, whose number greatly increases after birth in control and operated rats, reaching the adult level around the 20th postnatal day (4).

In the present study, special attention was paid to the distribution patterns and the interrelations of SIF cells in axotomized ganglia. In this respect, little difference was found between the control and axotomized ganglia 2, 5, or 10 days after operation in spite of the drastic changes in the principal neurons. However, dramatic SIF cell changes occurred later. Large, irregularly outlined SIF cell clusters appeared 20 days after axotomy, and the SIF cell clusters became very large 60 days after axotomy, clearly exceeding the size the largest control clusters ever seen.

Since these changes were late, temporary operation stress was clearly not responsible for the observed SIF cell hyperplasia, even if it may have caused an elevation of blood corticosteroid concentration. Although the development of large SIF cell clusters was a relatively slow process which was apparent first long after the disappearance of nerve cells from the ganglion, it may yet be causally connected with the loss of neurons. The SIF cells, whose number greatly increases also after axotomy, may seek proximity to related cells and, because of the lack of principal neurons, form close contacts with other SIF cells. Such a process would gradually result in the formation of large SIF cell conglomerates.

Axotomy in Adult Rats

In the superior cervical ganglion of adult rats, the peak reaction of nerve cell bodies to postganglionic nerve division has been reported to develop between the 5th and 20th postoperative day (5,11) or 1 to 3 weeks after operation (8). Partial recovery takes place slowly, and the ultimate degree of cell death

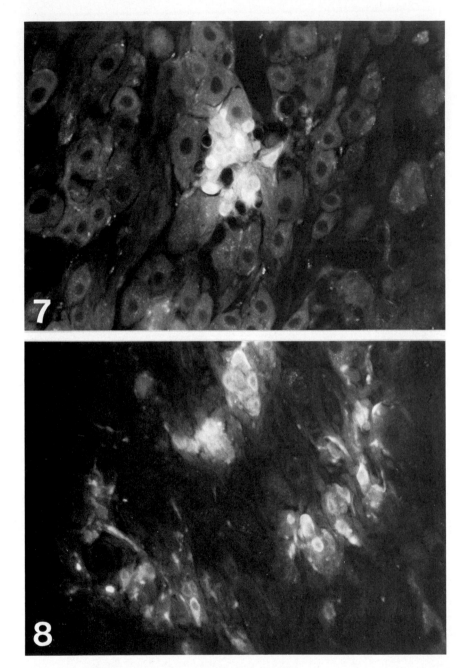

FIG. 7. An SIF cell cluster in an adult control ganglion. ×456.
FIG. 8. Several SIF cell clusters and solitary SIF cells of variable fluorescence intensity in an axotomized ganglion of an adult rat 20 days after axotomy. ×456.

depends on the type and closeness of the lesion (8). Axotomy in adult rats also causes temporary dissociation of synapses from the injured postganglionic neurons (10). This re-

action reaches its maximum intensity between the 3rd and 7th postoperative day, and gradual recovery takes place to a great extent by the 40th postoperative day. Furthermore, after axotomy "apparent synapses *between* small granule-containing cells, rarely seen in normal ganglia, become more numerous" (1).

The present study demonstrates that, in the adult rat, neither the total SIF cell number per ganglion nor their number per cluster is significantly affected by postganglionic nerve division. However, 20 days after operation, when the axon reaction of the neurons is still in intensive progress, the SIF cells of the axotomized ganglia showed a peculiar, "restless" pattern as if the clusters were disintegrating. It is possible that detachment of efferent SIF cell synapses from axotomized postganglionic neurons results in a tendency to seek contact with nonreacting or recovered neurons or other SIF cells.

It is not clear whether SIF cells later establish new contacts or reestablish their efferent synapses with adrenergic neurons. In any case, on the 80th postoperative day the size and form of SIF cell clusters in axotomized ganglia are indistinguishable by fluorescence microscopy from control clusters.

ACKNOWLEDGMENT

This work was supported by a grant from Sigrid Jusélius Foundation to O.E.

REFERENCES

1. Case, C.P., and Matthews, M.R. (1976): Effects of postganglionic axotomy on synaptic connexions of small granule-containing (SG) cells in the rat superior cervical ganglion. *J. Anat.*, 122:732.

2. Diem, K., and Leitner, C., editors (1973): *Scientific Tables,* Ciba-Geigy, Basel.

3. Eränkö, O. (1955): *Quantitative Methods in Histology and Microscopic Histochemistry,* pp. 76–78. Little Brown, Boston.

4. Eränkö, O., and Soinila, S. (1980): Effect of early postnatal division of the postganglionic nerves on the development of principal cells and small intensely fluorescent (SIF) cells in the rat superior cervical ganglion. *J. Neurocytol. (in press).*

5. Härkönen, M. (1965): Carboxylic esterases, oxidative enzymes and catecholamines in the superior cervical ganglion of the rat and the effect of pre- and postganglionic nerve division. *Acta Physiol. Scand. [Suppl. 237],* 63.

6. Hendry, I.A. (1975): The effects of axotomy on the development of the rat superior cervical ganglion. *Brain Res.,* 86:483–487.

7. Hendry, I.A., and Campbell, J. (1976): Morphometric analysis of rat superior cervical ganglion after axotomy and nerve growth factor treatment. *J. Neurocytol.,* 5:351–360.

8. Lieberman, A.R. (1971): The axon reaction: A review of the principal features of perikaryal responses to axon injury. *Int. Rev. Neurobiol.,* 14:49–124.

9. Matthews, M.R. (1976): Synaptic and other relationships of small granule-containing cells (SIF cells) in sympathetic ganglia. In: *Chromaffin, Enterochromaffin and Related Cells,* edited by R.E. Coupland and T. Fujita, pp. 131–146. Elsevier, Amsterdam.

10. Matthews, M.R., and Nelson, V.H. (1975): Detachment of structurally intact nerve endings from chromatolytic neurones of rat superior cervical ganglion during the depression of synaptic transmission induced by postganglionic axotomy. *J. Physiol. (Lond),* 210:11–14P.

11. Matthews, M.R., and Raisman, G. (1972): A light and electron microscopic study of the cellular response to axonal injury in the superior cervical ganglion of the rat. *Proc. R. Soc. Lond. [Biol.],* 181:43–79.

Histochemistry and Cell Biology of Autonomic Neurons, SIF Cells, and Paraneurons,
edited by O. Eränkö et al.
Raven Press, New York © 1980.

Effects of Cell Culture Conditions, Nerve Growth Factor, Dexamethasone, and Cyclic AMP on Adrenal Chromaffin Cells In Vitro

Klaus Unsicker, Bibiana Rieffert, and Wolfgang Ziegler

Department of Anatomy and Cell Biology, Philipps-Universität Marburg, Marburg, FRG

Work in the field of the development of peripheral autonomic neurons has centered on the problem of how nerve cells gain the competence for producing a certain type of transmitter, either acetylcholine (ACh) or norepinephrine (NE) (see ref. 18 for review). Environmental influences from nonneuronal cells and stimuli provided by preganglionic innervation have been shown to channel neural-crest-derived cells into the ACh (parasympathetic) or NE (sympathetic) producing line. Considerable attention has also been paid to the signals which govern the differentiation of the immature adrenergic neuron into its mature form (2,4,5), whereas the differentiation of the endocrine derivatives of the sympathetic system (e.g., the adrenal chromaffin cell) has largely been neglected. We performed *in vitro* studies on the morphological and functional characteristics of postnatal adrenal chromaffin cells which aimed at obtaining information on the factors that keep adrenal chromaffin cells differentiated and eventually cause these cells to express an endocrine instead of a neuronal phenotype. Furthermore, we tried to analyze the conditions which allow the transformation of adrenal medullary endocrine cells into neuron-like cells. This chapter briefly summarizes our results obtained with adrenal chromaffin cells from rats, guinea pigs, and cattle in various tissue culture systems.

METHODOLOGY

Animals

Adrenal glands from young (second postnatal week) Hanover-Wistar rats, newborn guinea pigs, and adult cattle were used for cell and tissue culture.

Preparation Procedures

Adrenal glands were freed from adhering cortical tissue (Fig. 1), and the medulla was dissociated into single cells with collagenase and trypsin or collagenase alone (for details see refs. 15,16). Alternatively, small pieces of medullary tissue were used for *in vitro* explants (14).

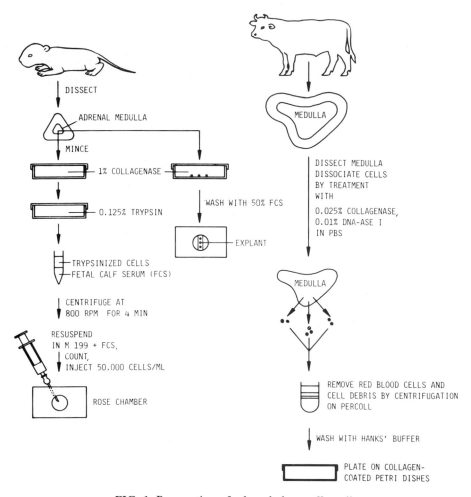

FIG. 1. Preparation of adrenal chromaffin cells.

Tissue Culture Devices

Petri dishes (NUNC, Falcon, Greiner), slide chambers (Labtec®), microwell plates, and modified Rose chambers, as well as glass, plastic, and collagen (prepared from rat tail tendons) surfaces were tested. Rose chambers offered optimum conditions for phase contrast observations and subsequent processing for electron microscopy. They were used for most of the experiments with explants and adrenal medullary cells from young rats and a few experiments with bovine chromaffin cells. The surface area of our

Rose chamber model was approximately 5 cm² and the volume 2 ml. Cells from at least four rat medullas had to be pooled to obtain a sufficient cell density. Thus the number of parallel experiments was limited to about four to six. This disadvantage could be overcome with a slide chamber model and microwell plates, which could be filled with 10,000 to 50,000 cells per well but did not offer adequate possibilities for microcopic examination. Petri dishes (35 to 50 mm) were suitable tools for cultures with cells from bovine adrenal medullas, which yielded 8 to 20 × 10⁶ cells per gland. However, Petri dishes had serious limitations regarding the quality of phase contrast micrographs obtained.

Adrenal chromaffin cells from all species studied showed excellent adherence to and long-term survival on collagen-coated surfaces but did not survive on glass for more than a week.

ADRENAL CHROMAFFIN CELL CULTURES

Effects of Various Sera and Serum Concentrations

For routine studies medium 199 was supplemented with 20% fetal calf serum (FCS; Flow Laboratories). Cultivation of rat and bovine chromaffin cells in serum-free medium resulted in shortened periods of survival (less than 1 week). Whether large doses of NGF may substitute for serum in the culture medium is under investigation. In addition, bovine adrenal medullary cells were grown with human placental serum (HPS) and chick hole embryo extract (CEE) (10% with 10% FCS). HPS and CEE improved the flattening out of chromaffin cells and the extension of processes (Fig. 2), which, however, did not acquire the length and appearance of processes formed by rat chromaffin cells. Omitting FCS and raising the concentration of HPS or CEE to 20% caused massive cell death.

Explants Versus Dissociated Cells

Chromaffin cells in explants of adrenal medullary tissue taken from rats during the second postnatal week spontaneously developed processes within 1 to 2 days (Fig. 3a, b). The cells and their processes showed catecholamine (CA) specific fluorescence after incubation with glyoxylic acid and exhibited ultrastructural features similar, but not identical, to their *in vivo* counterparts (Fig. 3c). They contained the specific chromaffin storage vesicles with cores of high electron densi-

ties, indicating the presence of a primary amine, probably NE or dopamine (DA). Cells with storage granules typical of epinephrine (E) disappeared within 1 to 2 weeks, although they constitute approximately 60% of the chromaffin cells in rat adrenal medulla at this age. There were no indications of a selective death of E-storing cells in culture. Rather, it appears that E-storing cells lost their capacity to form the secondary amine and gradually shifted to the production of a primary amine. Processes of chromaffin cells contained two populations of vesicles within their varicose regions: small clear (50 nm) vesicles with occasional dark, eccentrically located cores, and large granular vesicles (80 to 120 nm) (Fig. 3d). Outgrowth of neurite-like processes from explants of newborn guinea pig adrenal medulla was relatively poor compared to that in explants taken from young rats.

Chromaffin cells dissociated from young rat adrenal medulla with collagenase and trypsin did not extend processes (Fig. 4a) unless they were kept in the presence of large amounts of adrenal fibroblast-like cells.

Chromaffin cells from adult bovine medullas were kept in cultures for up to 4 weeks. They retained a large number of ultramorphological features typical of differentiated chromaffin cells (Fig. 2d) but lost about 50% of their original amine content and stored equal amounts of NE and E from day 9 onward compared to an initial NE/E ratio of approximately 1:6. About 50% of bovine chromaffin cells spontaneously formed short, broad processes (maximum length about 60 μm).

Effects of NGF and NGF Antiserum

Administration of 2.5S NGF 20ng/ml to cultures of young rat adrenal chromaffin cells elicited the formation of processes within 1 week (Fig. 4b–e). A few cells, some of which displayed neuronal characteristics (particularly large nuclei and cell bodies) responded

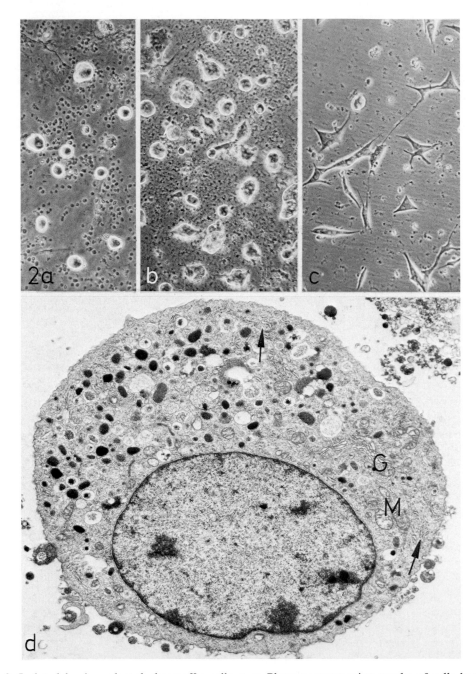

FIG. 2. Isolated bovine adrenal chromaffin cells. **a–c**: Phase contrast micrographs of cells kept in culture for 9 days in medium supplemented with 20% fetal calf serum (**a**), 10% fetal calf serum and 10% chick hole embryo extract (**b**), and 10% fetal calf serum and 10% human placenta serum (**c**). Note the differences in flattening out. ×280. **d**: Ultramorphology of a typical epinephrine-storing cell displaying chromaffin granules, Golgi cisternae (G), mitochondria (M), and profiles of rough endoplasmic reticulum *(arrows)*. ×7,200.

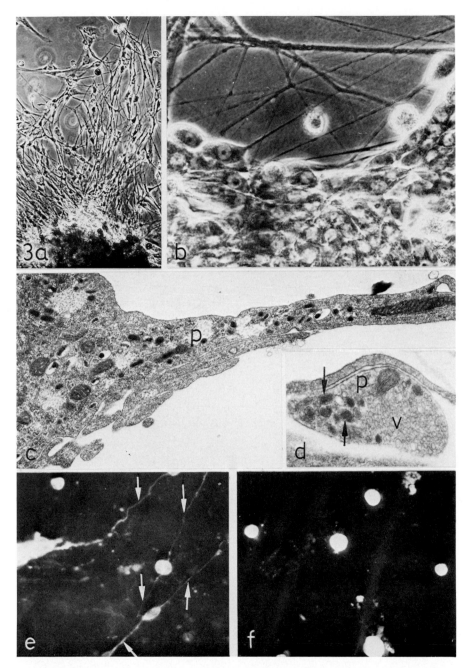

FIG. 3. Rat (second postnatal week) adrenal chromaffin cells in culture. **a, b:** Fiber outgrowth from explants of adrenal medullary tissue. **(a)** ×140. **(b)** ×480. **c, d:** Ultramorphology of chromaffin cell processes (P) containing large chromaffin storage granules *(arrows)* and small clear vesicles (v). **(c)** ×15,000. **(d)** ×36,000. **e, f:** Catecholamine-specific fluorescence of isolated chromaffin cells treated with NGF **(e)** or NGF and cholera toxin **(f)** for 9 days. NGF-treated cells show fluorescent processes *(arrows),* which are absent in cultures treated with NGF and cholera toxin. ×400.

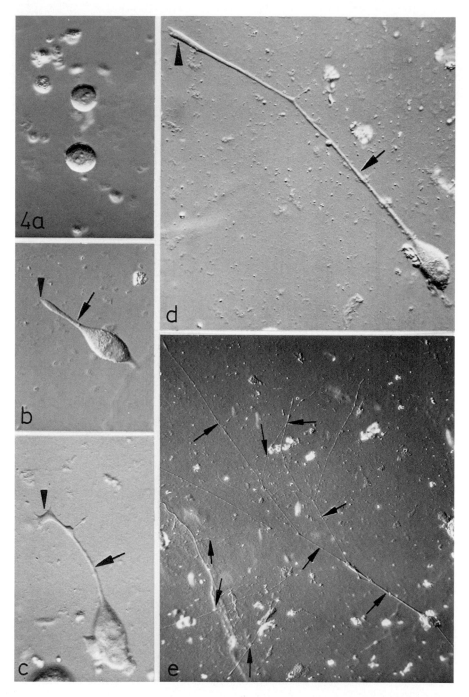

FIG. 4. Isolated adrenal chromaffin cells (rat, second postnatal week) kept under control conditions (**a**) and in the presence of NGF 20 ng/ml (**b–e**) for 4 (**a**), 12 (**b–d**, and 16 (**e**) days. NGF- treated cells extend neurite-like processes *(arrows)* with typical growth cones *(arrowheads)*. Nomarski optics. (**a–d**) × 960. (**e**) ×390.

to NGF within 2 days, but most cells did not extend processes until 5 to 7 days or later. These neurite-like processes grew to several hundred microns (Fig. 4e) and displayed varicosities, extensive ramifications, typical growth cones, and CA-specific histofluorescence even within their growth cones. In addition, NGF apparently facilitated adhesion of chromaffin cells to the collagen surface and increased flattening out. Higher concentrations of NGF (100 to 1,000 ng/ml) enhanced rather than inhibited the process of fiber formation, in contrast to what would have been expected on the basis of results obtained with chicken dorsal root ganglia (6) and sympathetic neurons (8).

Another characteristic response to NGF, which also can be observed under *in vivo* conditions (10), was an increase in tyrosine hydroxylase (TH) activity ($145 \pm 21\%$) after 7 days' treatment with NGF (12 ng/ml), compared to control cultures.

Effects of Dexamethasone and cAMP

Dexamethasone administered to dissociated chromaffin cells from young rats in combination with NGF (12 ng/ml) inhibited fiber outgrowth in a dose-dependent fashion. Whereas concentrations of 10^{-5} and 10^{-6} M completely abolished process formation, 10^{-7} M dexamethasone significantly reduced sprouting of axons. Cells that had been treated with 10^{-5} M dexamethasone alone or in combination with NGF (12 ng/ml) did not contain storage granules typical of E as would have been expected if dexamethasone had led to an induction of the methylating enzyme phenylethanolamine N-methyltransferase. However, the preferential storage of primary amines in dexamethasone-treated chromaffin cells might be interpreted in terms of an unsufficient supply with an important cofactor of the enzyme dopamine-β-hydroxylase, ascorbic acid.

Treatment of explants of young rat adrenal medulla with dibutyrylic cyclic AMP (dbcAMP, 10^{-3} to 1 mM) together with equimolar amounts of theophylline inhibited axon outgrowth in a dose-dependent fashion. Virtually no outgrowth was obtained with 0.1 and 1 mM dbcAMP (13). In addition, cholera toxin, which has been shown to increase intracellular cAMP (7), was administered to fibroblast-enriched or NGF (25 ng/ml) treated cultures of rat chromaffin cells for 7 days (Fig. 3e, f). Cholera toxin completely abolished fiber outgrowth under these conditions. This result favors the concept that cAMP efficiently antagonizes the neuronal transformation of chromaffin cells. Cholera toxin also kept the few adrenocortical cells present in the culture in a differentiated state. Hence it cannot be ruled out that corticosteroids from these cells exert a similar axon growth-inhibiting effect.

CONCLUSIONS

The following conclusions can be drawn from the results obtained by now with chromaffin cells from postnatal mammals in cultures.

1. Adrenal chromaffin cells may undergo spontaneous or NGF-mediated neuronal transformation. A similar transformation can occur in the adrenal medulla *in situ*, as shown by Aloe and Levi-Montalcini (1), if high doses of NGF are administered to fetal rats. Neuronally transformed adrenal chromaffin cells display features of sympathetic adrenergic neurons or small granule-containing cells, respectively. Some effects observed with NGF (e.g., increased substratum adhesiveness, fiber outgrowth, and TH induction) are in agreement with the documented effects of NGF on adrenergic neurons and the PC 12 pheochromocytoma cell line (2,8,12). In contrast, NGF apparently does not affect survival in short-term cultures and, in higher doses, does not inhibit fiber outgrowth. Interestingly, fibroblast-like cells seem to be able to substitute for NGF.

Adrenal chromaffin cells from adult bovine medullas do not adopt a neuronal phenotype under the experimental conditions employed by us. Comparing our results obtained with chromaffin cells from *young* rats and *adult* cattle, it might be argued that chromaffin cells can be induced to express neuronal characteristics only during a limited period of prenatal and early postnatal life. However, investigations with transplants of adult rat and guinea pig adrenal medulla into the anterior chamber of the eye have clearly shown that even adult chromaffin cells may form a substantial number of axons (9,17).

2. Our results suggest that glucocorticoids, apart from their well-known role in the induction of various CA-synthesizing enzymes (3,18), contribute to the maintenance of differentiation of adrenal chromaffin cells by preventing them from expressing a neuronal phenotype. Other hitherto unknown factors might have similar effects by acting through cAMP-dependent processes.

SUMMARY

Adrenal chromaffin cells from young rats, newborn guinea pigs, and adult cattle were grown under various cell culture conditions. Nerve growth factor (NGF) elicited the outgrowth of neurite-like processes from rat, but not from bovine and guinea pig, chromaffin cells. Dexamethasone, dibutyryl cAMP, theophylline, and cholera toxin, which are known to augment intracellular cAMP, specifically inhibited neurite outgrowth. We propose that glucocorticoid hormones and mechanisms which act through an increase of cAMP are prerequisites for the acquisition and maintenance of differentiation and expression of an endocrine instead of a neuronal phenotype of adrenal chromaffin cells.

ACKNOWLEDGMENTS

This work was supported by a grant from Deutsche Forschungsgemeinschaft (Un 34/6). Cholera toxin was a generous gift to Dr. W. Ziegler from Dr. H.J. Stärk, Behringwerke A.G., Marburg. We thank Mr. C. Fiebiger, Mr. W. Lorenz, Mrs. H. Reichert, Mrs. C. Riehl, and Mrs. H. Schneider for skillful technical assistance and editorial help.

REFERENCES

1. Aloe, L., and Levi-Montalcini, R. (1979): Nerve growth factor-induced transformation of immature chromaffin cells in vivo into sympathetic neurons: Effect of antiserum to nerve growth factor. *Proc. Natl. Acad. Sci. USA,* 76:1246–1250.
2. Black, I.B. (1978): Sympathetic neurone development. *TINS,* 1:101–104.
3. Ciaranello, R.D., Wooten, F.G., and Axelrod, J. (1976): Regulation of rat adrenal dopamine β-hydroxylase. II. Receptor interaction in the regulation of enzyme synthesis and degradation. *Brain Res.,* 113:349–362.
4. Coughlin, M.D., Boyer, D.M., and Black, I.B. (1977): Embryologic development of a mouse sympathetic ganglion in vivo and in vitro. *Proc. Natl. Acad. Sci. USA,* 74:3438–3442.
5. Coughlin, M.D., Dibner, M.D., Boyer, D.M., and Black, I.B. (1978): Factors regulating development of an embryonic mouse sympathetic ganglion. *Dev. Biol.,* 66:513–528.
6. Fenton, E.L. (1970): Tissue culture assay of nerve growth factor and of the specific antiserum. *Exp. Cell Res.,* 59:383–392.
7. Gill, D.M., and King, C.A. (1977): The mechanism of action of cholera toxin. *J. Biol. Chem.,* 250:6424–6432.
8. Hill, C.E., and Hendry, I.A. (1976): Differences in sensitivity to nerve growth factor of axon formation and tyrosine hydroxylase induction in cultured sympathetic neurons. *Neuroscience,* 1:489–496.
9. Olson, L., and Malmfors, T. (1970): Growth characteristics of adrenergic nerves in the adult rat. *Acta Physiol. Scand. [Suppl. 348],* 1–112.
10. Otten, U., Schwab, M., Gagnon, C., and Thoenen, H. (1977): Selective induction of tyrosine hydroxylase and dopamine β-hydroxylase by nerve growth factor: Comparison between adrenal medullae and sympathetic ganglia of adult and newborn rats. *Brain Res.,* 133:291–301.
11. Patterson, P.H. (1978): Environmental determination of autonomic neurotransmitter functions. *Ann. Rev. Neurosci.,* 1:1–17.
12. Schubert, D., Lacorbiere, M., Whitlock, C., and Stallcup, W. (1978): Alterations in the surface properties of cells responsive to nerve growth factor. *Nature,* 273:718–722.
13. Unsicker, K., and Chamley, J.H. (1976): Effects of dbcAMP and theophylline on rat adrenal medulla grown in tissue culture. *Histochemistry,* 46:197–201.
14. Unsicker, K., and Chamley, J.H. (1976): Growth characteristics of postnatal rat adrenal medulla in culture. *Cell Tissue Res.,* 177:247–268.
15. Unsicker, K., Griesser, G-H., Lindmar, R., Löffelholz, K., and Wolf, U. (1979): Establishment and characterization of isolated bovine adrenal chro-

maffin cells in long-term cultures. *Neuroscience (submitted).*

16. Unsicker, K., Krisch, B., Otten, U., and Thoenen, H. (1978): Nerve growth factor-induced fiber outgrowth from isolated rat adrenal chromaffin cells: Impairment by glucocorticoids. *Proc. Natl. Acad. Sci. USA,* 75:3498–3502.

17. Unsicker, K., Tschechne, B., and Tschechne, D. (1978): Formation of cholinergic synapses on adrenal chromaffin cells in anterior eye chamber transplants. *Brain Res.,* 152:334–340.

18. Wurtmann, R.J., and Axelrod, J. (1965): Adrenaline synthesis: Control by the pituitary gland and adrenal glucocorticoids. *Science,* 150:1464–1465.

Histochemistry and Cell Biology of Autonomic Neurons, SIF Cells, and Paraneurons,
edited by O. Eränkö et al.
Raven Press, New York © 1980.

Phenotypic Plasticity of Pheochromocytoma and Normal Adrenal Medullary Cells

*Arthur S. Tischler and †Lloyd A. Greene

*Department of Pathology, Tufts University School of Medicine, Boston, Massachusetts 02111; and
†Department of Pharmacology, New York University Medical Center, New York, New York 10016

Traditional models of the histogenesis of the adrenal medulla and its tumors posit divergent paths of development for chromaffin cells and neurons. According to these models, stem cells known as "sympathogonia" or "primitive sympathetic cells" migrate into the adrenal medullary region during embryogenesis and thereafter become committed to mature into either sympathetic neurons or chromaffin cells (3). Maturation is accompanied by increasing cell size, and increasing quantity and complexity of cytoplasm. Maturing neuronal cells form abundant filaments, tubules, and aggregates of rough endoplasmic reticulum (RER); extend processes; and develop two types of amine-storing vesicles: large granular vesicles (LGV) about 70 to 120 nm in diameter and small synaptic-like vesicles about 30 to 60 nm in diameter. Maturing chromaffin cells remain smaller than neurons, lack processes, contain less RER, and develop characteristic catecholamine-storing granules larger and more numerous than LGV in neurons. These granules, the most distinctive ultrastructural feature of chromaffin cells, range in diameter from about 50 to 350 nm, with a mean of about 200 nm in adult rats and humans (4,21). Tumors showing increasing morphological evidence of neuronal maturation are termed neuroblastomas, ganglioneuroblastomas, and ganglioneuromas, whereas tumors with chromaffin cell features are grouped as pheochromocytomas (21). The preceding concepts may also be applied to sympathetic ganglia. In ganglia, however, the chromaffin-like small intensely fluorescent (SIF) or small granule-containing (SGC) cells are a minority of the total population, and typically have granules about half the size of those in adrenal chromaffin cells (7). Neuronal and chromaffin-cell-like tumors arise in extra-adrenal locations as well as in the adrenal medulla.

Recent ultrastructural studies of normal tissues suggest that such dichotomous developmental models are simplistic. The adrenal medulla of several species has been reported to include a small population of cells which contain smaller granules than those of epinephrine (E) or norepinephrine (NE) storing chromaffin cells, and which resemble SIF cells (14). It has also been reported that sympathetic ganglia may contain at least two types of SIF cell, one of which has granules in the same size range as those in adrenal chromaffin cells (7). Further, SIF cells in the adrenal and in ganglia may contain synaptic-like vesicles in addition to larger granules, and may form short processes *in vivo* and *in vitro* (7,14).

During the past 4 years it has become apparent that cultured cells from human and rat pheochromocytomas can acquire morphological and functional characteristics of neurons. This chapter discusses evidence indicating that pheochromocytoma plasticity is a reflection of pluripotency retained by at least some normal chromaffin cells.

PLASTICITY OF PHEOCHROMOCYTOMA CELLS

Phenotypic plasticity of pheochromocytoma cells was initially reported in primary cultures of five human adrenal pheochromocytomas (25) and of a transplantable rat adrenal pheochromocytoma (22). In these studies, addition of nerve growth factor (NGF) to the culture medium promoted extensive neurite outgrowth. A small percentage of cells in cultures of the rat tumor (5,22) and of some but not all human tumors (13,25) also formed processes without added NGF. It was not determined whether the latter finding resulted from endogenous NGF production by fibroblasts or by the tumor cells themselves, or other mechanisms. These observations were rapidly followed by establishment of a clonal cell line, designated PC12, from the rat pheochromocytoma, which had been employed in the initial studies (10). The PC12 line was subsequently investigated intensively, and its characteristics were recently reviewed in detail (9,11).

In the absence of NGF, PC12 cells have no processes. They contain fluorogenic stores of NE and dopamine (DA), and varying numbers of membrane-bound cytoplasmic granules measuring up to about 350 nm. After glutaraldehyde and osmium fixation, granules of varying electron density are often seen in the same cells (10,23). In this regard PC12 cells resemble immature rather than fully developed chromaffin cells (4). In the adult rat adrenal medulla, "dark" granules have been correlated with NE and "light" granules with E content, and the two types are located in separate cells. PC12 cells do

not contain E. Their lighter granules might contain DA or NE in small amounts, or other substances. Granules resembling those in PC12 cells are also seen in SIF cells (7).

Unlike many normal chromaffin cells, PC12 cells do not show induction of phenylethanolamine-N-methyltransferase by corticosteroids (10). Corticosteroids do, however, increase specific activity of tyrosine hydroxylase (TH) (6). A number of other influences on synthesis and storage of catecholamines have also been demonstrated in PC12 cells. Specific activity of TH is dependent on culture density and is three- to fourfold higher in crowded than in sparse cultures (6,9). Effects on TH also appear to mediate short-term stimulation of catecholamine synthesis by dibutyryl cyclic AMP (dbcAMP) and depolarizing concentrations of K^+ (9). In addition, synthesis of NE by PC12 cells appears to be dependent on an adequate supply of reduced ascorbate, a cofactor for dopamine-β-hydroxylase. Thus in contrast to normal chromaffin cells and to PC12 cells growing *in vivo* as tumors, cultured PC12 cells contain about three times as much DA as NE (10). Addition of reduced ascorbate to the culture medium, however, is followed by a rapid 10-fold increase in the rate of NE accumulation (9). This finding is of interest in view of the high concentration of ascorbate, as well as steroids, in the adrenal cortex.

Addition of NGF to PC12 cultures is followed by inhibition of cell division, an increase in cell size, and the formation of neurites. The latter contain variable numbers of LGV (measuring up to about 120 nm) and synaptic-like vesicles, often in clusters. In early process formation intense formaldehyde-induced fluorescence (FIF) can be observed in cell bodies and processes. Later there is a marked reduction of FIF and a decreased number of large "chromaffin" granules in process-forming cells. The cell bodies as well as processes of NGF-treated cells continue to contain small numbers of LGV measuring up to about 120 nm, often in clusters with synaptic-like vesicles (9,10,23).

Concomitant with the acquisition of neuronal morphological features, PC12 cells treated with NGF become more "neuron-like" physiologically. They show increased specific activity of choline acetyltransferase (CAT) (6,9), as do normal neonatal rat sympathetic neurons in culture (19), and they develop an ability to generate all-or-nothing sodium action potentials. The latter are demonstrable by microelectrode recording and release of catecholamines in response to the sodium action potential ionophore activator veratridine (9). Quantitative alterations of catecholamine uptake and release (9), cell surface changes (2), and increased synthesis of NILE, a specific glycoprotein antigen (11), also occur.

Induction of neurite outgrowth in PC12 cells is a highly specific effect of NGF and is not mimicked by other substances (10). We have observed occasional cells with short cytoplasmic extensions in the presence of dbcAMP and related compounds (11). These extensions, however, are seldom longer than 20 to 30 μm, whereas NGF-induced neurites typically are hundreds of micrometers long (10). In contrast to catecholamine storage, neurite outgrowth from PC12 cells correlates negatively with plating density. In sparse cultures virtually all cells form processes. At higher densities, particularly when cells are in contact with one another on all sides, neurite outgrowth appears to be inhibited (9). Intensely fluorescent cells and cells containing many large granules can often be found within large cell clusters after prolonged NGF treatment (23). High concentrations of glucocorticoids do not inhibit neurite outgrowth, either by PC12 cells or by cultured cells of the tumor from which the PC12 line was derived (22), but they do inhibit NGF-induced increases in CAT (6). Removal of NGF from PC12 cultures causes degeneration of processes, but the cells remain viable and resume division (10).

One possible interpretation of the preceding findings is that pheochromocytoma cells exhibit properties of chromaffin cells and neurons as a manifestation of neoplasia, having partially regained the pluripotency of a primitive precursor to both cell types. Another possibility is that the plasticity of pheochromocytoma cells reflects pluripotency normally retained by differentiated chromaffin cells. Several studies in general now favor the latter possibility.

PLASTICITY OF NORMAL ADRENAL MEDULLARY CELLS

Immature Adrenal Medullary Cells In Vivo and In Vitro

Explants of mouse embryonic (18), human fetal (12), and rat postnatal (26) adrenal medullas in organ culture have been reported to contain cells which form processes even in the absence of exogenous NGF. Addition of NGF to cultures of explanted (26) or enzymatically dissociated (27) chromaffin cells from postnatal rats, however, markedly enhances neurite outgrowth. Further, injection of pre- and postnatal rats with NGF causes transformation of immature chromaffin cells *in vivo* into sympathetic neurons (1). Electron microscopic study of the cultured mouse and rat cells reveals that in addition to forming processes these cells have smaller granules than normal chromaffin cells and also contain synaptic-like vesicles (18,26,27). They thus resemble SIF cells of sympathetic ganglia. NGF-treated chromaffin cells *in vivo* also have ultrastructural features intermediate between chromaffin cells and neurons (1). NGF-induced neurite outgrowth from cultured postnatal rat chromaffin cells has been reported to be inhibited by high concentrations of corticosteroids (27), an effect consistent with the suggestion by Kuntz (15) and experimental evidence by Lempinen (16) that hormones from the adrenal cortex might promote differentiation of immature sympathetic cells into chromaffin cells.

In contrast to PC12 cells, immature rat chromaffin cells *in vivo* appear to require NGF for survival (1). Cultured postnatal rat chromaffin cells, on the other hand, have not been reported to require NGF for survival. Neurite outgrowth from these cultured cells in the absence of exogenous NGF, however, is reportedly inhibited by anti-NGF, suggesting the presence of endogenous NGF-like substances. Inhibition of neurite outgrowth and maintenance of chromaffin granules of typical size and morphology have been reported in these cultures in response to dbcAMP (26).

Adult Adrenal Medullary Cells In Vivo and In Vitro

Without exogenous NGF, transplants of adult guinea pig adrenal medulla to the kidney (28) and rat adrenal medulla to the anterior chamber of the eye (20) reportedly contain many cells which show plasticity comparable to that of immature chromaffin cells *in vivo* and *in vitro*. We are presently studying dissociated adult human and rat chromaffin cells in culture and report here on spontaneous and NGF-induced phenotypic changes.

In our current studies we utilize portions of adult human adrenal glands from radical nephrectomies for renal tumors or endocrine-ablative surgery for breast carcinoma. Only glands free of tumor are studied. After diag-

nostically appropriate histologic sections are taken, the remaining medulla is dissected from the cortex and dissociated in collagenase followed by trypsin plus collagenase. Rat chromaffin cells are prepared similarly. Methodological details are published elsewhere (24). Four adult human adrenals have been studied to date. In all of these, NGF consistently increased the percentage of cells with processes and the mean process length (24). Cell yields and proportional NGF responses, however, have varied by more than an order of magnitude. Our most extensive data have been obtained from the adrenal of a 67-year-old man with a kidney tumor. A total yield of 6×10^4 viable cells was plated on collagen in 35 mm tissue culture dishes at 4×10^3 cells per dish. Medium in replicate dishes was supplemented with 2.5S mouse salivary gland NGF (50 ng/ml), dexamethasone (10^{-6} M), NGF plus dexamethasone, or anti-NGF. In some dishes NGF was added at 11 days *in vitro* (d.i.v.) or removed at 13 d.i.v. Adrenal medullary cells were recognized by phase contrast microscopy as polygonal phase dark cells, and their identity was confirmed by intense FIF and electron microscopy. Percentages of process-forming cells and mean process lengths were calculated by strip counts (8,10) on consecutive

TABLE 1. *Effects of NGF, dexamethasone and anti-NGF on dissociated cells from an adult human adrenal medulla, 11 d.i.v.*[a]

	Cells with processes (%) (mean ± SEM)	Process length (μm) (mean ± SEM)	Total cells scored (No.)
Control medium[b]	5.5 ± 1.5	43.1 ± 5.6	101
NGF[c]	66.8 ± 10.8	185.5 ± 8.2	158
Dexamethasone[d]	7.5 ± 2.5	56.3 ± 8.3	200
NGF + dexamethasone	25.5 ± 3.5	62.4 ± 8.4	200
Anti-NGF[e]	5.4 ± 0.9	45.0 ± 15	76

[a] Means are derived from pooled data from duplicate dishes, with an approximately equal number of cells scored per dish. Only cells with processes longer than 24 μm are scored.

[b] McCoy's 5A medium with 20% fetal calf serum.

[c] 2.5S Mouse salivary gland NGF, 50 ng/ml.

[d] Dexamethasone disodium phosphate in aqueous solution, 10^{-6} M.

[e] Rabbit anti-NGF. The concentration employed was in excess of that required to totally inhibit neurite outgrowth inducible by NGF 10 ng/ml in the PC12 cell bioassay (8). Conditioned medium from the control cultures produced no neurite outgrowth, either neat or diluted 1:2, in the PC12 cell bioassay.

days in culture. Results are summarized in Table 1.

As in other studies of primary cultures, some process outgrowth was observed in cultures without added NGF. Increased neurite outgrowth began to be evident in NGF-treated cultures by about 1 week, and was marked by 11 to 12 d.i.v. Cultures with 10^{-6} M dexamethasone showed about 60% inhibition of NGF-induced process outgrowth, but not of spontaneous outgrowth. In contrast to the reported effects in rat cultures (27), anti-NGF also failed to inhibit spontaneous outgrowth. The latter finding might reflect differences between immature and mature chromaffin cells, immunologic differences between human and rodent NGF, or other influences. Cultures with NGF added at 11 d.i.v. showed about the same degree of process outgrowth by day 19 as cultures exposed to NGF at day 0 (75% versus 85% of cells with processes), indicating that process-forming cells did not require added NGF for survival. Cultures with NGF removed at day 13 showed only a small and possibly insignificant decrease in the percentage of cells with processes by day 18.

Many ultrastructural changes were comparable to those reported in PC12 cells and in immature normal chromaffin cells. Granule size ranged widely, from 30 to 350 nm, but in cultures with or without dexamethasone the granules tended to become smaller than in adrenal chromaffin cells *in vivo*. Granules of varying electron density were often seen in the same cell, in the bodies and the processes. In contrast to reported findings on cultured rat chromaffin cells (27), microtubule-containing processes could be observed emerging from cells with E- as well as NE-type granules. Additional findings in some cells were aggregates of RER in cells with large chromaffin granules (24) and accumulations of 30 to 50-nm agranular synaptic-like vesicles, often associated with tubular reticulum, as in early PC12 cell processes (17) (Fig. 1). Similar vesicles in cholinergic neurons and in acetylcholine (ACh) producing PC12 cells cannot be labeled with the false neurotransmitter 5-hydroxydopamine (5-OHDA) (23). Comparable unlabeled vesicles occur in adult guinea pig chromaffin cells transplanted to the kidney (28), although ACh synthesis has not been reported in these transplants. Although 5-OHDA labeling and CAT activity have not yet been studied in cultured adult chromaffin cells, present evidence is consistent with the possibility that ACh synthesis can be induced in these cells under appropriate environmental conditions.

In preliminary studies, dissociated adult rat medullary cells show spontaneous and NGF-induced process outgrowth comparable to that of human cells, but a smaller proportion of cells (about 15%) respond to NGF by 11 d.i.v. (24).

SUMMARY

Much of the phenotypic plasticity exhibited by pheochromocytoma cells is shown by normal immature chromaffin cells. Considerable plasticity is also retained by human adrenal medullary cells and probably by adrenal medullary cells of other species in adult life. Many of the cells which exhibit this plasticity have or acquire characteristics of SIF cells. There are a number of unexplained differences among pheochromocytoma, immature and adult chromaffin cells *in vivo* and *in vitro,* particularly with regard to NGF requirements for cell survival and maintenance of processes, the role of endogenous NGF-like substances in promoting neurite outgrowth, glucocorticoid modulation of NGF responsiveness, and selective enzyme induction by NGF and steroids. Continued parallel studies of pheochromocytoma lines and normal chromaffin cells might help to clarify morphological and functional interrelationships within and between classes of sympathetic cells, and to elucidate mechanisms which regulate growth and phenotypic expression in these cells in normal and neoplastic states.

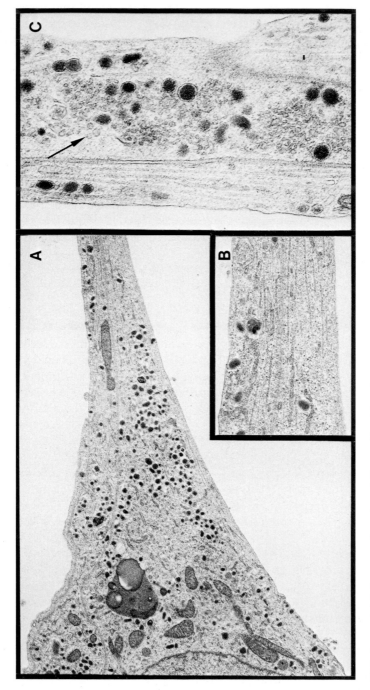

FIG. 1. **a**: Adult human adrenal medullary cells with NGF + 10⁻⁶ M dexamethasone, 14 d.i.v. In the lower cell many pleomorphic granules of varying electron density, 40 to 200 nm in diameter, are accumulated near the origin of a process. ×10,400. **B**: Higher magnification of the extreme right of **A**, where processes begin to show prominent parallel arrays of microtubules. ×18,800. **C**: Fascicle of processes from cells with NGF 14 d.i.v. One process contains agranular synaptic-like vesicles (*arrow*), 60 to 150 nm granules, and some tubular reticulum. ×31,000.

ADDENDUM

In an abstract, process outgrowth has been reported from cultured bovine adrenal medullary cells. These processes may contain either E or NE-type granules (Livett, B.G., Deanne, M., and Bray, G.M. [1978]: *Neurosci. Abstr.*, 4:592).

ACKNOWLEDGMENTS

Supported by grant 1 R23 CA27808-01 PTHA, awarded by the National Cancer Institute, DHEW (AST); NIH grant 5-S07-RR-005589-13 (AST), USPHS grant NS 1157, and The National Foundation, March of Dimes (L.A.G.) and NIH grant CA47389 (H. Wolfe and R.A. DeLellis). The authors thank Drs. Louise Edds, Ronald DeLellis, and Hubert Wolfe, and Mr. Bernard Biales, for helpful discussion.

REFERENCES

1. Aloe, L., and Levi-Montalcini, R. (1979): Nerve growth factor-induced transformation of immature chromaffin cells in vivo into sympathetic neurons: Effect of antiserum to nerve growth factor. *Proc. Natl. Acad. Sci. USA*, 76:1246-1250.

2. Connolly, J.L., Greene, L.A., Viscarello, R.R., and Riley, W.D. (1979): Rapid sequential changes in surface morphology of PC12 pheochromocytoma cells in response to nerve growth factor. *J. Cell Biol.*, 82:820-827.

3. Coupland, R.E. (1965): *The Natural History of the Chromaffin Cell*, pp. 47-87. Longmans, London.

4. Coupland, R.E. (1971): Observations on the form and size distribution of chromaffin granules and on identity of adrenaline and nonadrenaline-storing chromaffin cells in vertebrates and man. *Mem. Soc. Endocrinol.*, 19:611-635.

5. DeLellis, R.A., Merk, F.B., Deckers, P., Warren, S., and Balogh, K. (1973): Ultrastructure and in vitro growth characteristics of a transplantable rat pheochromocytoma. *Cancer*, 32:227-235.

6. Edgar, D.H., and Thoenen, H. (1978): Selective enzyme induction in a nerve growth factor-responsive pheochromocytoma cell line. *Brain Res.*, 154:186-190.

7. Eränkö, O., editor (1976): *SIF Cells. Structure and Function of the Small Intensely Fluorescent Sympathetic Cells.* Fogarty International Center Proceedings No. 30. DHEW Publication No. (NIH)

76-942. Government Printing Office, Washington, D.C.

8. Greene, L.A. (1977): A quantitative bioassay for nerve growth factor (NGF) employing a clonal pheochromocytoma cell line. *Brain Res.*, 133:350-353.

9. Greene, L.A. (1978): NGF-responsive clonal PC12 pheochromocytoma cells as tools for neuropharmacologic investigations. *Adv. Pharamacol. Ther.*, 10:197-206.

10. Greene, L.A., and Tischler, A.S. (1976): Establishment of a noradrenergic clonal line of rat adrenal pheochromocytoma cells which respond to nerve growth factor. *Proc. Natl. Acad. Sci. USA*, 73:2424-2428.

11. Greene, L.A., Burstein, D.E., McGuire, J.C., and Black, M. (1979): Cell culture studies on the mechanism of action of nerve growth factor. In: *Society for Neuroscience Symposia*, Vol. 4, edited by J.A. Ferrendelli; pp. 153-171. Society for Neuroscience, Bethesda.

12. Hervonen, A., and Kanerva, L. (1973): Neuronal differentiation in human fetal adrenal medulla. *Int. J. Neurosci.*, 5:43-46.

13. Kadin, M.E., and Bensch, K.G. (1971): Comparison of pheochromocytes with ganglion cells and neuroblasts grown in vitro. *Cancer*, 27:1148-1160.

14. Kobayashi, S. (1977): Adrenal medulla: Chromaffin cells as paraneurons. *Arch Histol. Jpn. (Suppl.)*, 40.61-79.

15. Kuntz, A. (1912): The development of adrenals in the turtle. *Am. J. Anat.*, 13:71-89.

16. Lempinen, M. (1964): Extra adrenal chromaffin tissue of the rat and the effect of cortical hormones on it. *Acta Physiol. Scand. [Suppl.]*, 62:231.

17. Luckenbill-Edds, L., Van Horn, C., and Greene, L.A. (1979): Fine structure of initial outgrowth of processes induced in a pheochromocytoma cell line (PC12) by nerve growth factor. *J. Neurocytol*, 8:493-511.

18. Manuelidis, L., and Manuelidis, E. E. (1975): Synaptic boutons and neuron-like cells in isolated adrenal gland cultures. *Brain Res.*, 96:181-186.

19. O'Lague, P.H., Obata, K., Claude, P., Furshpan, E.J., and Potter, D.D. (1974): Evidence for cholinergic synapses between dissociated rat sympathetic neurons in culture. *Proc. Natl. Acad. Sci. USA*, 71:3602-3606.

20. Olson, L. (1970): Fluorescence histochemical evidence for axonal growth and secretion from transplanted adrenal medullary tissue. *Histochemie*, 22:1-7.

21. Tannenbaum, M. (1970): Ultrastructural pathology of adrenal medullary tumors. In: *Pathology Annual,* edited by S.C. Sommers, pp. 145-171. Appleton-Century-Crofts, New York.

22. Tischler, A.S., and Greene, L.A. (1975): Nerve growth factor-induced process formation by cultured rat pheochromocytoma cells. *Nature*, 258:341-342.

23. Tischler, A.S., and Greene, L.A. (1978): Morphologic and cytochemical properties of a clonal line

of rat adrenal pheochromocytoma cells which respond to nerve growth factor. *Lab. Invest.*,
39:77–89.

24. Tischler, A.S., DeLellis, R.A., Biales, B., Nunne-
macher, G., Carraba, V., and Wolfe, H.J. (1979):
Nerve growth factor-induced neurite outgrowth
from normal human chromaffin cells. Submitted
for publication.

25. Tischler, A.S., Dichter, M.A., Biales, B., DeLellis,
R.A., and Wolfe, H.J. (1976): Neural properties of
cultured human endocrine tumor cells of proposed
neural crest origin. *Science,* 192:902–904.

26. Unsicker, K., and Chamley, J.H. (1977): Growth
characteristics of postnatal rat adrenal medulla in
culture. *Cell Tissue Res.,* 177:247–268.

27. Unsicker, K., Krisch, B., Otten, U., and Thoenen,
H. (1978): Nerve growth factor-induced fiber out-
growth from isolated rat adrenal chromaffin cells:
Impairment by glucocorticoids. *Proc. Natl. Acad.
Sci. USA,* 75:3498–3502.

28. Unsicker, K., Zwarg, U., and Habura, O. (1977):
Electron microscopic evidence for the formation of
synapses and synaptoid contacts in adrenal medullary grafts. *Brain Res.,* 120:533–539.

Histochemistry and Cell Biology of Autonomic Neurons, SIF Cells, and Paraneurons,
edited by O. Eränkö et al.
Raven Press, New York © 1980.

Factors Influencing Transmitter Type in Sympathetic Ganglion Cells

Caryl E. Hill, Ian A. Hendry, and Ian S. McLennan

Department of Pharmacology, John Curtin School of Medical Research, Australian National University, Canberra City ACT 2601, Australia

The superior cervical ganglion (SCG) of the adult rat contains mainly adrenergic neurons. Indications of cholinergic neurons in sympathetic ganglia stem from the description of cholinergic sweat gland responses in the cat (17) and an acetylcholinesterase and catecholamine histofluorescence study of adjacent sections of the SCG in adult rats (19). In the latter study, the number of cholinergic neurons was estimated to be as few as 5%.

When SCG from newborn rats are dissociated and single neurons cultured in the presence of nonneuronal cells or medium conditioned by them, there is a surprisingly large increase in acetylcholine production (14) and cholinergic synapses form between as many as 50% of the total number of neurons (7,10,-11). Similarly, in explants of intact SCG, which contain both neurons and nonneuronal cells, the activity of choline acetyltransferase (CAT), the enzyme involved in the biosynthesis of acetylcholine, increases 30-fold after 14 days in tissue culture (4). The growth of SCG neurons in the absence of nonneuronal cells results in negligible acetylcholine synthesis (9); with increasing concentrations of conditioned medium, however, there is an increase in acetylcholine synthesis and an associated decrease in norepinephrine synthesis (15).

A CULTURE ARTIFACT?

The change in transmitter type seen *in vitro* was restricted to developing neurons (4,16), and the period of greatest sensitivity coincided with the time when the neurons were receiving synapses from the preganglionic spinal fibers and were themselves making synapses with their prospective target organs (1,2). A possible source of artifact, however, is that the effect is due to a substance released by nonneuronal cells only when in culture. Nerve growth factor (NGF) has been reported to be synthesized and released in tissue culture by cells which *in vivo* do not contain it (3). Since medium conditioned by heart cells was a potent source of the "cholinergic" factor *in vitro,* we tested the ability of neurons of SCG from 3-day-old rats to become cholinergic under the influence of a homogenate of the heart. A 10% (weight/volume) cardiac extract from adult rats doubled the CAT activity in the ganglion but had no effect on the activity of tyrosine hydroxylase (TH), the enzyme involved in the biosynthesis of norepinephrine (Table 1). Thus the CAT/TH ratio, which provides a measure of the degree of cholinergic character of the ganglion, was doubled following exposure of the neurons to cardiac extract,

TABLE 1. *Effect of cardiac extracts and dialysis on acetylcholine synthetic ability in SCG*

Medium	CAT (nmoles/hr /ganglion)	TH (pmoles/hr /ganglion)
Control medium	1.9 ± 0.26 (6)	121 ± 11.8
10% Cardiac extract	3.9 ± 0.14 (6)	111 ± 9.0
Control medium after dialysis	2.5 ± 0.21 (6)	135 ± 7.0
Cardiac extract after dialysis	3.5 ± 0.04 (6)	89 ± 6.6

The activities of choline acetyltransferase (CAT) and tyrosine hydroxylase (TH) were determined after 14 days' culture of SCG from 3-day-old rats. Control medium was dialyzed against cardiac extract for 24 hr prior to the culture period, and cardiac extract was similarly dialyzed against control medium. Values are means and SEMs, with the number of ganglia in parentheses.

demonstrating the presence of the cholinergic factor *in vivo*.

A CALCIUM EFFECT?

Since muscle cells sequester calcium (8), it is possible that the effects of conditioned medium or cardiac extract were due simply to changes in the concentration of free calcium. We studied calcium binding in the presence of heart extract in an equilibrium dialysis experiment (3 days at 4°C) using ^{45}Ca and found that, for a starting calcium concentration of 2.5 mM (equivalent to the calcium concentration in the tissue culture medium used in the above experiments), the free calcium in a 10% cardiac extract with serum was 2.1 mM. Thus extracts of rat heart do bind some calcium and reduce the concentration of free calcium in the culture medium. We then studied the effect of reducing the free calcium by simply altering the concentration of calcium in control media and growing SCG from 3-day-old rats in this modified medium for 14 days. For concentrations of NGF greater than 0.1 μg/ml, decreased calcium increased the activities of both CAT and TH, however, and had no effect on the CAT/TH ratio (Fig. 1). Hence selective stimulation of CAT by cardiac extract is not caused by a reduction in calcium.

Furthermore, the potency of the cardiac extract appears to depend on a nondialyzable macromolecule, since dialysis of cardiac extract against control medium (three changes of a 10-fold larger volume of control medium for 24 hr) did not significantly alter its ability to induce CAT activity in SCG after 14 days in culture (Table 1). Similarly, control medium was not significantly enriched by dialysis against cardiac extract (Table 1). Thus the stimulation by cardiac extract of CAT activity in sympathetic neurons in tissue culture is due to a nondialyzable macromolecule and not to alterations in the free calcium of the medium.

WHAT HAPPENS *IN VIVO?*

During the development of the SCG *in vivo,* the intrinsic CAT activity remains at a low level despite the potential presence of "cholinergic" factor in nonneuronal cells in the ganglion and in tissues to which nerve terminals may grow (4). Why do the neurons not become cholinergic *in vivo*? There are three factors which may affect the differentiation of the neurons *in vivo* but not *in vitro:* (a) preganglionic synapses from the spinal cord; (b) postganglionic synapses of the sympathetic axons with effector organs; and (c) the blood supply of the ganglia.

PREGANGLIONIC INFLUENCE

One possibility is that the influence of the preganglionic synapses is to prevent the neurons of the SCG from becoming cholinergic. Ko et al. (7), however, found that cholinergic synapses formed between more than 25% of the sympathetic neurons cultured in the presence of spinal cord explants. Although these results suggest that the preganglionic nerves have no effect on the differentiation of cholinergic neurons, it should be noted that only 20% of the sympathetic neurons in these cultures were actually synaptically coupled with the spinal nerves (6). In other experiments,

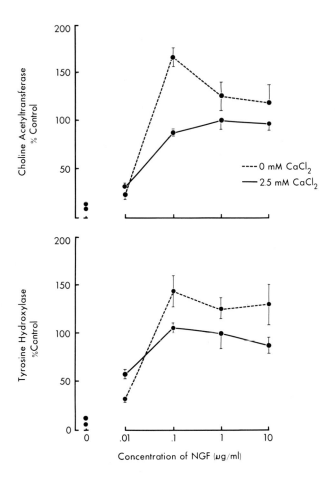

FIG. 1. Effect of calcium on the changes in CAT and TH activities following alterations in the concentration of NGF in SCG from 3-day-old rats. Enzyme activities are expressed as a percentage of the activity in SCG grown in 1 μg/ml NGF and 2.5 mM calcium chloride. Values are means and SEMs.

prolonged depolarization of all the sympathetic neurons using high potassium, veratridine, or direct electrical stimulation prevented the large increase in acetylcholine synthesis caused by conditioned medium from nonneuronal cells (18). Hence *in vitro* the bulk of evidence is in favor of a role for the preganglionic fibers in the selection of transmitter type in SCG neurons.

We tested the influence of the preganglionic fibers *in vivo* by sectioning the preganglionic nerve trunk to the right SCG in rats 1, 3, 7, and 14 days after birth and assaying the right ("decentralized") and the unoperated left SCG for CAT and TH at 21 days of age.

To compare changes in intrinsic CAT between the operated and the unoperated ganglion, it was necessary to "decentralize" the left (unoperated) ganglion 4 days prior to enzyme assay. This was the time required for the complete elimination of CAT activity present in the preganglionic cholinergic nerve terminals. The operated ganglion was also checked for reinnervation at the time of the second operation. There was no significant difference between the intrinsic CAT activity of the operated ganglion and its control "unoperated" ganglion for any day of operation (Table 2). In addition, there was no significant difference between the CAT activi-

TABLE 2. *Effect of preganglionic nerve section on acetylcholine synthetic ability in SCG of 21-day-old rats*

Postnatal day of operation	CAT activity (nmole/hr/ganglion)	
	Operated ganglion	Unoperated ganglion
1	0.19 ± .021 (5)	0.26 ± .018 (5)
3	0.23 ± .069 (5)	0.16 ± .032 (5)
7	0.27 ± .056 (7)	0.21 ± .056 (7)
14	0.12 ± .016 (8)	0.16 ± .036 (8)

The preganglionic nerve trunk to the right SCG was sectioned in rats 1, 3, 7, or 14 days old and the activity of choline acetyltransferase (CAT) compared with the intrinsic CAT activity of the control (left) ganglion when the rats were 21 days old. Values are means and SEMs, with the number of animals in parentheses.

ties of the operated and its control ganglion when ganglia were "decentralized" on day 3 and the animals injected daily with 10 μg/g NGF until biochemical assay on day 14. Thus *in vivo,* even as early as 1 day postnatal-

ly, the vast majority of the neurons of the SCG are differentiating as adrenergic neurons and this development is not altered following prolonged deprivation of the preganglionic input.

Although not directly affecting the choice of transmitter, the electrical activity of the preganglionic nerve fibers could determine or fix the transmitter type in the SCG neurons and hence make them unresponsive to further developmental influences. In the absence of activity *in vivo,* then, the neurons would retain the ability to become cholinergic when placed in culture at any time during their development. We decentralized the right SCG of 3-day-old rats and, at various times up to 21 days of age, the right and its contralateral unoperated ganglion were removed and cultured for a further 14 days. Assay for CAT revealed that there was a progressive reduction in the ability to become cholinergic as both the left and right ganglia matured (Fig. 2). Thus the neurons of the SCG become determined as adrenergic with

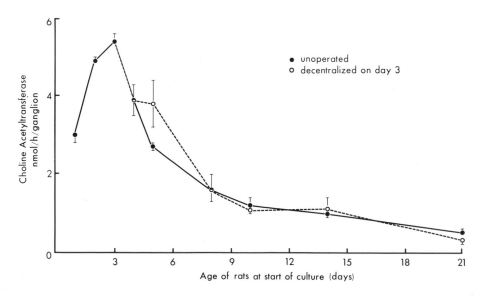

FIG. 2. CAT activity after 14 days' culture of paired left (●, unoperated) and right (O, preganglionic nerve sectioned at 3 days of age) SCG from rats of variable age up to 21 days. Note that the ability to induce CAT activity in culture decreases with the same time course in the presence or absence of preganglionic activity.

the same time course even in the absence of the preganglionic input.

POSTGANGLIONIC INFLUENCE

The role of the postganglionic synapses on the choice of transmitter in the SCG *in vitro* is equivocal. Explants of the SCG formed functional cholinergic synapses *in vitro* with one of its normal target organs, the iris. This occurred in 50% of cases (5), suggesting that there was a large number of cholinergic neurons within the explants. In contrast, cholinergic synapses formed between only 7% of the sympathetic neurons when they were grown in the presence of another target organ, brown fat cells, but between 30% of the neurons in control cultures (7).

In vivo there was less than a 1.5-fold increase in CAT activity following pre- and postganglionic axotomy (4), whereas the corresponding period of time *in vitro* led to a 50- to 1,000-fold increase in CAT activity (4,15).

BLOOD SUPPLY

The circulation of blood could influence the transmitter type in the SCG neurons in two ways. Firstly, it could remove the factor released by the nonneuronal cells within the ganglion. As a concentration-dependent response of the neurons to the factor released by heart cells has been described (15), the levels of factor remaining in the ganglion may be inadequate to produce a detectable effect. Secondly, humoral agents within the circulation may directly influence the ultimate cell type. Glucocorticoids augment the induction of adrenergic neurotransmitter synthetic enzymes in the SCG (12,13). We investigated the effect of glucocorticoids on the activities of CAT and TH in SCG from 3-day-old rats grown for 14 days *in vitro*. We found that the synthetic glucocorticoid, dexamethasone, and the naturally occurring glucocorticoid in rats, corticosterone, prevented

TABLE 3. *Effect of glucocorticoids on acetylcholine synthetic ability in SCG of 3-day-old rats*

Treatment	CAT/TH
14-Day culture	
Control	13.6 ± 1.8 (6)
10^{-7} M Dexamethasone	1.0 ± 0.2 (6)
10^{-6} M Corticosterone	0.5 ± 0.1 (6)
14-Day decentralization *in vivo*	0.6 ± 0.1 (5)

Choline acetyltransferase (CAT) and tyrosine hydroxylase (TH) activities (nmoles/hr/ganglion) were determined after either 14 days' culture of SCG from 3-day-old rats or 14 days after preganglionic nerve section of SCG in 3-day-old rats; i.e., SCG were removed from 17-day-old rats. Values are means and SEMs, with the number of ganglia in parentheses.

the development of a predominantly cholinergic SCG. The CAT/TH ratio decreased with increasing glucocorticoid concentration, reaching a level comparable to that of decentralized SCG *in vivo* when the concentration of corticosterone was equivalent to that in plasma (10^{-6} M) (Table 3). Thus physiological levels of glucocorticoids can influence the transmitter type in the developing SCG *in vitro*. Preliminary results, however, showed that reduction of circulatory glucocorticoids by adrenalectomy does not lead to an increase in CAT in the SCG *in vivo*. Thus although glucocorticoids can prevent the SCG from becoming cholinergic *in vitro*, it seems that adrenalectomy is insufficient to prevent the neurons from becoming adrenergic *in vivo*. In order to detect a change in the ultimate cell type *in vivo*, it may be necessary to combine some of the altered conditions which individually have no effect.

ACKNOWLEDGMENTS

We wish to thank Professor D.R. Curtis and Dr. K.G. Hill for criticizing the manuscript. C.E.H. is supported by an Overseas Research Fellowship and a Grant-in-Aid from the National Heart Foundation of Australia.

REFERENCES

1. Black, I.B., Hendry, I.A., and Iversen, L.L. (1971): Trans-synaptic regulation of growth and development of adrenergic neurons in a mouse sympathetic ganglion. *Brain Res.,* 34:229-240.
2. De Champlain, J., Malmfors, T., Olson, L., and Sachs, C. (1970): Ontogenesis of peripheral adrenergic neurons in the rat: Pre- and postnatal observations. *Acta Physiol. Scand.,* 80:276-288.
3. Harper, G.P., Pearce, F.L., and Vernon, C.A. (1976): Production of nerve growth factor by the mouse adrenal medulla. *Nature,* 261:251-253.
4. Hill, C.E., and Hendry, I.A. (1977): Development of neurons synthesizing noradrenaline and acetylcholine in superior cervical ganglion of the rat in vivo and in vitro. *Neuroscience,* 2:741-749.
5. Hill, C.E., Purves, R.D., Watanabe, H., and Burnstock, G. (1976): Specificity of innervation of iris musculature by sympathetic nerve fibers in tissue culture. *Pfluegers Arch.,* 361:127-134.
6. Ko, C-P., Burton, H., and Bunge, R.P. (1976): Synaptic transmission between rat spinal cord explants and dissociated superior cervical ganglion neurons in tissue culture. *Brain Res.,* 117:437-460.
7. Ko, C-P., Burton, H., Johnson, M.I., and Bunge, R.P. (1976): Synaptic transmission between rat superior cervical ganglion neurons in dissociated cell cultures. *Brain Res.,* 117:461-485.
8. MacLennan, D.H., and Wong, P.T.S. (1971): Isolation of a calcium-sequestering protein from sarcoplasmic reticulum. *Proc. Natl. Acad. Sci. USA,* 68:1231-1235.
9. Mains, R.E., and Patterson, P.H. (1973): Primary cultures of dissociated sympathetic neurons. *J. Cell Biol.,* 59:329-366.
10. O'Lague, P.H., Obata, K., Claude, P., Furshpan, E.J., and Potter, D.D. (1974): Evidence for cholinergic synapses between dissociated rat sympathetic neurons in cell culture. *Proc. Natl. Acad. Sci. USA,* 71:3602-3606.
11. O'Lague, P.H., Obata, K., Claude, P., Furshpan, E.J., and Potter, D.D. (1978): Studies on rat sympathetic neurons developing in cell culture. II. Synaptic mechanisms. *Dev. Biol.,* 67:404-423.
12. Otten, U., and Thoenen, H. (1976): Selective induction of tyrosine hydroxylase and dopamine-β-hydroxylase in sympathetic ganglia in organ culture: Role of glucocorticoids as modulators. *Mol. Pharmacol.,* 12:353-362.
13. Otten, U., and Thoenen, H. (1976): Modulatory role of glucocorticoids on NGF-mediated enzyme induction in organ cultures of sympathetic ganglia. *Brain Res.,* 111:438-441.
14. Patterson, P.H., and Chun, L.L.Y. (1974): The influence of non-neuronal cells on catecholamine and acetylcholine synthesis and accumulation in cultures of dissociated sympathetic neurons. *Proc. Natl. Acad. Sci. USA,* 71:3607-3610.
15. Patterson, P.H., and Chun, L.L.Y. (1977): The induction of acetylcholine synthesis in primary cultures of dissociated rat sympathetic neurons. *Dev. Biol.,* 56:263-280.
16. Ross, D., Johnson, M., and Bunge, R. (1977): Evidence that development of cholinergic characteristics in adrenergic neurons is age dependent. *Nature,* 267:536-539.
17. Sjöqvist, F. (1963): The correlation between the occurrence and localization of acetylcholinesterase-rich cell bodies in the stellate ganglion and the outflow of cholinergic sweat secretory fibers to the forepaw of the cat. *Acta Physiol. Scand.,* 57:339-351.
18. Walicke, P.A., Campenot, R.B., and Patterson, P.H. (1977): Determination of transmitter function by neuronal activity. *Proc. Natl. Acad. Sci. USA,* 74:5767-5771.
19. Yamauchi, A., Lever, J.D., and Kemp, K.W. (1973): Catecholamine loading and depletion in the rat superior cervical ganglion. *J. Anat.,* 114:271-282.

Fine Structure of SIF Cells and Modulation of Ganglionic Function

*Histochemistry and Cell Biology of
Autonomic Neurons, SIF Cells, and
Paraneurons,*
edited by O. Eränkö et al.
Raven Press, New York © 1980.

Ultrastructural Studies Relevant to the Possible Functions of Small Granule-Containing Cells in the Rat Superior Cervical Ganglion

Margaret R. Matthews

Department of Human Anatomy, University of Oxford, Oxford OX1 3QX, England

It is now more than a decade since the first accounts were published of the ultrastructure of the catecholamine-rich small granule-containing (SG) cells in sympathetic ganglia (11,18,39,50,51,56,59). From the beginning (18) the likely correlation was noted, and has now been repeatedly confirmed (e.g., refs. 19,57), between these and the small intensely fluorescent (SIF) cells, which had then recently been discovered by the powerful Falck-Hillarp technique of formaldehyde-induced fluorescence for catecholamines (15,42). It happened that most of the early ultrastructural studies were made in the superior cervical ganglion (SCG) of the rat, and the various groups of authors observed four cardinal features of the intraganglionic relationships of the SG cells they studied: first, that they received "afferent" synaptic inputs, from nerve terminals resembling the preganglionic nerve endings; second, that areas of their surfaces were free of satellite wrapping and were separated only by basal lamina from the tissue spaces of the ganglion; third, that they were closely associated with fenestrated small blood vessels, unusual elsewhere in the ganglion; and fourth, that the SG cells also gave rise to outgoing (or "efferent") synapses, which were directed toward other nerve elements in the ganglion, including dendrite-like profiles. The hypothesis was therefore advanced (56) that these cells might be interneurons, interposed in neuronal circuits within the ganglion; it was also evident that they might act in a neuroendocrine sense by release of granule contents into the intraganglionic tissue spaces and toward the associated small blood vessels (see above references).

SG CELLS IN GANGLIA OF VARIOUS SPECIES; POSSIBILITY OF TWO TYPES OF SG CELLS

Further information rapidly accumulated about similar cells in other sites and in other ganglia, both in the rat and in a number of other species (e.g. refs. 6,14). It has become clear that among sympathetic ganglia the superior cervical ganglion of the rat may be unusual in having relatively large numbers of SG cells which give efferent synapses. SG efferent synapses have been found also in the pelvic ganglion, a mixed sympathetic-parasympathetic ganglion, in the male rat (8), though not in the corresponding paracervical (uterine) ganglion of the female in the same species. Examples of similar synapses have been described in the SCG of the rhesus monkey (5). In most other sympathetic ganglia in which their ultrastructure has been investigated, however, SG cells have been found to receive an afferent innervation but do not seem to give efferent synapses (e.g. refs. 1,12, 17,45,54,55). In the rabbit SCG, although the principal neurons receive synapses of adrenergic type, it is not clear whether (or to what extent) these come from SG cells (7). The association of SG cells with fenestrated small blood vessels or with glomera of highly vesiculated small vessels (24) has been observed rather consistently, however, as has the incomplete nature of their satellite cell wrapping (e.g., refs. 6,14). Chiba and Williams (4) and Williams et al. (57,58) observed in a number of species (cow, cat, rabbit, rat, guinea pig) that some SG cells are solitary and have relatively long processes which run among the ganglionic neurons, whereas other SG cells lie in clusters near small blood vessels and have few or no processes (and if there are any, they are directed toward the vessels). They therefore advanced the suggestion that intraganglionic SG cells are of two types. They propose that the solitary cell is a true interneuron, forming efferent synapses (type I), and that the clustered SG cells are paraneuronic or neuroendocrine in character, secreting directly into the small vessels with which they are associated (type II). Certainly, in further support of a possible type II function, occasional profiles suggesting exocytosis from the large granular vesicles of the SG cells have been reported for ganglionic SG cells (12,17,35). In the rat and guinea pig, Chiba and Williams (4) observed that a type I–type II distinction was not possible: The two types of SG cells appeared to be mixed in the clusters found in these ganglia. The author has observed that the same SG cell process in the rat SCG can in fact give an efferent synapse and exhibit an area of its membrane directly applied to the basal lamina. Heym and Williams (24) have reconstructed the thin-walled vessels associated with large clusters of SG cells at the proximal

end of the SCG of the tree shrew *(Tupaia)* and have shown that these vessels arise from arteries supplying the ganglion and "re-enter large vessels before the latter are deployed to form the principal capillary network of the ganglion." Since these SG cells receive an afferent innervation, such an arrangement would clearly allow them to act as "paraneurons," in a neuroendocrine role.

SIF (SG) CELLS IN EXTRAGANGLIONIC SITES: PARAGANGLIA AND GLOMERA

As in ganglia, SIF cells elsewhere in the body are typically associated with thin-walled small blood vessels, which are usually fenestrated. In respect of their synaptic relationships, the striking characteristic of SIF (SG) cells in extraganglionic sites is their variability and apparent malleability in adapting to differing situations (e.g., ref. 34). SIF cells form "paraganglia," some of which lie in the posterior abdominal and pelvic regions, whereas others are associated with the vagus nerve and with its nodose ganglion; these paraganglionic SG cells either receive no innervation or receive afferent synapses apparently of cholinergic type, but they have not in general been found to give efferent synapses (3,19,31,32); one of the vagal "paraganglia" recently studied by Kondo (26) was of a different, glomus-like type. Cells of SIF (SG) type form the chief cells of the carotid body and the aortic and subclavian glomera; and morphologically similar (although 5-hydroxytryptamine-containing) cells constitute the neuroendocrine cells of glomera associated with the bronchial epithelium [e.g., in the rabbit (27)] and with the amphibian lung (48). In all these glomera the SG cells enter into synaptic relationships with sensory nerve endings, but the precise pattern varies according to site and probably also to species. In the carotid body the chief cells (SG in type) form reciprocal synapses with chemoreceptor sensory nerve endings of the glossopharyngeal nerve and are thought to receive a minor input from sympathetic preganglionic nerve fibers; in addition, synaptic connections may be found between chief cells (22,23,40,41,53). In the neuroepithelial bodies of the rabbit lung, there could be similar relationships, although these SIF (SG) cells appear to contain 5-hydroxytryptamine, not catecholamine (27). In the rabbit aortic body and subclavian glomus, the synaptic relationship of catecholamine-containing SIF (SG) cells with sensory nerve endings does not appear to be reciprocal, nor do there seem to be synaptic connections between the SG cells (20,21). [Synapses, including reciprocal synapses, have been seen, however, between some SG cells of autonomic ganglia (34).] In cardiac (parasympathetic) ganglia of the rat and turtle, the SG cells enter into a feedback synaptic relationship, receiving inputs of cholinergic type not from the preganglionic vagal nerve fibers but, it seems, from the postganglionic neurons, and giving reciprocal (adrenergic) efferent synapses to these same neurons (60–62).

It is thus evident that SIF (SG) cells may be deployed in widely differing roles in different sites and different species according to the environment in which they develop, in the context of the peripheral nervous system. Such a diversity of roles is not surprising in view of the great lability not only as to synapse formation but also as to transmitter type, which is shown during development by closely related cell types: thus, sympathetic neurons are capable of being powerfully influenced during development by environmental factors (28,43,46). Even differentiated chromaffin tissue can show considerable plasticity. Adrenal medullary tissue, when transplanted to the anterior eye chamber, gives rise to catecholamine-containing nerve fibers which include fibers derived from groups of highly fluorescent cells (i.e., not from neurons in the transplant), and these fibers can grow into the host iris and form networks there (44). Chromaffin cells transplanted under the kidney capsule form synapses on each other and on neurons in the

graft (52). SIF cells during development are highly susceptible to the influences of hydrocortisone (13,16,O. Eränkö, *this volume*) and of nerve growth factor (NGF) (Levi-Montalcini and Aloe, *this volume*). In the adult, their unifying general characteristics of size and catecholamine content should not be allowed to mislead one into ignoring their variations or assuming homogeneity of function; rather, the particular effects of the SG cells may prove to be at least as numerous as the various methods of recruitment, types of transmitter or neurohumoral secretion [including possibly neuroactive peptides (e.g., ref. 49)], postsynaptic sites or neuroendocrine targets, and varieties of receptor mechanisms would permit. These require separate study in each situation in which these cells occur.

SYNAPTIC RELATIONSHIPS OF SG CELLS: RAT SCG

As more detailed and extensive studies are undertaken and new species are examined, the various patterns of relationships formed by SG cells are more fully revealed. In the rat SCG, studies of synapses have produced new information which adds significantly to the picture. It has been observed that preganglionic denervation removes all the incoming (afferent) nerve endings which form synapses on the SG cells (33,38), and that the outgoing (efferent) synapses are given to nerve processes which include dendrites and dendritic spines of principal neurons (34), or occasionally the soma of a principal neuron (36) or a somatic spine (63). These observations reinforce the theory of an interneuron-like role for some SG cells. Yokota (63) examined serial sections throughout a group of four SG cells and counted their synapses. She found that each cell received one to four afferent synapses but that only two of them gave rise to efferent synapses (two each), whereas the other two had more extensive areas exposed to the basal lamina. This finding indicates that not all the SG cells of a cluster necessarily play the same role.

of SG cell efferent synapses became reduced after axotomy, but to a lesser extent and more transiently than in the axotomized ganglia. There was here no question of local damage to postganglionic nerve bundles or to neighboring cranial nerves. In axotomized ganglia in which the preganglionic trunk was cut at a second operation 2 days before removal of the ganglia for analysis, the incidence of SG cell efferent synapses did not differ from that seen in other axotomized ganglia at similar survival intervals; thus the profiles postsynaptic to the SG cells were not derived from nerve fibers entering the ganglion along the preganglionic trunk.

In contrast to the behavior of the efferent synapses, the incidence of afferent synapses to the SG cells increased postoperatively, from a mean level of approximately 0.45 ± 0.06 per SG cell nuclear profile in normal ganglia to approximately 0.66 ± 0.09 over the period 4 to 7 days, and 1.4 ± 0.26 per SG cell nuclear profile by 34 to 42 days postoperatively. SG cell groups in contralateral control ganglia showed comparable increases in the incidence of their afferent synapses over the first 2 to 3 weeks postoperatively; however, by 27 days the ratio of afferent to efferent SG cell synapses in control ganglia had returned to normal. In the axotomized ganglia in which the preganglionic trunks had also been cut 2 days beforehand, no SG cell afferent synapses remained; i.e., all the newly acquired afferent synapses, as well as the original ones, were derived from the preganglionic nerve fibers. It is not known whether the new synapses had arisen by sprouting from the original inputs to the SG cells, or whether the SG cells had become innervated by nerve fibers which would normally terminate on the principal neurons. Sham operations in which the ganglia were exposed but no nerves were cut did not lead to any synaptic changes in either afferent or efferent synapses.

The marked dissociation of the behavior of afferent and efferent synapses shown by these experiments, especially after preganglionic

axotomy, makes it highly improbable that both could be derived from the same axons in more than a very small proportion of cases. The loss of efferent synaptic contacts from SG cells in the axotomized ganglia could be due to the reaction of the injured neurons, and this seems a very likely cause. Other possible causes would be that the SG cells have been injured directly (this is unlikely because all observations so far indicate that any processes which they possess are relatively short, and there is no evidence that they send projections outside the ganglia), or that the postsynaptic structures themselves are axons entering the ganglion along the injured postganglionic trunks, in which case they would have been separated from their cell bodies by the lesion and thus would undergo direct degeneration. There was no evidence for the latter possibility; e.g., no degenerating nerve profiles were found adjacent to the SG cells at short postoperative intervals. The occurrence of changes in afferent synapses on the SG cells in the axotomized ganglia, and in afferent and efferent synapses in ganglia contralateral to these, suggests some central or general cause for these wider reactive changes. Since sham axotomy did not lead to synaptic changes, it seems more likely that they are mediated via nervous than via general hormonal influences. These possibilities are being explored in further experiments.

REACTIVE CHANGES IN SG CELLS

Various changes in the SG cells themselves accompanied these major disturbances of their synaptic relationships. During the first postoperative week, the incidence of profiles of SG cell processes increased sharply, up to fivefold, suggesting an increase either in their number or in their length or tortuosity; it was found that this could happen without gross change in the total area of cytoplasm per SG cell nuclear profile. In addition, synapses were observed between the SG cells: Such synapses are rare in normal rat SCG but are increased transiently after postganglionic axotomy, reaching a peak at about 5 days postoperatively before declining. Unsicker et al. (52) observed similar changes in rat adrenal medullary cells transplanted to beneath the renal capsule, i.e., to a site where afferent innervation is not immediately available to them; the ganglionic SG cells may here be behaving in an analogous manner at a stage when there is apparently reduced access to their normal postsynaptic targets, and despite a level of afferent innervation which is beginning to increase above the normal. It might be the increase in their afferent input which causes the synapses between the SG cells to disappear again. Later, the SG cells of some groups become enlarged; the increase in size involves both nucleus and cytoplasm, and the cytoplasm-nucleus ratio may even be increased. A linear correlation has been found between the mean nuclear diameter for an SG cell group and the incidence of afferent synapses; thus the observed hypertrophy of SG cells may be directly or indirectly related to the increase in incidence of afferent synapses.

CONCLUSIONS; POSSIBLE FUNCTIONS OF SG CELLS IN THE SCG

It appears that the majority of the groups of SG cells of the rat SCG are well integrated into intraganglionic neuronal circuits after the fashion of interneurons, morphologically (by way of their afferent and efferent synapses) and in a metabolically dependent sense, while still retaining the neuroendocrine characteristics, which in most other ganglionic SG cell sites are the predominant ones. Others, a minority, are like chemoreceptor glomera included within the ganglion but able also to interact with ganglionic elements. We still do not know, however, quite what any of them do, nor what could be the range of selective functions of the intraganglionic SG cells. In the rat, most of the SG cells of the SCG, like the chief cells of the carotid body,

seem to contain dopamine (e.g., ref. 14). Their preganglionic nerve inputs place them under central nervous control, but they are variable in extent (Matthews and Case, *in preparation*). It is likely that they may be recruited in various ways, according to differing nerve connections, differing patterns of nervous activity, and different functional requirements. It seems at least possible that some of them, perhaps including the glomus-like groups, might act in a chemosensitive capacity and perhaps might assist in regulating and preserving the blood flow to vital structures, even possibly acting through a relationship with appropriate postganglionic neurons as part of a local short-loop effector circuit. For instance, groups associated with the internal carotid nerve territory might act on the neurons (or on their axons or even their presynaptic inputs) which innervate the stems of the cerebral blood vessels (as well as the ophthalmic territory); and groups associated with the external carotid nerve territory might have similar functions with respect to the vessels of the important oronasal and cutaneous territories of the external carotid artery. As yet, not only is the functional significance of the SG cells unknown but also the precise manner and site of their action(s). It would be of special interest to know what particular advantage is conferred by the direct focal synaptic pathway to the principal neurons in the rat SCG over the more diffuse and probably slower neurohumoral pathway that appears to coexist with it.

ACKNOWLEDGMENT

The author thanks the Medical Research Council of Great Britain for research support.

REFERENCES

1. Becker, K. (1972): Paraganglienzellen im ganglion cervicale uteri der maus. *Z. Zellforsch. Mikrosk. Anat.*, 130:249-261.
2. Case, C.P., and Matthews, M.R. (1976): Effects of postganglionic axotomy on synaptic connexions of

small granule-containing (SG) cells in the rat superior cervical ganglion. *J. Anat.*, 122:732.

3. Chen, I-L., and Yates, R.D. (1970): Ultrastructural studies of vagal paraganglia in Syrian hamsters. *Z. Zellforsch. Mikrosk. Anat.*, 108:309–323.

4. Chiba, T., and Williams, T.H. (1975): Histofluorescence characteristics and quantification of small intensely fluorescent (SIF) cells in sympathetic ganglia of several species. *Cell Tissue Res.*, 162:331–342.

5. Chiba, T., Black, A.C., Jr., and Williams, T.H. (1977): Evidence for dopamine-storing interneurons and paraneurons in rhesus monkey sympathetic ganglia. *J. Neurocytol.*, 6:441–453.

6. Coupland, R.E., and Fujita, T., editors (1976): *Chromaffin, Enterochromaffin and Related Cells.* Elsevier, Amsterdam.

7. Dail, W.G., and Evan, A.P. (1978): Ultrastructure of adrenergic nerve terminals and SIF cells in the superior cervical ganglion of the rabbit. *Brain Res.*, 148:469–477.

8. Dail, W.G., Jr., Evan, A.P., Jr., and Eason, H.R. (1975): The major ganglion in the pelvic plexus of the male rat: A histochemical and ultrastructural study. *Cell. Tissue Res.*, 159:49–62.

9. Dunant, Y., and Dolivo, M. (1967): Relations entre les potentiels synaptiques lents et l'excitabilité du ganglion sympatique chez le rat. *J. Physiol. (Paris)*, 59:281–294.

10. Eccles, R.M., and Libet, B. (1961): Origin and blockade of the synaptic response of curarized sympathetic ganglia. *J. Physiol. (Lond)*, 157:484–503.

11. Elfvin, L-G. (1968): A new granule-containing nerve cell in the inferior mesenteric ganglion of the rabbit. *J. Ultrastruct. Res.*, 22:37–44.

12. Elfvin, L-G., Hökfelt, T., and Goldstein, M. (1975): Fluorescence microscopical, immunohistochemical and ultrastructural studies on sympathetic ganglia of the guinea-pig, with special reference to the SIF cells and their catecholamine content. *J. Ultrastruct. Res.*, 51:377–396.

13. Eränkö, L., and Eränkö, O. (1972): Effect of hydrocortisone on histochemically demonstrable catecholamines in the sympathetic ganglia and extra-adrenal chromaffin tissue of the rat. *Acta Physiol. Scand.*, 84:125–133.

14. Eränkö, O., editor (1976): *SIF Cells: Structure and Function of the Small Intensely Fluorescent Sympathetic cells.* Fogarty International Center Proceedings No. 30. DHEW Publication No. (NIH) 76–942. Government Printing Office, Washington, D.C.

15. Eränkö, O., and Härkönen, M. (1965): Monoamine-containing small cells in the superior cervical ganglion of the rat and an organ composed of them. *Acta Physiol. Scand.*, 63:511–512.

16. Eränkö, O., Eränkö, L., Hill, C.E., and Burnstock, G. (1972): Hydrocortisone-induced increase in the number of small intensely fluorescent cells and their histochemically demonstrable catecholamine content in cultures of sympathetic ganglia of the newborn rat. *Histochem. J.*, 4:49–58.

17. Furness, J.B., and Sobels, G. (1976): The ultrastructure of paraganglia associated with the infe-rior mesenteric ganglia in the guinea-pig. *Cell Tissue Res.*, 171:123–139.

18. Grillo, M.A. (1966): Electron microscopy of sympathetic tissues. *Pharmacol. Rev.*, 18:387–399.

19. Grillo, M.A., Jacobs, L., and Comroe, J.H., Jr. (1974): A combined fluorescence histochemical and electron microscopic method for studying special monoamine-containing cells (SIF cells). *J. Comp. Neurol.*, 153:1–14.

20. Hansen, J.T. (1979): An ultrastructural stereological analysis of the aortic body chief cell of adult rabbits. *Cell Tissue Res.*, 196:511–518.

21. Hansen, J.T., and Yates, R.D. (1975): Light, fluorescence and electron microscopic studies of rabbit subclavian glomera. *Am. J. Anat.*, 144:477–490.

22. Hess, A. (1975). The significance of the ultrastructure of the rat carotid body in structure and function of chemoreceptors. In: The *Peripheral Arterial Chemoreceptors,* edited by M.J. Purves, pp. 51–68. Cambridge University Press, Cambridge, England.

23. Hess, A., and Zapata, P. (1972): Innervation of the cat carotid body: Normal and experimental studies. *Fed. Proc.*, 31:1365–1382.

24. Heym, C., and Williams, T.H. (1979): Evidence for autonomic paraneurons in sympathetic ganglia of a shrew *(Tupaia glis). J. Anat.*, 129:151–164.

25. Kanerva, L., and Teräväinen, H. (1972): Electron microscopy of the paracervical (Frankenhäuser) ganglion of the adult rat. *Z. Zellforsch. Mikrosk. Anat.*, 129:161–177.

26. Kondo, H. (1977): Innervation of SIF cells in the superior cervical and nodose ganglia: An ultrastructural study with serial sections. *Biol. Cell.* 30:253–264.

27. Lauweryns, J.M., and Cokelaere, M. (1973): Hypoxia-sensitive neuroepithelial bodies: Intrapulmonary secretory neuroreceptors, modulated by the CNS. *Z. Zellforsch. Mikrosk. Anat.*, 145:521–540.

28. Le Douarin, N.M., and Teillet, M.M. (1974): Experimental analysis of the migration and differentiation of neuroblasts of the autonomic nervous system and of neuroectodermal mesenchymal derivatives using a biological cell marking technique. *Dev. Biol.*, 41:162–184.

29. Libet, B. (1976): The SIF cell as a functional dopamine-releasing interneurone in the rabbit superior cervical ganglion. In: *SIF Cells, Structure and Function of the Small, Intensely Fluorescent Sympathetic Cells,* edited by O. Eränkö, pp. 163–177. Fogarty International Center Proceedings No. 30. DHEW Publication No. (NIH) 76–942. Government Printing Office, Washington, D.C.

30. Libet, B. (1977): The role SIF cells play in ganglionic transmission. *Adv. Biochem. Psychopharmacol.*, 16:541–546.

31. Mascorro, J.A., and Yates, R.D. (1970): Microscopic observations on abdominal sympathetic paraganglia. *Texas Rep. Biol. Med.*, 28:363–372.

32. Mascorro, J.A., and Yates, R.D. (1971): Ultrastructural studies of the effects of reserpine on mouse abdominal sympathetic paraganglia. *Anat. Rec.*, 120:269–280.

33. Matthews, M.R., (1971): Evidence from degeneration experiments for the preganglionic origin of

afferent fibres to the small granule-containing cells of the rat superior cervical ganglion. *J. Physiol. (Lond)*, 218:95-96P.

34. Matthews, M.R. (1976): Synaptic and other relationships of small granule-containing cells (SIF cells) in sympathetic ganglia. In: *Chromaffin, Enterochromaffin and Related Cells*, edited by R. Coupland and T. Fujita, pp. 131-146. Naito Foundation Symposium. Elsevier, Amsterdam.

35. Matthews, M.R. (1978): Ultrastructural evidence for discharge of granules by exocytosis from small granule-containing cells of the superior cervical ganglion in the rat. In: *Peripheral Neuroendocrine Interaction*, edited by R.E. Coupland and W.G. Forssman, pp. 80-85. Springer-Verlag, Berlin.

36. Matthews, M.R., and Nash, J.R.G. (1970): An efferent synapse from a small granule-containing cell to a principal neurone in the superior cervical ganglion. *J. Physiol. (Lond)*, 210:11-14P.

37. Matthews, M.R., and Nelson, V.H. (1975): Detachment of structurally intact nerve endings from chromatolytic neurones of rat superior cervical ganglion during the depression of synaptic transmission induced by postganglionic axotomy. *J. Physiol. (Lond)*, 245:91-135.

38. Matthews, M.R., and Ostberg, A. (1973): Effects of preganglionic nerve section upon the afferent innervation of the small granule-containing cells in the rat superior cervical ganglion. *Acta Physiol. Pol.*, 24:215-224.

39. Matthews, M.R., and Raisman, G. (1969): The ultrastructure and somatic efferent synapses of small granule-containing cells in the superior cervical ganglion. *J. Anat.*, 105:255-282.

40. McDonald, D.M., and Mitchell, R.A. (1975): A quantitative analysis of synaptic connections in the rat carotid body. In: *The Peripheral Arterial Chemoreceptors*, edited by M.J. Purves, Cambridge University Press, Cambridge, England.

41. McDonald, D.M., and Mitchell, R.A. (1975): The innervation of glomus cells, ganglion cells and blood vessels in the rat carotid body: A quantitative ultrastructural analysis. *J. Neurocytol.*, 4:177-230.

42. Norberg, K-A., and Sjöqvist, F. (1966): New possibilities for adrenergic modulation of ganglionic transmission. *Pharmacol. Rev.*, 18:743-751.

43. O'Lague, P.H., MacLeish, P.R., Nurse, C.A., Claude, P., Furshpan, E.J., and Potter, D.D. (1975): Physiological and morphological studies on developing sympathetic neurons in dissociated cell culture. *Cold Spring Harbor Symp. Quant. Biol.*, 40:399-407.

44. Olson, L. (1970): Fluorescence histochemical evidence for axonal growth and secretion from transplanted adrenal medullary tissue. *Histochemie*, 22:1-7.

45. Ostberg, A. (1970): Granule-containing cells of the inferior mesenteric ganglion. *Proc. Aust. Physiol. Pharm. Soc.*, 1:72.

46. Patterson, P.H., Reichardt, L.F., and Chun, L.L.Y. (1975): Biochemical studies on the development of primary sympathetic neurons in cell culture. *Cold Spring Harbor Symp. Quant. Biol.*, 40:389-397.

47. Purves, D. (1975): Functional and structural changes of mammalian sympathetic neurones following interruption of their axons. *J. Physiol. (Lond)*, 252:429-463.

48. Rogers, D.C., and Haller, C.J. (1978): Innervation and cytochemistry of the neuroepithelial bodies in the ciliated epithelium of the toad lung (Bufo marinus). *Cell Tissue Res.*, 195:395-400.

49. Schultzberg, M., Hökfelt, T., Terenius, L., Elfvin, L-G., Lundberg, J.M., Brandt, J., Elde, R.P., and Goldstein, M. (1979): Enkephalin immunoreactive nerve fibres and cell bodies in sympathetic ganglia of the guinea pig and rat. *Neuroscience*, 4:249-270.

50. Siegrist, G., De Ribaupierre, F., Dolivo, M., and Rouiller, C. (1966): Les cellules chromaffines des ganglions cervicaux supérieurs du rat. *J. Microsc.*, 5:791-794.

51. Siegrist, G., Dolivo, M., Dunant, Y., Foroglou-Kerameus, C., De Ribaupierre, F., and Rouiller, C. (1968): Ultrastructure and function of the chromaffin cells in the superior cervical ganglion of the rat. *J. Ultrastruct. Res.*, 25:381-407.

52. Unsicker, K., Zwarg, U., and Habura, O. (1977): Electron microscopic evidence for the formation of synapses and synaptoid contacts in adrenal medullary grafts. *Brain Res.*, 120:533-539.

53. Verna, A. (1975): Observations on the innervation of the carotid body of the rabbit. In: *The Peripheral Arterial Chemoreceptors*, edited by M.J. Purves. Cambridge University Press, Cambridge, England.

54. Watanabe, H. (1971): Adrenergic nerve elements in the hypogastric ganglion of the guinea-pig. *Am. J. Anat.*, 130:305-330.

55. Watanabe, H., and Burnstock, G. (1976): A special type of small granule-containing cell in the abdominal para aortic region of the frog. *J. Neurocytol.*, 5:465-478.

56. Williams, T.H. (1967): Electron microscopic evidence for an autonomic interneuron. *Nature*, 214:309-310.

57. Williams, T.H., Black, A.C., Jr., Chiba, T., and Bhalla, R.C. (1975): Morphology and biochemistry of small, intensely fluorescent cells of sympathetic ganglia. *Nature*, 256:315-317.

58. Williams, T.H., Chiba, T., Black, A.C., Jr., Bhalla, R.C., and Jew, J. (1976): Species variation in SIF cells of superior cervical ganglia: are there two functional types? In: *SIF Cells. Structure and Function of the Small, Intensely Fluorescent Sympathetic Cells*, edited by O. Eränkö, pp. 143-162. Fogarty International Center Proceedings No. 30. DHEW Publ. No. (NIH) 76-942. Government Printing Office, Washington, D.C.

59. Williams, T.H., and Palay, S.L. (1969): Ultrastructure of the small neurons in the superior cervical ganglion. *Brain Res.*, 15:17-34.

60. Yamauchi, A. (1976): Ultrastructure of chromaffin-like interneurons in the autonomic ganglia. In: *Chromaffin, Enterochromaffin and Related Cells*, edited by R. Coupland and T. Fujita, pp. 117-130.

Naito Foundation Symposium. Elsevier, Amsterdam.

61. Yamauchi, A., Fujimaki, Y., and Yokota, R. (1975): Reciprocal synapses between cholinergic postganglionic axon and adrenergic interneuron in the cardiac ganglion of the turtle. *J. Ultrastruct. Res.,* 50:47–57.

62. Yamauchi, A., Yokota, R., and Fujimaki, Y. (1975): Reciprocal synapses between cholinergic axons and small granule-containing cells in the rat cardiac ganglion. *Anat. Rec.,* 181:195–210.

63. Yokota, R. (1973): The granule-containing cell somata in the superior cervical ganglion of the rat, as studied by a serial sampling method for electron microscopy. *Z. Zellforsch. Mikrosk. Anat.,* 141:331–345.

Histochemistry and Cell Biology of Autonomic Neurons, SIF Cells, and Paraneurons,
edited by O. Eränkö et al.
Raven Press, New York © 1980.

Catecholamines in Paraganglionic Cells of the Rat Superior Cervical Ganglion: Functional Aspects

Christine Heym, *Klaus Addicks, Norbert Gerold, Hannsjörg Schröder, Rainer König, and Christa Wedel

*Department of Anatomy, University of Heidelberg, D-6900 Heidelberg, GFR; and *Department of Anatomy, University of Köln, D-5000 Köln 41, GFR*

A considerable body of evidence supports the modulatory significance of paraganglionic [small granule-containing or small intensely fluorescent (SIF)] cells in sympathetic ganglia, suggesting that the intracellular catecholamine (CA) stores may act as neuroregulators. The functional modes of paraganglionic cells have been subject to controversy because of a great species diversity in their distribution and biochemistry (4). It is claimed that paraganglionic cells characterized as interneurons (20,23) contain dopamine (DA) (18) and correlate with isolated cells with long ramifying processes (3). However, a very small number of interneurons have been counted (25) which comprise no more than a small proportion of the multitude of principal neurons. On the other hand, the close vascular relation to paraganglionic cell clusters possessing no or only short processes suggests the possibility that amines released from these cells into regional capillaries may influence ganglionic activity in an endocrine fashion (12,24), their synaptic effects being restricted to specific pathways (20).

Whereas in the superior cervical ganglion (SCG) of some species morphological criteria of paraganglionic cell types were found which correspond to biochemical findings, such conformity was not obtainable in the paraganglionic cells of the rat SCG, which occur in clusters (24) but also show characteristics of interneurons. To further elucidate their functional implication, a great deal of attention has been given to the nature of the transmitter released by paraganglionic cells.

IDENTIFICATION OF CATECHOLAMINES IN PARAGANGLIONIC CELLS

Microspectrofluormetric investigations indicate that paraganglionic cells in the rat SCG contain most of the DA present in the ganglion (2). Moreover, from quantitative mass fragmentography following pharmacological manipulations (16), it was concluded that all of the epinephrine (E) and a large percentage of the DA found in the rat SCG was present in paraganglionic cells, whereas most of the norepinephrine (NE) was located in the principal neurons. The ultrastructural diversity of CA-storing granules within different paraganglionic cells of the rat SCG (22), being particularly visible following alteration of the CA metabolism (11), as well as heterogenous responses of these cells to

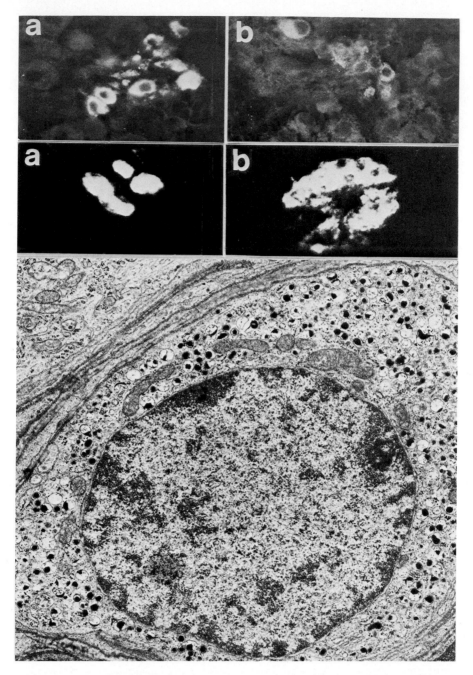

FIG. 1. (Top) Strongly DBH-positive cell clusters in the rat SCG. (**a:** From ref. 14. **b:** From *Verh. Anat. Ges., in press.*) **a:** Adult. **b:** Three days of age. × 300.
FIG. 2. (Middle) Formaldehyde-induced fluorescence of SIF cell clusters in the rat SCG. **a:** Control, fluorescent areas covering nuclei. **b:** Following 12 periods of immobilization; dark nuclei are clearly visible. × 350.
FIG. 3. (Bottom) Paraganglionic perikaryon in the rat SCG following 18 immobilization periods. A large number of large dcv (160 to 280 nm) with cores of varying size and shape are visible. × 12,000.

formaldehyde vapor or to glutaraldehyde-di-chromate treatment (17), led to the assumption of two paraganglionic cell populations in the rat SCG containing different CAs (19).

We confirmed these findings by applying a combined method for demonstration of glyoxylic acid-induced CA fluorescence (6) and evaluation of indirect FITC-labeled antibodies to dopamine-β-hydroxylase (DBH) (10) to serial sections of the rat SCG (14). A small proportion of paraganglionic cells (1 to 5%) stained distinctly DBH-positive, indicating their ability to synthesize NE, not only in the adult but also in the neonatal rat (Fig. 1). However, the presence of one CA does not necessarily rule out the presence of other transmitters, e.g., other CAs in paraganglionic cells.

Since immunohistochemical reactions obviously depend on a sufficient level of enzyme being present in the cell, it was desirable to apply a more specific method for identification of the CA in paraganglionic cells. Yet because of their irregular distribution in the ganglion, measurement of the neurotransmitter content within paraganglionic cells has been hampered by the lack of appropriate methodology. The BTG micro-laser unit enabled the dissection of single clusters of paraganglionic cells from the surrounding neuropil following their identification in alternate sections treated with glyoxylic acid (13). After processing the samples for biochemical assay, the content of the primary CAs (NE and E) could be determined by gas chromatography/mass fragmentography. To verify correct measurement of these two CAs, the technique of multiple ion detection (16) was applied. The analysis showed a high degree of specificity. In about 20 paraganglionic cells from two large clusters, an E content of 5 pmoles and significantly less NE (Table 1) were measured.

The finding agrees with the E content found in the ganglion (150 pmoles) when related to the mean total number of paragan-

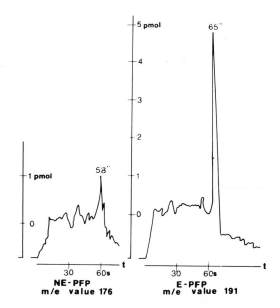

TABLE 1. Gas chromatographic/mass fragmentographic analysis of two SIF cell clusters consisting of about 20 cells, as prepared by the BTG micro-laser unit. **Left:** NE-PFP (pentafluorpropionic acid) fragment; retention time 58 sec. **Right:** E-PFP fragment; retention time 65 sec. (From ref. 13.)

glionic cells (388 to 986) counted per SCG (3,7). Moreover, it is supported by measurements from the entire ganglion (15), which showed a comparable pool of E in the SCG. Our data do not necessarily infer that all paraganglionic cells in the rat SCG contain E, particularly since x-ray microanalysis has indicated (17) that a small number of these cells which may have been missed in our samples store NE. In any case, the majority of paraganglionic cells, besides the main transmitter DA, seems to store E and a very small amount of NE.

CATECHOLAMINES IN PARAGANGLIONIC CELLS FOLLOWING IMMOBILIZATION

Having detected more than one CA within paraganglionic cells of the rat SCG, it was our aim to determine how the CAs in this cell population react to functional strain of the aminergic system. Paraganglionic cells are known to be susceptible to the influence of glucocorticoids by way of the blood current (8). Since they are also capable of being influenced by pre- and postganglionic sympathetic neurons (21), alteration of either of these influences is thought to affect their CA content.

Applying a familiar model for the enhanced activity of the vegetative nervous system (1) we subjected young adult Wistar rats (200 g body weight) to immobilization in perforated metallic cylinders. The duration of restraint was 12 hr, with intermittent 12 hr rest, repeated up to 23 times.

After treatment with fluorescence-labeled antibodies to DBH, the ratio of 5% NE-storing cells to 95% DA-storing cells in controls did not visibly change after any period of immobilization, indicating that the beta-hydroxylation mechanisms in most paraganglionic cells had not altered to any noticeable extent. Using constant photography exposure times, in comparison with controls (Fig. 2a) the intensity of the paraformaldehyde-induced fluorescence in many of the paraganglionic cells was slightly impaired following restraint. The fluorescent areas were confined to the cell periphery (Fig. 2b). In correlation with the restraint duration, an increasing diminution of CA fluorescence could be verified by microscope photometry, showing a close resemblance to the reduced fluorescence product as measured in type I cells of the carotid body and adrenal medullary cells (Table 2; 1). The similar reaction of paraganglionic cells and other paraganglionic tissues

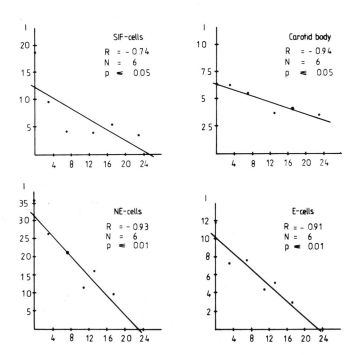

TABLE 2. Fluorescence cytophotometric evaluation of formaldehyde-induced fluorescence intensity of SIF cells in the SCG, type I cells in the carotid body, and NE- and E-containing cells in the adrenal medulla, compared to a uranyl glass standard GG 17/090 with 100%. Ordinate: Fluorescence intensity (I). Abscissa: Periods of immobilization.

tempts us to assume that there are analogous functional mechanisms. Assuming a participation of paraganglionic cells in a general paraganglionic system, the stress-induced CA depletion in the SCG and the carotid body may follow similar mechanisms. This hypothesis also fits into the paraneuronal concept of Fujita (9). It does not exclude the additional possibility of synaptic transmission. However, it should be mentioned that in all experiments in which there was comparison with controls a few cells exhibited no distinct decrease in fluorescence intensity.

The fluorescence histochemical observations were supported by electron microscopic findings. Following repeated immobilization, compared with controls the paraganglionic cells exhibited an intensified Golgi vesiculation, augmented parallel rough endoplasmic reticulum, and an increased number of lysosomes as signs of increased cytoplasmic activity. The number of dense-cored vesicles (dcv; 106 ± 20.5 nm in size) in the perikarya was in general decreased. Whereas the degranulation following long-term stress was mainly observed in the perikarya (Fig. 4), specific granules often accumulated in the cell periphery (Fig. 5a). The cell processes in addition exhibited dense bodies of varying shapes and irregular tubular profiles (Fig. 5b). They were regarded as degenerating or regenerating features of enhanced cell metabolism. Highly degranulated cells were found to exhibit synaptic contact with presumed dendritic spines of principal neurons (Fig. 4). Since synaptic transmission and neurosecretion are not necessarily mutually exclusive (24), this finding supports previous suggestions (22,24) that paraganglionic cells in the rat SCG could exert their inhibitory influence by a dual mechanism.

Again, a few cells packed with large irregularly shaped opaque dcv (226 ± 70 nm in size) (Fig. 3) were observed reminiscent of NE storage sites in the adrenal medulla (5). Previous studies (17,22) disclosed more than one paraganglionic cell type in the rat SCG. On the basis of the present observations, few cells, if not representing a functional stage of DA/E-storing elements such as exhaustion, may belong to the NE-storing cells of type II (17). These cells then did not participate in the degranulation or decrease in fluorescence intensity, thus suggesting a different functional implication.

CONCLUSIONS AND SUMMARY

In the rat SCG two types of paraganglionic cells have been identified immunohistochemically in a ratio of about 5% DBH-positive to 95% DBH-negative cells. Measurements of CAs in SIF cell clusters separated from principal neurons with the microlaser and using the technique of quantitative mass fragmentography disclosed the presence of E within paraganglionic cells, thus confirming the proposal that, besides DA, E also is stored in most cells. Following immobilization of adult rats, the majority of SIF cells revealed a time-dependent loss of formaldehyde-induced fluorescence intensity that could be ultrastructurally correlated with an increasing degranulation of small granule-containing perikarya, presumed to store DA and E. Similar results were obtained in type I cells of the rat carotid body and adrenal medullary cells, thus suggesting a mutual endocrine function. An additional interneuronal mode of modulation is assumed, since degranulated cells are capable of displaying efferent synaptic contact with postganglionic neurons. A minor cell population, storing granular vesicles similar to those in the NE-containing cells of the adrenal medulla, did not respond with degranulation to functional strain of the vegetative nervous system. Our results strongly corroborate the hypothesis that a functional dualism exists: Most paraganglionic cells can act as an interneuron and as an endocrine cell.

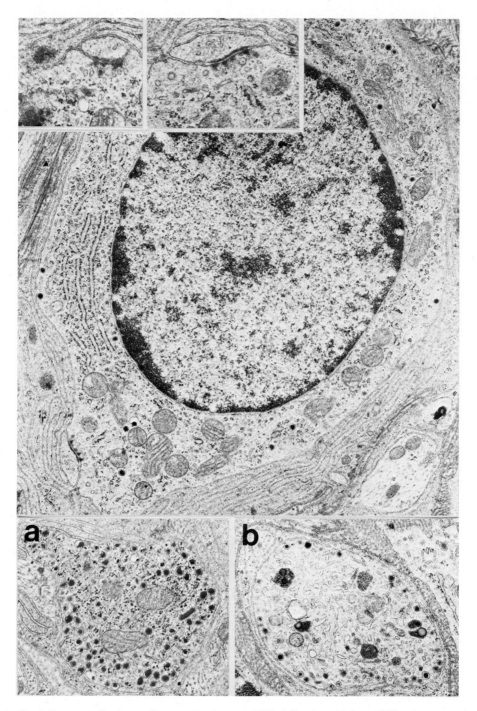

FIG. 4. (Top) Paraganglionic perikaryon in the rat SCG following 15 immobilization periods, with parallelly arranged cisternae of granular ER. Note the single dcv (80 to 120 nm) and synaptic contact **(insets)** with a presumed dendritic spine. × 12,000. **Insets** × 45,000.
FIG. 5. (Bottom) Paraganglionic cell processes. **A:** With mitochondria and numerous dcv. **b:** With dcv, electron-dense inclusions, as well as irregular vesicular and tubular profiles. × 18,000.

ACKNOWLEDGMENTS

This study has been supported by the DFG grants He 919/4, Ad 54/1 and by a trainee scholarship of the Sandoz-AG, Department of Preclinical Research. We wish to thank Miss Hannelore Ehlers for writing the manuscript. The BTG micro-laser unit was kindly made available by Sandoz, Department of Preclinical Research.

REFERENCES

1. Addicks, K., Heym, Ch., and Schindelmeiser, J. (1979): Morphologie paraganglionärer Zellsysteme der Ratte nach Immobilisation. *Proc. 1. Arbeitstagung Anat. Ges.,* 7.-8.9.79 Würzburg.
2. Björklund, A., Cegrell, L., Falck, B., Ritzén, M., and Rosengren, E. (1970): Dopamine-containing cells in sympathetic ganglia. *Acta Physiol. Scand.,* 78:334-338.
3. Chiba, T., and Williams, T.H. (1975): Histofluorescence characteristics and quantification of small intensely fluorescent (SIF) cells in sympathetic ganglia of several species. *Cells Tissue Res.,* 162:331-341.
4. Chiba, T., Black, A.C., Jr., and Williams, T.H. (1977): Evidence for dopamine storing interneurons and paraneurons in rhesus monkey sympathetic ganglia. *J. Neurocytol.* 6:441-453.
5. Coupland, R.E., and Hopwood, D. (1966): The mechanism of differential staining reaction for noradrenalin- and adrenalin-storing granules in tissues fixed in glutaraldehyde. *J. Anat.,* 100:227-243.
6. De la Torre, J.C., and Surgeon, J.W. (1976): A methodological approach to rapid and sensitive monoamine histofluorescence using a modified glyoxylic acid technique: The SPG method. *Histochemistry,* 49:81-93.
7. Eränkö, O., and Eränkö, L. (1971): Small intensely fluorescent granule-containing cells in the sympathetic ganglion of the rat. *Brain Res.,* 34:39-51.
8. Eränkö, L, and Eränkö, O. (1972): Effect of hydrocortisone on histochemically demonstrable catecholamines in the sympathetic ganglia and extraadrenal chromaffin tissue of the rat. *Acta Physiol. Scand.,* 84:125-133.
9. Fujita, T. (1977): Concept of paraneurons. *Arch. Histol. Jpn., (Suppl.)* 40:1-12.
10. Hartman, B.K. (1973): Immunofluorescence of dopamine-β-hydroxylase: Application of improved methodology to the localization of the peripheral and central noradrenergic nervous system. *J. Histochem. Cytochem.,* 21:321-331.
11. Heym, Ch. (1978): Functional morphology of monoamine-storing cells in the rat superior cervical ganglion. In: *Peripheral Neuroendocrine Interaction,* edited by R.E. Coupland and W.G. Forssmann, pp. 70-79. Springer-Verlag, Heidelberg.
12. Heym, Ch., and Williams, T.H. (1979): Evidence for autonomic paraneurons in sympathetic ganglia of a shrew (Tupaia glis). *J. Anat.,* 129:151-164.
13. Heym, Ch., König, R., Schröder, H.J., and Gerold, N. (1979): Immunofluorescent and biochemical studies on dopamine-β-hydroxylase and catecholamines. *Acta Histochem. (in press).*
14. König, R., and Heym, Ch. (1978): Immunofluorescent localization of dopamine-β-hydroxylase in small intensely fluorescent cells of the rat superior cervical ganglion. *Neurosci. Lett.,* 10:187-191.
15. Koslow, S.H. (1976): Mass fragmentographic analysis of SIF cell catecholamines of normal and experimental rat sympathetic ganglia. In: *SIF Cells. Structure and Function of the Small, Intensely Fluorescent Sympathetic Cells,* edited by O. Eränkö, pp. 82-88. Fogarty International Center Proceedings No. 30. DHEW Publ. No. (NIH) 76-942. Government Printing Office, Washington, D.C.
16. Koslow, S.H., Racagni, G., and Costa, E. (1974): Mass fragmentographic measurement of norepinephrine, dopamine, serotonin and acetylcholine in seven discrete nuclei of the rat teldiencephalon. *Neuropharmacology,* 13:1122-1129.
17. Lever, J.D., Santer, R.M., Lu, K.S., and Presley, R. (1977): Electron probe x-ray microanalysis of small granulated cells in rat sympathetic ganglia after sequential aldehyde and dichromate treatment. *J. Histochem. Cytochem.,* 25:275-279.
18. Libet, B., and Owman, Ch. (1974): Concomitant changes in formaldehyde-induced fluorescence of dopamine interneurons and in slow inhibitory postsynaptic potentials of the rabbit superior cervical ganglion induced by stimulation of the preganglionic nerve or by a muscarinic agent. *J. Physiol. (Lond),* 237:635-662.
19. Lu, K.S., Lever, J.D., Santer, R.M., and Presley, R. (1976): Small granulated cell types in rat superior cervical and coeliac-mesenteric ganglia. *Cell Tissue Res.,* 172:331-343.
20. Matthews, M.R., and Raismann, G. (1969): The ultrastructure and somatic efferent synapses of small granule-containing cells in the superior cervical ganglion. *J. Anat.,* 105:255-282.
21. Pearson, J.D.M., and Sharman, D.F. (1974): Increased concentration of acidic metabolites of dopamine in the superior cervical ganglion following preganglionic stimulation in vivo. *J. Neurochem.,* 22:547-550.
22. Siegrist, G., Dolivo, M., Dunant, Y., Foroglou-Kerameus, C., de Ribaupierre, F., and Rouiller, Ch. (1968): Ultrastructure and function of the chromaffin cells in the superior cervical ganglion of the rat. *J. Ultrastruct. Res.,* 25:381-407.
23. Williams, T.H. (1967): Electron microscopic evidence for an autonomic interneuron. *Nature,* 214:309-310.
24. Williams, T.H., Chiba, T., Black, A.C., Jr., Bhalla,

R.C., and Jew, J. (1976): Species variation in SIF cells of superior cervical ganglia: Are there two functional types? In: *SIF Cells. Structure and Function of the Small Intensely Fluorescent Sympathetic Cells,* edited by O. Eränkö, pp. 143-162. Fogarty International Center Proceedings No. 30.

DHEW Publ. No. 76-942. Government Printing Office, Washington, D.C.

25. Yokota, R. (1973): The granule-containing cell somata in the superior cervical ganglion of the rat, as studied by a serial sampling method for electron microscopy. *Z. Zellforsch.,* 141:331-345.

Histochemistry and Cell Biology of
Autonomic Neurons, SIF Cells, and
Paraneurons,
edited by O. Eränkö et al.
Raven Press, New York © 1980.

SIF Cells in Cat Sympathetic Ganglia and Associated Paraganglia

A. Autillo-Touati, and R. Seite

Groupe de Neurocytobiologie, Laboratoire de Biologie Cellulaire et Histologie (Secteur Sud), Faculté de Médecine, 13385 Marseille cedex 4, France

Small intensely fluorescent (SIF) cells were first described in the rat superior cervical ganglion (SCG) by Eränkö and Härkönen (7) and were considered afterwards as inhibitory dopaminergic interneurons (11). Many studies concerning these cells (2,3,5,9,12,15,17) have shown important differences in their number, morphology, and amine content according to species and ganglia. Because of this quantitative and qualitative heterogeneity, the role of SIF cells in modulating ganglionic transmission remains to be elucidated.

SIF CELL HETEROGENEITY

As a contribution to this important problem, serial sections of the SCG and celiac ganglion (CG) in the cat were investigated by fluorescence and electron microscopy (1). A combined method (9) for formaldehyde-induced fluorescence and electron microscopy was used. This study showed that the CG contains five times more SIF cells than the SCG in the cat. Furthermore, two types of SIF cells were identified in the CG and SCG (Table 1). Type I cells, presenting large polymorphous dense-cored vesicles, are arranged as clusters around capillaries and are surrounded by a dense network of collagen fibers (Figs. 1 and 5). Type II is composed of isolated cells which possess small dense-cored vesicles and long varicose processes. These cells establish synaptic junctions with preganglionic fibers and dendrites of principal neurons (Figs. 2 and 5). Less than 10% of the total SIF cells are of type II, which could be considered as interneurons. On the other hand, type I cells largely predominate in the CG and SCG of the cat, representing more than 90% of total SIF cells observed by fluorescence microscopy (Table 1). The question naturally arises as to the endocrine and paracrine role of the type I cells and their influence on modulation of neural transmission.

Our study of the SIF cells in cat SCG and CG point out two main facts: The number of these cells varies greatly according to the ganglionic territory examined, and two distinct cell types are present in different ratios. Such differences have been observed in other ganglia in other species. In the cat SCG there are three times fewer SIF cells than in the rat SCG (16,17). The rabbit SCG is very rich in type II cells compared with the SCG and CG in the cat (16,17). In guinea pig SCG, no distinction can be made between isolated and clustered cells (16,17). Moreover, Eränkö and Eränkö (6) reported that in the rat SCG the clustered cell type increases with age as the isolated type decreases. It is well known that

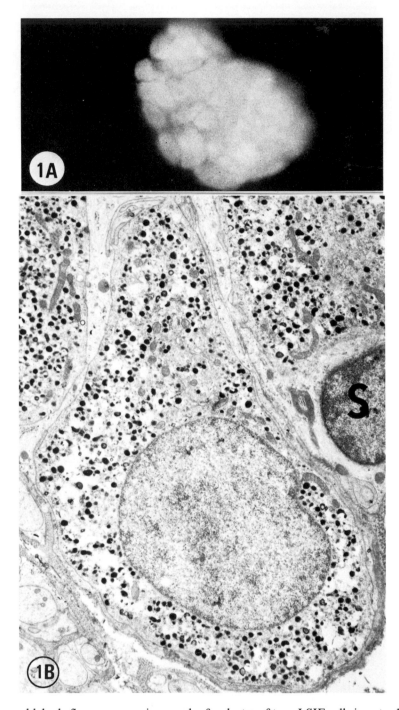

FIG. 1. A: Formaldehyde fluorescence micrograph of a cluster of type I SIF cells in cat celiac ganglion. ×900. **B:** Electron micrograph of type I SIF cell. The cytoplasm is filled with dense-cored vesicles. Satellite cell (S) is associated with SIF cell. × 10,000.

TABLE 1. *SIF cell characteristics in cat celiac ganglion.*

Cell	Formalde-hyde fluorescence	Organization	Distribution	Synapses		Dense cored vesicles	Frequency (%)	Functional interpretation
				Afferent	Efferent			
Type I (av. diameter 8-12 μm)	Intense yellow	Clusters centered around capillaries	Ganglion periphery	—	—	Polymorphous, 100-300 nm	90	Endocrine Paracrine
Type II (av. diameter 8-12 μm)	Intense yellow	Isolated cells with varicose processes	Ganglionic mass	?	+	Regular, 50-150 nm	10	Neurocrine

cells of the adrenal medulla, SIF cells, and sympathetic neurons share the same neuroectodermal origin. It has been shown that fluorescent cells in cultured chicken sympathetic ganglia develop morphological characteristics of sympathetic neurons (10). In addition, rat adrenal medullary cells transplanted in the anterior chamber of the eye are able to reinnervate the iris (15). Thus the common origin of SIF cells, sympathetic neurons, and chromaffin cells of the adrenal medulla and the plasticity of the adrenal sympathetic system could account for the coexistence of two cell types in different ratios. One of the cell types would be closer to the endocrine system, and the other would resemble nerve cells. Further studies are required to elucidate the significance of such quantitative and qualitative heterogeneity according to species and ganglionic territories.

ASSOCIATED PARAGANGLIA

In the present work special attention was given to the existence of very large fluorescent clusters of small cells associated only to the CG and which form small organs called paraganglia (PG) (Figs. 3, 4, and 5).

The PG cells display a connective tissue capsule, one side of which is in continuity with the ganglion capsule. The other side separates the PG cells from the principal neurons, the former appearing as distinct aggregates (Fig. 3). The PG cells can be easily distinguished from the noradrenergic neurons by their intense yellow fluorescence (Fig. 3), their cytoplasm filled with large polymorphous dense-cored vesicles (Figs. 3, 4, and 5), and their subcapsular clustered organization (Fig. 3). The PG cells are located close to large vessels. Expansions from the PG cells end close to the capillaries. The satellite cells surrounding the PG cells are identical to those of the principal neurons and do not completely cover the surface of these cells. Serial sections were used to establish the general morphological features of the PG cells as well as their relationship to one another and among the various ganglionic constituents. The PG cells do not exhibit processes and do not establish contact with the principal neurons or the pre- and postganglionic fibers. Accordingly, we feel that the PG cells could belong to the same cell category as the type I SIF cells. According to some authors, PGs are considered as SIF cells of the endocrine type (5,8). PGs are frequently associated with the abdominal ganglia in various species (5,8,13,14) and, like the adrenal medulla, may be innervated by cholinergic fibers (8,14).

SIF CELL FUNCTION IN NERVOUS TRANSMISSION OF CAT CG

Our observations underline three points. First, the cat CG has few SIF cells relative to the total number of principal neurons. Second, two types of SIF cell are present, with the endocrine type strongly predominating.

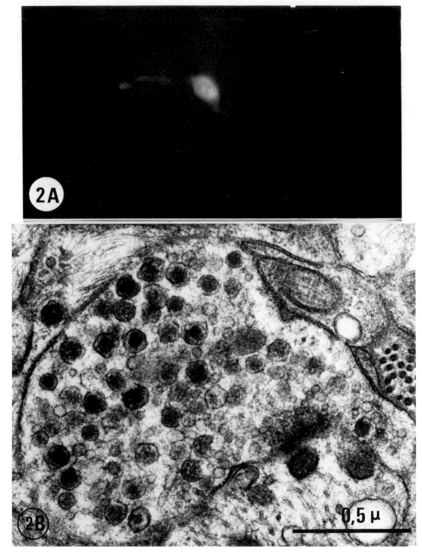

FIG. 2. A: Formaldehyde fluorescence micrograph of type II isolated SIF cell in cat celiac ganglion. Varicose processes are visible. ×900. **B:** Synapse. The presynaptic element contains dense-cored vesicles. × 60,000.

Third, PGs are associated with the CG in the cat. Additionally, as stated previously, the cat CG displays five times more SIF cells than the SCG. These observations raise the following question: Does modulation of neural transmission exist in the cat CG, and, if so, what is its importance and by what mechanisms does it operate?

Recent biochemical and pharmacological findings (4) suggest the existence of an inhibitory-type modulation of transmission in the cat CG. Thus, according to cell type, SIF cells of the cat CG might modulate ganglionic transmission as true interneurons, as endocrine cells secreting directly into the blood stream, or as paracrine cells secreting directly into neighboring cells. The last two mechanisms would probably act more slowly than

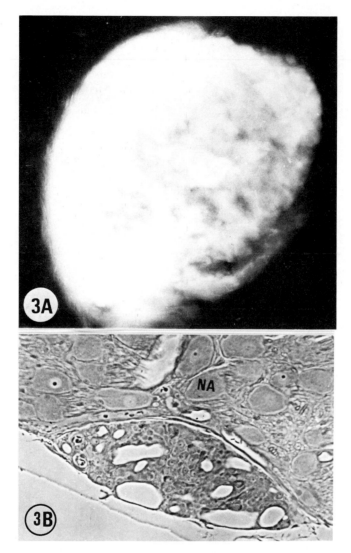

FIG. 3. A: Formaldehyde fluorescence micrograph of paraganglion cells. × 300. **B:** Paraganglion (PG), observed on phase contrast, is surrounded by a connective tissue capsule. The cell cytoplasm is filled with granulations. PG cells are smaller than the noradrenergic neurons (NA). × 1,000.

the first. It is known that under physiological conditions the sympathetic system transmits very few messages. However, with intense stimulation there is a great volume of transmission, during which the SIF cells could act as a "brake" to avoid over exerting the system. This explanation could account for the small number of these cells encountered in certain ganglia.

The high degree of vascularization and the absence of afferent and efferent synapses point out that the PG associated with the CG are small endocrine organs. Although anatomical proof is lacking, the PG capillaries could possibly carry blood to the celiac ganglion where catecholamines would be released into the principal neurons, thereby modulating their activity (8).

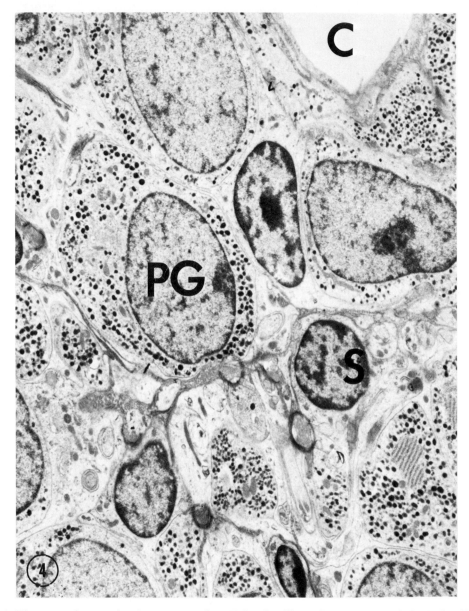

FIG. 4. Electron micrograph of a paraganglion (PG). Satellite cells (S) surround the PG cells. One capillary (C) is visible. × 5,000.

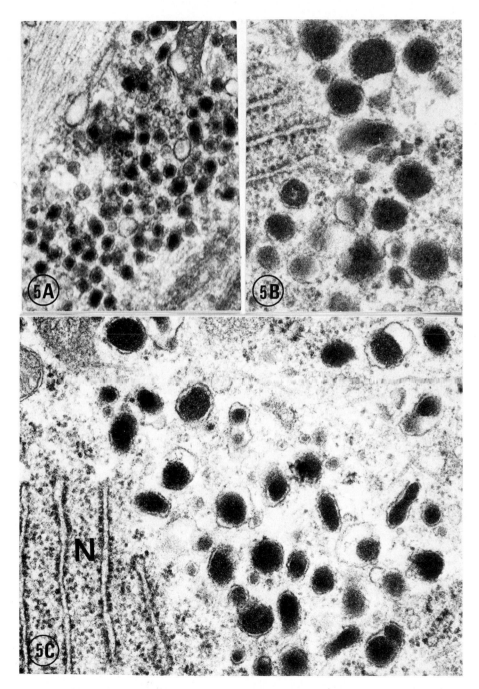

FIG. 5. Note the differences in size and aspect of the dense-cored vesicles among a type II SIF cell **(A)**, a type I SIF cell **(B)** and a paraganglion cell **(C)**. A Nissl body (N) is visible. × 60,000.

CONCLUSIONS

The data presented herein are not incompatible with the notion of modulation of ganglionic transmission. The latter may be based not only on mechanisms of neurotransmission but on forms of action embodied in the broader conception of the neuromodulation phenomenon.

ACKNOWLEDGMENTS

We wish to thank Mrs. J. Vuillet, Mrs. Vio, Mrs. J. Bottini, and Mr. C. Cataldo for their expert technical assistance.

REFERENCES

1. Autillo-Touati, A., and Seite, R. (1979): A cytochemical and ultrastructural study of the "S.I.F." cells in cat sympathetic ganglia. *Histochemistry,* 60:189-223.
2. Chiba, T. (1977): Monoamine-containing paraneurons in the sympathetic ganglia of mammals. In: *Paraneurons. New Concepts on Neuroendocrine Relatives,* edited by S. Kobayashi, and T. Chiba, *Arch. Histol. Jpn.,* 40:163-176.
3. Dail, W. (1977): Histochemical and fine structural studies of SIF cells in the major pelvic ganglion of the rat. In: *SIF Cells, Structure and Function of Small, Intensely Fluorescent, Sympathetic Cells,* edited by O. Eränkö, pp. 8-17. Fogarty International Center Proceedings No. 30. Government Printing Office, Washington, D.C.
4. Della Bella, D., and Benelli, G. (1977): Evidence of a modulatory role of SIF cells on cat celiac ganglion synaptic transmission. In: *Advances in Biochemical Psychopharmacology,* edited by E. Costa and G.L. Gessa, 16:513-517. Raven Press, New York.
5. Elfvin, L.G., Hökfelt, T., and Goldstein, M. (1975): Fluorescence microscopical, immunohistochemical and ultrastructural studies on sympathetic ganglia of the guinea-pig with special reference to the S.I.F. cells and their catecholamine content. *J. Ultrastruct. Res.,* 51:377-396.
6. Eränkö, O., and Eränkö, L. (1971): Small, intensely fluorescent granule-containing cells in the sympathetic ganglion of the rat. In: *Progress in Brain Re-*search, edited by O. Eränkö, 34:39-51. Elsevier, Amsterdam.
7. Eränkö, O., and Härkönen, M. (1963): Histochemical demonstration of fluorogenic amines in the cytoplasm of sympathetic ganglion cells of the rat. *Acta Physiol. Scand.,* 58:285-286.
8. Furness, J.B., and Sobels, G. (1976): The ultrastructure of paraganglia associated with the inferior mesenteric ganglia in the guinea-pig. *Cell Tissue Res.,* 171:123-139.
9. Grillo, M.A., Jacobs, L., and Comroe, J.H. (1974): A combined fluorescence histochemical and electron microscopic method for studying special monoamine containing cells. *J. Comp. Neurol.,* 153:1-14.
10. Hervonen, A., and Eränkö, O. (1975): Fluorescence histochemical and electron-microscopical observations on sympathetic ganglia of the chick embryo cultured with and without hydrocortisone. *Cell. Tissue Res.,* 156:145-166.
11. Libet, B., and Owman, Ch. (1974): Concomitant changes in formaldehyde-induced fluorescence of dopamine interneurons, and in slow-inhibitory postsynaptic potentials of the rabbit, induced by stimulation of the pre-ganglionic nerve or by a muscarinic agent. *J. Physiol. (Lond),* 237:635-662.
12. Lu, K.S., Lever, J.D., Santer, R.M., and Presley, R. (1976): Small granulated cell types in rat superior cervical and celiac-mesenteric ganglia. *Cell. Tissue Res.,* 172:331-343.
13. Mascorro, J.A., and Yates, R.D. (1970): Microscopic observations on abdominal sympathetic paraganglia. *Texas Rep. Biol.,* 28:59-68.
14. Mascorro, J.A., and Yates, R.D. (1975): A review of abdominal paraganglia: ultrastructure, mitotic cells, catecholamine release, innervation, light and dark cells, vascularity. In: *Electron microscopic concepts of secretion,* edited by M. Hess, P.M. Heidger, R.F. Dyer, and J.R. Ruby, pp. 435-450. Wiley, New York.
15. Olson, L. (1970): Fluorescence histochemical evidence for axonal growth and secretions from transplanted adrenal medullary tissue. *Histochemie,* 22:1-7.
16. Williams, T.H., Black, A.C., Jr., Chiba, T., and Jew, J. (1976): Interneurons/SIF cells in sympathetic ganglia of various mammals. In: *Chromaffin, Enterochromaffin and Related Cells,* edited by R.E. Coupland and T. Fujita, pp. 95-116. Elsevier, Amsterdam.
17. Williams, T.H., Black, A.C., Jr., Chiba, T., and Jew, J.Y. (1977): Species differences in mammalian SIF cells. In: *Advances in Biochemical Psychopharmacology,* edited by E. Costa and G.L. Gessa, 16:505-511. Raven Press, New York.

Histochemistry and Cell Biology of
Autonomic Neurons, SIF Cells, and
Paraneurons,
edited by O. Eränkö et al.
Raven Press, New York © 1980.

Is the SIF Cell Truly an Interneuron in the Superior Cervical Ganglion?

Hisatake Kondo

Department of Anatomy, College of Medicine, University of Illinois, Chicago, Illinois 60612

Electron microscopic studies of the innervation of SIF cells in the superior cervical ganglion were first reported by Matthews and Raisman (13) and Williams and Palay (15) during the late 1960's. Two types of neuronal profiles were described; the first contained small clear vesicles and the second one clusters of granular vesicles characteristic of SIF cells. The former type of neuronal profile was regarded as presynaptic to SIF cells and presumed to come from the preganglionic neurons, whereas the latter type was regarded as postsynaptic to SIF cells and was believed to synapse with the postganglionic principal neuron. This interpretation was favorable to the physiological idea of dopaminergic interneuron-generating slow inhibitory postsynaptic potential (s-IPSP) proposed by Libet and his group (11). Since then, most physiological as well as morphological studies concerning the mechanism of s-IPSP in the autonomic ganglia have been based on the interpretation that the SIF cell is synaptically interposed between pre- and postganglionic neurons. However, this interpretation has two major problems which remain to be confirmed: (a) Is the identification of synapses at sites of contact between these two neuronal profiles and SIF cells well compatible

with the present morphological criteria for chemical synapses? (b) Are these two types of neuronal profiles truly of different origin, the first preganglionic and the second postganglionic?

By electron microscopy, the chemical synapses have been recognized as sites at which clusters of small vesicles are in close association with the cytoplasmic surface of the presynaptic membrane (14). At such sites, the pre- and postsynaptic membranes lie parallel to each other with a gap of about 20 nm. The cytoplasmic surface of one or both membranes is coated with electron-dense material. This coating may be either symmetrical or asymmetrical. When neuronal profiles on both sides contain vesicles, the profiles with a vesicle accumulation close to the synaptic membrane is considered to be the presynaptic component. In this regard, it should be noted that in studies so far published on the innervation of the SIF cells most electron micrographs of neuronal profiles which were regarded as presynaptic to SIF cells presented only junctional sites with a symmetrical coating of dense material similar to the desmosomes. At these sites no clustered vesicles were seen close to the junctional membrane, although many vesicles were scattered in the neuronal profile. Only a few reports showed typical incoming synapses to SIF cells fulfilling the morphological criteria for the chemical synapses described above. Desmosomes are known to be ubiquitous in the peripheral as well as the central nervous system.

The origin of these two neuronal profiles is not fully confirmed by experimental findings, although only one study (12) was reported in which the severance of the preganglionic trunk results in degeneration of most neuronal profiles apposing SIF cells. In an attempt to solve these problems, a study was undertaken to re-examine the innervation of SIF cells using long series of thin sections (4).

SERIAL SECTION STUDY ON THE INNERVATION OF THE CHIEF CELL OF THE CAROTID BODY

Before presenting the data on SIF cells, it is useful to describe the results of a similar serial section study on the innervation of the carotid body chemoreceptor because the chief cell of the carotid body shows ultrastructural characteristics similar to those of SIF cells (7).

In many studies of the carotid body using random sections, the presence of two types of synapse have been noted on the chief cells. Many vesicles and clusters of vesicles were seen inside the type 1 (incoming to cell) synapse, and vesicle clusterings were found inside the chief cells at the type 2 (outgoing from cell) synapse. In addition to these two membrane specializations, desmosomes were often seen at apposition sites between nerves and chief cells. The type 1 synapse is considered to be formed by motor fibers of the glossopharyngeal nerve, and the type 2 synapse by sensory fibers of the same nerve. However, the serial section study by the present author (6) revealed that a single nerve fiber forms both types of synapse, with the type 2 predominating (Fig. 1). Thus the possibility was strongly suggested that most of type 1 synapses are not formed by motor fibers but solely by sensory fibers also having many type 2 synapses.

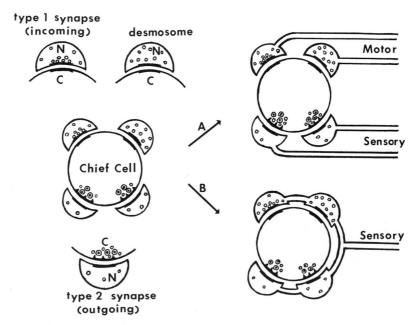

FIG. 1. Comparison of the results from random section studies hitherto published (A) and those from my serial section study (B) of the innervation of the chief cell of the carotid body.

SERIAL SECTION STUDY ON THE INNERVATION OF SIF CELLS

Within the reconstructed area (60 × 40 × 40 μm), using 400 serial ultrathin sections, there were 19 SIF cells forming groups (15 complete and 4 incomplete), each of which was given a number (1 to 19), and one ganglion cell (Fig. 2). SIF cells had several cytoplasmic processes, but all these processes were confined within the reconstructed cell groups and did not extend beyond the boundary of the cell group. Most neuronal profiles apposed to the reconstructed SIF cells were found to originate from axons (Fig. 3). Two types of synapse (types 1 and 2) were formed at apposition sites between SIF cells and neuronal profiles. SIF cells were postsynaptic elements in type 1 synapses and formed presynaptic elements in type 2 synapses. A single axon made *en passant* and bou-

ton synaptic contact with many SIF cells and formed both types of synapse, with type 2 predominating (Figs. 2 through 4). In other words, type 1 synapses occurred in a small number on axons also forming numerous type 2 synapses. Thus the main direction of transmission appears to be from the SIF cell to axons. Many desmosomes were found along the trajectory of the axons where no clustered vesicles were found, although abundant vesicles were scattered inside the axons. A few neuronal profiles apposed to SIF cells were derived from dendrites of adjacent ganglion cells. Dendrites formed a few type 2 synapses with SIF cells.

Where do these axons come from? As described above, the severance of the preganglionic trunk results in the degeneration of a considerable number of neuronal profiles apposed to SIF cells. Therefore it is possible that the axons originate from the ones travel-

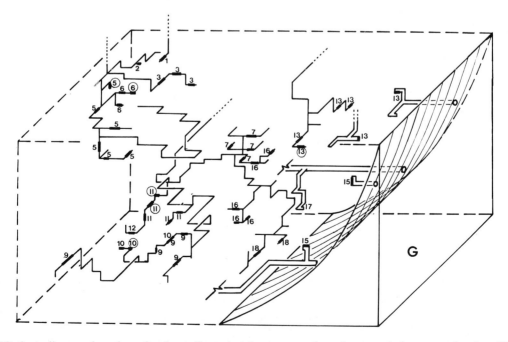

FIG. 2. A diagram based on the three-dimensional reconstruction of neuronal elements related to SIF cells. A hexahedron indicated by solid and broken lines represents the reconstructed area containing 19 SIF cells. Thin solid lines represent axons, and thick solid lines represent sites of synaptic contacts. Dendrites of a ganglion cell (G) are indicated by double solid lines. Dotted lines represent the portion of axons and dendrites extending beyond the reconstructed area. Circled numbers indicate SIF cell numbers facing type 1 synapses. Uncircled numbers depict SIF cell numbers facing type 2 synapses.

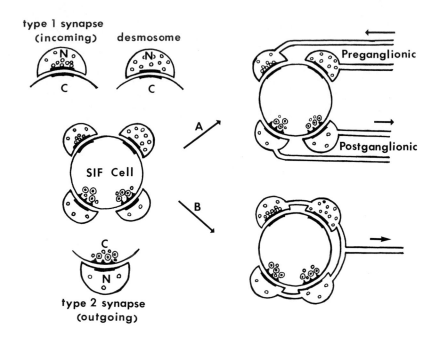

FIG. 3. Comparison of the results from random section studies hitherto published (A) and those from my serial section study (B) of the innervation of the SIF cells in the SCG.

ing in the preganglionic trunk (Fig. 4). This implies that some axons in the trunk might conduct centripetal or sensory impulses from SIF cells, although the possibility cannot be ruled out that these axons might be branches of the ordinary preganglionic nerve fibers.

POSSIBLE FUNCTIONS OF THE SIF CELLS

This striking similarity between SIF cells and chief cells of the carotid body in terms of synaptic relation implies that SIF cells function as chemoreceptors in the ganglion. With regard to this speculation, it should be noted that another serial section study revealed SIF cells in the nodose ganglion to receive innervation quite similar to those of SIF cells in the superior cervical ganglion (SCG) and chief cells in the carotid body (8). In addition, some physiological studies have been reported in which the nodose ganglion has chemoreceptive properties, responding to various chemicals injected into its circulation (1,2,5).

There are two possible explanations for the data presented here in relation to the idea of dopaminergic interneuron. According to the first explanation, the stimulation of the preganglionic trunk arouses the antidromic stimulation of some basically centripetal nerve fibers. Then the stimulation affects the SIF cells through a few reciprocal type 1 synapses, influencing ganglion cells through very few type 2 synapses between SIF cells and dendrites of ganglion cells. Even if this is so, the number of such pathways is very small, and their significance is questionable. According to another more plausible explanation, the dopaminergic interneuron might be entirely different from SIF cells. Some observations favor this latter explanation: There are marked species differences in the population of SIF cells in the mammalian SCG (3), and the total number of SIF cells in the SCG of any mammals is too small compared to the number of ordinary ganglion cells. The rat ganglion contains rather numerous SIF cells, and the rabbit ganglion contains few. The monkey ganglion has very few SIF cells, and

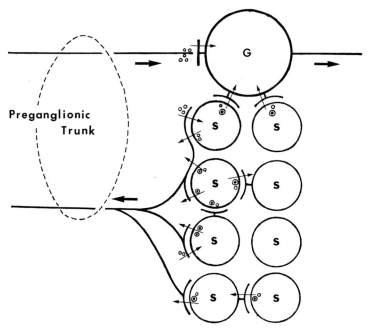

FIG. 4. Synaptic organization of SIF cells in the SCG. Arrows indicate directions of nerve conduction and synaptic transmission. (G) Ordinary ganglion cells. (S) SIF cells.

no typical SIF cells have been found in the postnatal human SCG (10). Interestingly, morphological findings on the SIF cells have been mainly obtained from studies with rat ganglia whereas most physiological studies on the s-IPSP have been done using rabbit ganglia containing few SIF cells.

MONOAMINERGIC NERVE FIBERS IN THE GANGLION—ANOTHER CANDIDATE FOR THE MEDIATOR OF s-IPSP

The SCG of rabbits was pretreated with 5-hydroxydopamine, which is known to enhance the appearance of the monoaminergic nerve fibers. After this treatment, thickened neuronal profiles packed with numerous small granular vesicles (45 nm in mean diameter) mixed with several large granular vesicles (80 to 140 nm in diameter) were found (4,9). They made incoming synapses onto ordinary ganglion cells, mostly on the processes but occasionally on the soma. The ratio of such synapses to those from the cholinergic

preganglionic nerve was about 3:19 in random sections. These monoaminergic neuronal profiles and cholinergic profiles were frequently enclosed by a common Schwann cell, and both formed synapses very close together on a single process of an ordinary ganglion cell (Fig. 5). Occasionally these two neuronal profiles were directly apposed to each other with slight membrane thickenings, but no synaptic membrane specializations were found at the sites (Fig. 6). These monoaminergic neuronal profiles survived preganglionic nerve division. Five monoaminergic neuronal profiles were followed up to 100 μm in 300 to 500 serial ultrathin sections (Fig. 7). These neuronal components first showed thickened vesicle-containing portions with a maximum diameter of 2 μm, changed their diameter abruptly into 0.1 to 0.2 μm, and remained thin except for several thickened portions along their trajectory. Some of these fibers had no branches, whereas others sent one branch and eventually extended beyond the reconstructed area. Thin portions contained several neurotubules and filaments

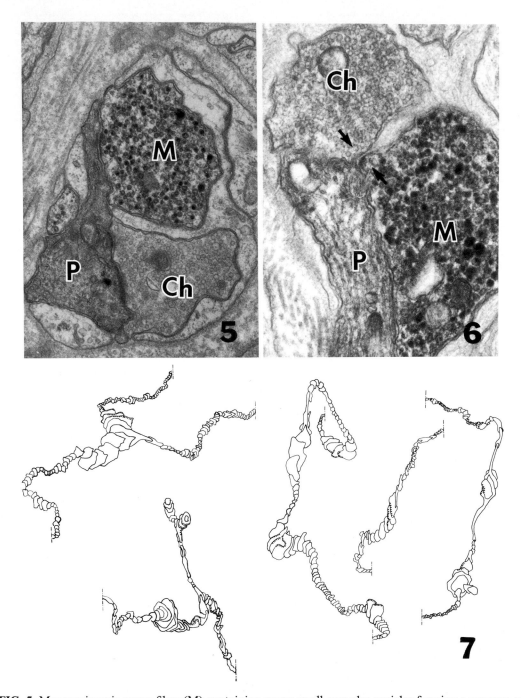

FIG. 5. Monoaminergic nerve fiber (M) containing many small granular vesicles forming a synapse on process (P) of an ordinary ganglion cell. (Ch) Cholinergic preganglionic nerve fiber. ×27,000.
FIG. 6. Occasional direct contact *(arrows)* between a monoaminergic and a cholinergic nerve fiber. Note the membrane densification without vesicle clustering. (P) Process of an ordinary ganglion cell. ×36,000.
FIG. 7. Three-dimensional reconstruction of monoaminergic nerve fibers using 300 to 500 serial sections. Stippled portions indicate synapses on ordinary ganglion cells.

but few if any granular vesicles. No ribosomes were encountered in any portions of the reconstructed monoaminergic nerve fibers. The nerve fibers received no synapses along their trajectory. No neuronal processes showing features of SIF cell processes were seen in the present material. This serial section study failed to disclose the perikaryal soma of the monoaminergic nerve fibers. However, it is unlikely from the present data that these fibers come from SIF cells. This strongly suggests that there is a special population of ganglion cells whose function is to provide monoaminergic synapses onto ordinary ganglion cells by sending nerve fibers that contain many small dense-cored vesicles.

CONCLUSION

Judging from the synaptic relation of SIF cells to other neuronal elements and the ultrastructural characteristics of the monoaminergic nerve fibers which frequently occur in the ganglion, the monoaminergic nerve fibers seem to be a more likely candidate for the mediator of s-IPSP than the SIF cells in the SCG. The SIF cell in the SCG might be a chemoreceptor with a similar mechanism for the origin of chemosensory discharge to that of the carotid body. Physiological studies in search for chemosensory centripetal impulses in the preganglionic trunk is crucial to verify the credibility of this possible function of the SIF cell.

ACKNOWLEDGMENT

The author wishes to thank Dr. T.Y. Yamamoto and C. Eyzaguirre for their encouragement throughout the course of this work.

REFERENCES

1. Borison, H.L., and Fairbanks, V.L. (1952): Mechanism of veratorum-induced emesis in the cat. *J. Pharmacol. Exp. Ther.,* 105:317-325.
2. Chai, C.Y., and Wang, S.C. (1966): Mechanism of sinus bradycardia induced by veratrum alkaloids-protoveratrine A. *J. Pharmacol. Exp. Ther.,* 154:546-557.
3. Chiba, T., and Williams, T.H. (1975): Histofluorescence characteristics and quantification of small intensely fluorescent (SIF) cells in sympathetic ganglia of several species. *Cell Tissue Res.* 162:331-341.
4. Dail, W.G., and Evan, A.P. (1978): Ultrastructure of adrenergic terminals and SIF cells in the superior cervical ganglion of the rabbit. *Brain Res.,* 148:469-477.
5. Jacobs, L., and Comroe, J.H., Jr., (1971): Reflex apnea, bradycardia, and hypotension produced by serotonin and phenyldiguanide acting on the nodose ganglia of the cat. *Circ. Res.,* 29:145-155.
6. Kondo, H. (1976): Innervation of the carotid body of the adult rat: A serial ultrathin section analysis. *Cell Tissue Res.,* 173:1-16.
7. Kondo, H. (1977): Innervation of the chief cells of the carotid body: An ultrastructural review. *Arch. Histol. Jpn.,* 40:221-230.
8. Kondo, H. (1977): Innervation of SIF cells in the superior cervical and nodose ganglia: An ultrastructural study with serial sections. *Biol. Cell.,* 30:253-264.
9. Kondo, H. (1978): A serial ultrathin section analysis of monoaminergic nerve fibers in the superior cervical ganglion of the rabbit. *Soc. Neurosci. Abstr.,* 8:276.
10. Kondo, H., and Fujiwara, S. (1979): Granule-containing cells in the human superior cervical ganglion. *Acta Anat. (Basel),* 103:192-199.
11. Libet, B. (1970): Generation of slow inhibitory and excitatory postsynaptic potentials. *Fed. Proc.,* 29:1945-1956.
12. Matthews, M.R., and Ostberg, A. (1973): Effects of preganglionic nerve section upon the afferent innervation of the small granule-containing cells in the rat superior cervical ganglion. *Acta Physiol. Pol.,* 26:215-223.
13. Matthews, M.R., and Raisman, G. (1969): The ultrastructure and somatic efferent synapses of small granule-containing cells in the superior cervical ganglion. *J. Anat.,* 105:255-282.
14. Pappas, G.D., and Waxman, S.G. (1972): Synaptic fine structure-morphological correlates of chemical and electronic transmission. In: *Structure and Function of Synapses,* edited by G.D. Pappas and D.P. Purpura, pp. 1-43. Raven Press, New York.
15. Williams, T.H., and Palay, S.L. (1969): Ultrastructure of the small neurons on the superior cervical ganglion. *Brain Res.,* 15:17-34.

Histochemistry and Cell Biology of
Autonomic Neurons, SIF Cells, and
Paraneurons,
edited by O. Eränkö et al.
Raven Press, New York © 1980.

Functional Roles of SIF Cells in Slow Synaptic Actions

Benjamin Libet

Department of Physiology, School of Medicine, University of California San Francisco, San Francisco, California 94143

Although small intensely fluorescent (SIF) cells have been found to be widely distributed in sympathetic ganglia and to exhibit considerable variations in their numbers, specific catecholamines, and morphological characteristics (16,17), positive evidence for specific functional roles of SIF cells is thus far chiefly based on studies with the rabbit superior cervical ganglia (SCG) (26,29) and the paravertebral ganglia of the frog (33,37, 42). It was indeed for rabbit SCG that the existence of and physiological role for a type of adrenergic interneuron in ganglia had been initially postulated on physiological and pharmacological grounds (14). The known functions relate entirely to two postsynaptic actions of catecholamines: (a) A hyperpolarizing action equivalent to that in the slow inhibitory postsynaptic potentials (s-IPSPs) (22,26); an example of s-IPSP and slow excitatory postsynaptic potential (s-EPSP) responses elicited orthodromically in curarized rabbit SCG is shown in Fig. 1. (b) A modulatory action in the rabbit SCG, in which a brief exposure to dopamine (DA) induces a long-lasting enhancement of the s-EPSP [or of the equivalent slow muscarinic depolarizing responses to acetylcholine (ACh) or its muscarinic agonists] (36) (Fig. 2). I review and clarify, briefly, present concepts of how SIF cells are involved in these functions, and suggest some possibilities for functions of various SIF cell types in other ganglia.

SIF CELLS AS MEDIATORS OF s-IPSP

Second Transmitter Required

Release of a second, noncholinergic transmitter is required in the preganglionic pathway between preganglionic release of ACh and the s-IPSP response for rabbit and frog ganglia. A muscarinic agonist elicits a biphasic response; an initially hyperpolarizing phase is followed by a more prolonged depolarizing one (36). These two phases correspond, respectively, to the s-IPSP and s-EPSP responses to orthodromic preganglionic input in curarized ganglia (26). When the Ca/Mg ratio in the bathing medium was lowered to a level that stopped presynaptic release of all transmitters, as judged by elimination of all PSP responses to preganglionic input, the hyperpolarizing (s-IPSP) phase of these responses to muscarinic agonists was selectively eliminated (26,33; see also 11a). This demonstrated that intraganglionic release of a second transmitter was necessary for production of the hyperpolarizing response to ACh, and it supported our hypoth-

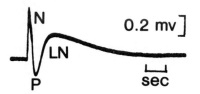

FIG. 1. Postsynaptic potentials of curarized rabbit SCG. Surface recordings with an air gap between one electrode on the ganglion and the reference electrode on the internal carotid branch (postganglionic nerve); response to stimulation of preganglionic nerve (cervical sympathetic) by a train of supramaximal pulses, 40 pps for 0.25 sec [as in Eccles and Libet (14)]. Surface-negative (N) potential is seen during stimulation; N represents the summated fast EPSPs, with their peaks sloping in the downward direction because of the developing surface-positive (P) or hyperpolarizing component. P outlasts, and reaches its peak after, the stimulus train; it is the s-IPSP. The "late-negative" (LN) depolarizing phase is the s-EPSP; since s-EPSP has a synaptic delay 200 to 300 msec longer than that for the s-IPSP, it would just begin to develop after the end of this brief stimulus train and could only minimally affect the peak amplitude of this s-IPSP. The form of this surface-recorded response is precisely similar to that of an intracellular recording from an impaled principal neuron [as in Libet and Tosaka (36)].

esis (14) that the intraganglionic pathway mediating the s-IPSP included an adrenergic interneuron (Fig. 3).

The intervening noncholinergic transmitter in the orthodromic pathway for the s-IPSP appears to be a catecholamine (26,33). DA is identified by various lines of evidence as the actual transmitter in the case of rabbit SCG (30,34,36). For frog ganglia, the direct transmitter is likely to be norepinephrine (NE) and/or epinephrine (E) rather than DA, chiefly based on the identification of these catecholamines in the SIF cells of frog ganglia (23). The s-IPSPs in rabbit and frog ganglia are blocked, relatively selectively, by α-adrenergic antagonists and enhanced by inhibitors of COMT (26,29,33).

Nicotine Blockade

It must be emphasized that the use of nicotine (instead of a curariform agent) to block nicotinic responses to ACh in frog ganglia

can lead to a different result and to misleading inferences about the nature of the s-IPSP. During the noncompetitive "late" blockade by nicotine, the hyperpolarizing response of frog ganglion to ACh is not eliminated by treatment with low Ca/Mg (33,45). That is, under the special and abnormal membrane conditions induced by nicotine, ACh becomes capable of eliciting a direct hyperpolarizing response without the intervention of a second transmitter. This direct response to ACh in nicotinized ganglia should not be confused with the physiological hyperpolarization or s-IPSP of nonnicotinized ganglia; the physiological s-IPSP of frog sympathetic ganglion does require release of a second transmitter, and it cannot be a direct response to ACh (33). A considerable amount of work was carried out with nicotinized ganglia of the frog before the distinction between their "s-IPSP" and the s-IPSP of normal (nonnicotinized) ganglia was fully recognized; the findings and conclusions from such work must obviously be reinterpreted. This includes the findings that s-IPSP of nicotinized ganglia was selectively sensitive to ouabain, leading to the proposal that the s-IPSP is generated by an activation of an electrogenic Na-K pump (40). However, the physiological s-IPSP of nonnicotinized frog ganglia is *not* selectively sensitive to ouabain (26). [The s-IPSP of rabbit ganglia (26), as well as the direct hyperpolarizing response to DA (39), are also not selectively sensitive either to ouabain or to Li^+ replacement of Na^+.] The further proposal that electrogenesis of the s-IPSP is mediated by a decrease ("inactivation") of Na^+ conductance was also based on studies with nicotinized frog ganglia (45). Regardless of any merits the "Na^+ inactivation" hypothesis (45) may have for explaining the direct ACh hyperpolarizing response of nicotinized ganglia, there is no reason at all to apply it to the case of physiological s-IPSPs of normal ganglia, as the latter are responses to another transmitter, probably a catecholamine, but not to ACh directly (33). Additionally, normal neu-

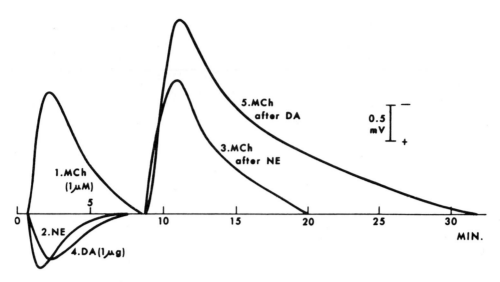

FIG. 2. Modulation of slow muscarinic depolarizations by DA. Surface-recorded responses of rabbit SCG (22°C) in a sucrose-gap superfusion chamber; the ganglion was pretreated with BCh to deplete intraganglionic DA and eliminate the initial hyperpolarizing component normally seen in the responses to methacholine (MCh) (34). The first test shown is the response to a single-bolus injection of 1 μmole MCh into the ganglionic superfusate. Test 2 is the response to 1 μg NE. Test 3 is a repeat of MCh, as in test 1, done shortly after the hyperpolarizing response to NE finished. Test 4 shows the response to 1 μg DA injected after the conclusion of response 3. Test 5 is a repeat of MCh, as in test 1, shortly after the hyperpolarizing response to DA finished. The substantial increase in amplitude and duration of the MCh response seen in test 5 after DA, but not in test 3 after NE, can be exhibited by succeeding tests with MCh for some hours even though no further DA is applied. (From ref. 29, based on experiments reported in ref. 36.)

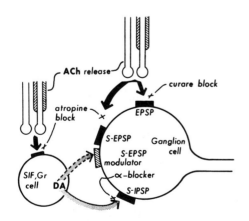

FIG. 3. Schema of intraganglionic synaptic connections for mediation of fast EPSP, s-IPSP, s-EPSP, and DA modulation of s-EPSP. (From ref. 28.)

rons in general have relatively low resting Na⁺ conductances; this does not provide a readily workable base from which to produce an s-IPSP generated by a proposed reduction ("inactivation") of the resting Na^+ conductance. However, treatment with nicotine has indeed been found to raise the resting Na^+ conductance substantially (43); thus treatment with nicotine can provide the more suitable conditions for a hyperpolarizing change that might depend on a reduced Na^+ conductance. It should be noted that in rabbit ganglion cells there is no detectable change in membrane conductance, up or down, during either an s-IPSP response or a direct catecholaminergic hyperpolarization (12,22).

SIF Cell Role in DA-Mediated s-IPSP

The SIF cell in the rabbit SCG exhibits the necessary and sufficient physiological and pharmacological features to constitute the functional interneuron in the intraganglionic pathway mediating the s-IPSP in the rabbit SCG. The fluorophore in these SIF cells has been identified as DA (34), as it has in some other ganglia [e.g., in rat SCG (2,3)] and, by indirect evidence, in SCG of monkey (6), cow, and possibly cat (4). Either orthodromic, preganglionic nerve impulses or direct action by the muscarinic agonist bethanechol can produce a release of DA from SIF cells (34). Depletion of DA content in SIF cells is accompanied by a reduction in the s-IPSP response; a complete loss of s-IPSP response developed when total DA concentration was lowered by about 50% (34). Depleted SIF cells could take up exogenous DA, and this reuptake was accompanied by partial restoration of the s-IPSP response. Exogenous NE and E were relatively ineffective for this kind of restoration by reuptake, even though exogenous NE and E can elicit the appropriate hyperpolarizing response; similarly, incubation with antagonists of dopamine-β-hydroxylase (diethyldithiocarbamate, fusaric acid, or FLA 63) did not depress the s-IPSP (30,36).

Morphological Features of SIF Cells, in Mediation of s-IPSP

At least some SIF cells have been shown to fulfill the classical morphological criteria for a role as an interneuron in that they provide specialized efferent synaptic terminals onto ganglion cells as well as receive afferent synaptic contacts from preganglionic cholinergic fibers (48). However, the relation of some overall morphological aspects to a role in mediating the s-IPSP is less obvious but explainable.

(a) The numbers of SIF cells appear to be quite low in many ganglia, including the rabbit SCG (47). However, in the case of perhaps analogous monoaminergic projections from nuclei in the brainstem to the rest of the central nervous system (CNS), where the general widespread pattern of their innervations is incontrovertible, there are also very small ratios (for the number of monoaminergic cells to the number of receiving neurons). In rabbit SCG the ubiquitous fluorescent network of fine beaded fibers that surrounds virtually all principal neurons was found to contain DA (34) [also shown indirectly in rat SCG, in that the networks did not exhibit any specifically NE fluorescence (2)]. This indicated that the relatively few DA-containing SIF cells could nevertheless supply ramifying projections that reached all ganglion cells. Other evidence supported this view. The numerous "adrenergic terminals" observable in the rabbit SCG (9) have been found *not* to contain horseradish peroxidase (HRP) when HRP was applied to postganglionic nerve (8); such terminals could not therefore originate from neurites of the principal neurons, which were invaded by the HRP. It was also shown that none of the adrenergic terminals appear to be supplied by the preganglionic nerve (10), and they must therefore arise from intraganglionic structures. Since these adrenergic terminals also were found to correspond to the varicose fibers (9) that were found to contain DA (34) in rabbit SCG, it seems likely that they represent ubiquitous DA-releasing sites by which the SIF cells can elicit s-IPSPs in most or all principal neurons (35).

(b) Although specialized efferent synaptic contacts between SIF cells and principal neurons have been detected in the ganglia of some species (48), such classical chemical-type junctions (i) are usually relatively rare, (ii) have not been detected in some ganglia, and (iii) are presumably completely absent for all those SIF cells that lack axonal projections, e.g., the "type II SIF cells" in mammalian ganglia (47,48) and types I to III in amphibian ganglia (19). However, specialized efferent contacts may not be essential. The relatively long synaptic delay of the s-IPSP (25,37) would permit junctional interactions via "loose synapses" (24,29,31) of a type more similar to those for *en passage* varicose terminations of peripheral sympathetic fibers near smooth muscle fibers (5) than to specialized junctions with 200 Å clefts; merely a close proximity of s-IPSP receptors to beaded, releasing sites of monoaminergic fibers could thus be sufficient. Again, the numerous "adrenergic terminals" in rabbit SCG (9), discussed above, appear to provide the kinds of efferent releasing sites envisioned in such a proposal. These adrenergic terminals, or contacts, in relation to ganglion cells appear to be generally less specialized than classical chemical synapses and seem, at least in the figures presented (9), to involve spaces between the pre- and postsynaptic membranes that are greater than 20 nm.

On the other hand, the axonless type II SIF cells in mammalian ganglia (4,6,48) could not serve as an interneuronal mediator of the s-IPSP in view of the relatively large transport distances to most principal neurons. However, the longer transport times would still be compatible with the other postsynaptic action by DA in rabbit SCG, that of modulatory enhancement of the s-EPSP (see further below). Indeed, there is morphological evidence for adrenergic release not only at specialized synapses but also at "loose" or nonspecialized contacts and at noncontact sites remote from any postsynaptic membrane. Three kinds of actual sites of catecholamine release by SIF cells, made apparent by either preganglionic or muscarinic stimulation, were recently demonstrated histologically in the rat SCG (18): Extruded material related to dense-cored vesicles appeared at (a) classical-type chemical synapses; (b) nonjunctional membrane specializations called "synaptoids"; and (c) "sheathless portions of an SIF cell surface bordering on the interstitial space and, in some cases, a fenestrated blood vessel".

In frog sympathetic ganglia, there is apparently a complete absence of specialized adrenergic efferent synapses between SIF cells and principal neurons (19,44,46). In addition, frog ganglia are reported to lack the adrenergic networks of beaded fibers (19,46) that surround principal neurons in mammalian ganglia [but compare a positive report (20)]. However, chromaffin cells in these ganglia were "often found to be closely applied to the surface of the principal neurons" (19);

such relatively close contacts could be compatible with their putative role as the adrenergic interneurons for the s-IPSP if they were located in relation to "C" ganglion cells (42), in view of the 100-msec synaptic delay for the s-IPSP in frog ganglia (37). Further allegedly negative anatomical evidence has been reported for frog sympathetic ganglia (46); in this report, no fluorescent processes were observed to emanate from SIF cells, and no afferent synapses onto chromaffin cells were found. This inability of one group (46) to find either of these structures is presumably of questionable significance in view of clearly positive findings for both by other investigators; numerous processes from at least some types of chromaffin cell in frog ganglia had been reported (19,20), and afferent synaptic contacts by apparently cholinergic terminals on chromaffin cells have been commonly observed (19).

In any case, it should be re-emphasized that there is clearcut physiological evidence that release of a second transmitter is a requirement for the orthodromic production of the s-IPSP in rabbit SCG and frog paravertebral ganglia (unless the frog ganglia are in a special abnormal state due to treatment with nicotine). Consequently, histological findings of a negative type cannot logically be used as a basis for denying the existence of a second transmitter; rather, the issue must be one of looking further for some suitable morphological substrates that can be shown to provide the delivery of the second transmitter in a manner consonant with the known physiological characteristics of the response.

SIF CELLS AS MEDIATORS FOR MODULATORY SYNAPTIC ACTIONS

Modulatory synaptic actions may be defined as those in which a transmitter alters effective synaptic transmission by mechanisms *other than* any postsynaptic potential (PSP) or direct change in excitatory level that may be produced by that transmitter (31). The modulatory action of DA in the rabbit

SCG, in which a brief exposure to DA induces a long-lasting enhancement of the muscarinic (s-EPSP) depolarizing responses to ACh (36), appears to have a latent period of 30 sec or more (29,30,38). (A more definitive investigation of this delay is needed and is in progress in our laboratory.) These very long delays are probably largely due to intracellular chemical processes mediated by cyclic AMP (32,38) which are required for the development of the persisting change in neuronal responsiveness to ACh. However, such large latencies could, in addition, easily accommodate time for transport of a transmitter from relatively remote releasing sites. This could include either diffusion over hundreds of micrometers and/or intraganglionic vascular transport via a possible "portal" system (6,48) from axonless type II SIF cells to the postsynaptic sites for modulatory action.

It seems possible that other kinds of slow synaptic potentials or modulatory actions may be discovered in sympathetic ganglia. Recent discoveries of neuroactive peptides in sympathetic ganglia include at least one example of a peptide (met-enkephalin) that can be found within some SIF cells (41). Such findings obviously raise interesting but still speculative additional possibilities for roles by SIF cells. However, we shall have to await physiological, neurochemical, and pharmacological investigations to establish the significance of any such newly available possibilities.

ROLES FOR MAMMALIAN SIF CELLS CONTAINING NE AND/OR E

SIF cells in sympathetic ganglia of some species have been found to contain NE (and E) rather than DA. The most striking example is the NE-containing SIF cell in the SCG of the guinea pig (15), especially in view of the fact that the homologous ganglion in the other rodents studied has DA-containing SIF cells. It should be recalled that, in contrast to the curarized SCG of the rabbit (14,26) and the rat (13; Ashe and Libet, *unpublished*), no s-IPSP is detectable in the curarized SCG of the guinea pig (11,26). An s-EPSP response has been elicitable in all mammalian sympathetic ganglia tested (including the guinea pig SCG) (11,26), but the s-EPSP is monosynaptically mediated by ACh without any requirement for another transmitter or for SIF cells (26).

A possible physiological role for the NE-containing SIF cells seen in guinea pig SCG and elsewhere may lie in another form of s-IPSP for which NE rather than DA is the transmitter. In rabbit SCG, exogenous NE and E are as, or even more, effective than DA (7,21,36) for eliciting a hyperpolarization. As with DA, the NE response is generated with the same absence of any change in membrane conductance that is seen for the s-IPSP itself (22). Despite the hyperpolarizing effectiveness of NE, it is DA and not NE (or E) that is the actual physiological transmitter for that s-IPSP which is mediated via a muscarinic activation of SIF cells in the rabbit SCG (26,30). More recently we gained evidence for the existence of another type of s-IPSP that is producible via a nonmuscarinic pathway in rabbit SCG; the response is not blocked by atropine. This s-IPSP appears to be mediated by NE, with the NE released intraganglionically by a nonmuscarinic or presumably nicotinic action of ACh (1,27). The intraganglionic structure that releases the NE transmitter in the rabbit SCG in response to the nonmuscarinic action of ACh is yet to be established. We are now investigating the possibility that guinea pig SCG may also exhibit an s-IPSP mediated by a nonmuscarinic release of NE; if such an s-IPSP is present in this ganglion, the NE-containing SIF cells would be a prime candidate for the intraganglionic source of the NE transmitter involved.

There is yet another possible role for NE-containing SIF cells in general. In rabbit SCG, NE and E were found to be relatively ineffective for mimicking the DA modulatory action on the s-EPSP type of response (36). However, the possibility that NE or E may be able to act as modulators of the s-EPSP in other ganglia (e.g., guinea pig SCG) has not yet been tested; such a possibility should not yet be excluded from consideration.

Clearly, there is still need for further investigation into the possible physiological roles of the various types of SIF cell. It is also desirable to conduct intracellular studies of the electrophysiological responses and characteristics of SIF cells themselves. Some morphological similarities of SIF cells to the monoaminergic neurons in the brainstem suggest that a fuller knowledge about functional roles and mechanisms for SIF cells may also help us better understand the important cerebral functions that arc mcdiatcd by thc monoaminergic nuclei in the brainstem.

ACKNOWLEDGMENTS

This research was supported by U.S. Public Health Service research grant NS-00884 from the National Institute of Neurological and Communicative Disorders and Stroke.

REFERENCES

1. Ashe, J.H., and Libet, B. (1979): A noradrenergic s-IPSP in mammalian sympathetic ganglion, elicited by a non-muscarinic action of preganglionic volleys. *Abstr. Soc. Neurosci.,* 5:743.
2. Baker, N.A., Redick, J.A., Schnute, W.J., and Van Orden, L.L., III (1974): Neurotransmitter identification in small, fluorescent cells of rat paracervical ganglia by microspectrofluorometry and immunohistochemistry. *Abstr. Soc. Neurosci.,* 126.
3. Björklund, A., Cegrell, L., Falck, B., Ritzén, M., and Rosengren, E. (1970): Dopamine-containing cells in sympathetic ganglia. *Acta Physiol. Scand.,* 78:334-338.
4. Black, A.C., Jr., Chiba, T., Wamsley, J.K., Bhalla, R.C., and Williams, T.H. (1978): Interneurons of sympathetic ganglia: Divergent cyclic AMP responses and morphology in cat and cow. *Brain Res.,* 148:389-398.
5. Burnstock, G., and Costa, M. (1975): *Adrenergic Neurons: Their Organisation, Function and Development in the Peripheral Nervous Sytem.* Chapman and Hall, London.
6. Chiba, T., Black, A.C., Jr., and Williams, T.H. (1977): Evidence for dopamine-storing interneurons and paraneurons in rhesus monkey sympathetic ganglia. *J. Neurocytol.,* 6:441-453.

7. Cole, A.E., and Shinnick-Gallagher, P. (1979): Characterization of a post-ganglionic catecholamine receptor in the rabbit superior cervical ganglion (RSCG). *Abstr. Soc. Neurosci.,* 5:738.
8. Dail, W.G., Barraza, C., Khoudary, S., and Murray, H. (1979): Horseradish peroxidase studies of an autonomic ganglion. *Abstr. Soc. Neurosci.,* 5:332.
9. Dail, W.G., and Evan, A.P. (1978): Ultrastructure of adrenergic terminals and SIF cells in the superior cervical ganglion of the rabbit. *Brain Res.,* 148:469-477.
10. Dail, W.G., Jr., and Wood, J. (1977): Studies on the possibility of an extraganglionic source of adrenergic terminals to the superior cervical ganglion. *Abstr. Soc. Neurosci.,* 3:248.
11. Dun, N., and Karczmar, A.G. (1977): The presynaptic site of action of norepinephrine in the superior cervical ganglion of guinea pig. *J. Pharmacol. Exp. Ther.,* 200:328-335.
11a. Dun, N., and Karczmar, A.G. (1978): Involvement of the interneuron in the generation of the slow inhibitory postsynaptic potential in mammalian sympathetic ganglia. *Proc. Natl. Acad. Sci. USA,* 75:4029-4032.
12. Dun, N., and Nishi, S. (1974): Effects of dopamine on the superior cervical ganglion of the rabbit. *J. Physiol. (Lond),* 239:155-164.
13. Dunant, Y., and Dolivo, M. (1967): Relations entre les potentiels synaptiques lents et l'excitabilite du ganglion sympathique chez le rat. *J. Physiol. (Paris),* 59:281-294.
14. Eccles, R.M., and Libet, B. (1961): Origin and blockade of the synaptic responses of curarized sympathetic ganglia. *J. Physiol. (Lond),* 157.484-503.
15. Elfvin, L.-G., Hökfelt, T., and Goldstein, M. (1975): Fluorescence microscopical, immunohistochemical and ultrastructural studies on sympathetic ganglia of the guinea pig, with special reference to the SIF cells and their catecholamine content. *J. Ultrastruct. Res.,* 51:377-396.
16. Eränkö, O. (1976): *SIF Cells: Structure and Function of the Small, Intensely Fluorescent Sympathetic Cells.* Government Printing Office, Washington, D.C.
17. Eränkö, O., and Eränkö, L. (1971): Small intensely fluorescent, granule-containing cells in the sympathetic ganglion of the rat. *Prog. Brain Res.,* 34:39-52.
18. Grillo, M.A. (1977): Ultrastructural studies on catecholamine release from SIF cells. *J. Cell Biol.,* 75:118a.
19. Hill, C.E., Watanabe, H., and Burnstock, G. (1975): Distribution and morphology of amphibian extra-adrenal chromaffin tissue. *Cell Tissue Res.,* 160:371-387.
20. Jacobowitz, D.M. (1970): Catecholamine fluorescence studies of adrenergic neurons and chromaffin cells in sympathetic ganglia. *Fed. Proc.,* 29:1929-1944.
21. Kobayashi, H. (1976): On the slow synaptic responses in autonomic ganglia. *Prog. Brain Res. (Jpn),* 20:186-198.

22. Kobayashi, H., and Libet, B. (1970): Actions of noradrenaline and acetylcholine on sympathetic ganglion cells. *J. Physiol. (Lond), 208*:353–372.

23. Kojima, H., Anraku, S., Onogi, K., and Ito, R. (1978): Histochemical studies on two types of cells containing catecholamines in sympathetic ganglia of the bullfrog. *Experientia, 34*:92–93.

24. Libet, B. (1965): Slow synaptic responses in autonomic ganglia. In: *Studies in Physiology,* edited by D.R. Curtis and A.K. McIntyre, pp. 160–165. Springer-Verlag, Berlin.

25. Libet, B. (1967): Long latent periods and further analysis of slow synaptic responses in sympathetic ganglia. *J. Neurophysiol., 30*:494–514.

26. Libet, B. (1970): Generation of slow inhibitory and excitatory postsynaptic potentials. *Fed. Proc., 29*:1945–1956.

27. Libet, B. (1977): Roles of intraganglionic catecholamines in slow synaptic actions. *Proc. Int. Union Physiol. Sci., 12*:558.

28. Libet, B. (1977): The role SIF cells play in ganglionic transmission. *Adv. Biochem. Psychopharmacol., 16*:541–546.

29. Libet, B. (1979): Slow synaptic actions in ganglionic functions. In: *Integrative Functions of the Autonomic Nervous System,* edited by C.McC. Brooks, K. Koizumi, and A. Sato, pp. 197–222. Tokyo University Press, Tokyo, and Elsevier/North Holland, Amsterdam.

30. Libet, B. (1979): Dopaminergic synaptic processes in the superior cervical ganglion: models for synaptic actions. In: *The Neurobiology of Dopamine,* edited by A. Horn, J. Korf, and B.H.C. Westerink. Academic Press, London.

31. Libet, B. (1979): Neuronal communication and synaptic modulation: experimental evidence vs. conceptual categories: Commentary to Dismukes, R.K.: New concepts of molecular communication among neurons. *Behav. Brain Sci. (in press).*

32. Libet, B. (1979): Which postsynaptic action of dopamine is mediated by cyclic AMP? *Life Sci., 24*:1043–1058.

33. Libet, B., and Kobayashi, H. (1974): Adrenergic mediation of the slow inhibitory postsynaptic potential in sympathetic ganglia of the frog. *J. Neurophysiol., 37*:805–814.

34. Libet, B., and Owman, Ch. (1974): Concomitant changes in formaldehyde-induced fluorescence of dopamine interneurons and in slow inhibitory postsynaptic potentials of rabbit superior cervical ganglion, induced by stimulation of preganglionic nerve or by a muscarinic agent. *J. Physiol. (Lond), 237*:635–662.

35. Libet, B., and Tosaka, T. (1969): Slow inhibitory and excitatory postsynaptic responses, in single cells of mammalian sympathetic ganglia. *J. Neurophysiol., 32*:43–50.

36. Libet, B., and Tosaka, T. (1970): Dopamine as a synaptic transmitter and modulator in sympathetic ganglia; a different mode of synaptic action. *Proc. Natl. Acad. Sci. USA, 67*:667–673.

37. Libet, B., Chichibu, S., and Tosaka, T. (1968): Slow synaptic responses and excitability in sympathetic ganglia of the bullfrog. *J. Neurophysiol., 31*:383–395.

38. Libet, B., Kobayashi, H., and Tanaka, T. (1975): Synaptic coupling into the production and storage of a neuronal memory trace. *Nature, 258*:155–157.

39. Libet, B., Tanaka, T., and Tosaka, T. (1977): Different sensitivities of acetylcholine-induced "after-HP" compared to dopamine-induced hyperpolarization, to ouabain or to lithium-replacement of sodium, in rabbit sympathetic ganglia. *Life Sci., 20*:1863–1870.

40. Nishi, S., and Koketsu, K. (1968): Early and late after-discharges of amphibian sympathetic ganglion cells. *J. Neurophysiol., 31*:109–121.

41. Schultzberg, M., Hökfelt, T., Lundberg, J.M., Terenius, L., Elfvin, L-G., and Elde, R. (1978): Enkephalin-like immunoreactivity in nerve terminals in sympathetic ganglia and adrenal medulla and in adrenal medullary gland cells. *Acta Physiol. Scand., 103*:475–477.

42. Tosaka, T., Chichibu, S., and Libet, B. (1968): Intracellular analysis of slow inhibitory and excitatory postsynaptic potentials in sympathetic ganglia of the frog. *J. Neurophysiol., 31*:396–409.

43. Wang, C.M., and Narahashi, T. (1972): Mechanisms of dual action of nicotine on end-plate membranes. *J. Pharmacol. Exp. Ther., 182*:427–441.

44. Watanabe, H., and Burnstock, G. (1978): Postsynaptic specializations at excitatory and inhibitory cholinergic synapses. *J. Neurocytol., 7*:119–133.

45. Weight, F.F., and Padjen, A. (1973): Acetylcholine and slow synaptic inhibition in frog sympathetic ganglion cells. *Brain Res., 55*:225–228.

46. Weight, F.F., and Weitsen, H.A. (1977): Identification of small intensely fluorescent (SIF) cells as chromaffin cells in bullfrog sympathetic ganglia. *Brain Res., 128*:213–226.

47. Williams, T.H., Black, A.C., Jr., Chiba, T., and Bhalla, R.C. (1975). Morphology and biochemistry of small intensely fluorescent cells of sympathetic ganglia. *Nature, 256*:315–317.

48. Williams, T.H., Chiba, T., Black, A.C., Jr., Bhalla, R.C., and Jew, J. (1976). Species variation in SIF cells of superior cervical ganglia: are there two functional types? In: *SIF Cells, Structure and Function of the Small, Intensely Fluorescent Sympathetic Cells,* edited by O. Eränkö, pp. 143–162. Government Printing Office, Washington, D.C.

*Histochemistry and Cell Biology of
Autonomic Neurons, SIF Cells, and
Paraneurons,*
edited by O. Eränkö et al.
Raven Press, New York © 1980.

Connections of Local Circuit Neurons in Guinea Pig and Rabbit Superior Cervical Ganglia

Jean Y. Jew

Department of Anatomy, College of Medicine, University of Iowa, Iowa City, Iowa 52242

Cajal contended that higher cerebral functions "did not depend on the sizes and number of [principal] cerebral neurons, but on the richness of the connective processes; in other words on the complexity of the association pathways to short and long distances." When considering the sympathetic ganglia, the presence of local circuit neurons or SIF cells invokes the same sort of probable conclusion; the influence exerted over transmission by the SIF cells may far outweigh their size and number.

Of the three basic neural constituents of sympathetic ganglia, the preganglionic input and the principal ganglionic neurons (PGNs) are in most respects similar in most mammalian species, whereas clear points of difference have been demonstrated between the SIF cells of different species (6). For example, SIF cells of the rabbit superior sympathetic ganglion contain dopamine (DA) as transmitter; those in the guinea pig contain norepinephrine (NE). Indeed, a DA-receptor-adenylate-cyclase complex, which is said to bear directly on the generation of the slow inhibitory postsynaptic potential (s-IPSP), is absent in the guinea pig as is the s-IPSP itself.[1]

In regard to the claim that SIF cells are concerned in s-IPSP production, the morphological feature that appears most worthy of attention is the SIF cell efferent, which is widely credited with transmitting this inhibitory signal to the PGNs. We hypothesize that this connection should be present in the rabbit superior cervical ganglion (SCG) and absent in guinea pig SCG. The more general objective was to seek an understanding (at present very far off) of local circuitry and the interdependence of SIF cells and PGNs in sympathetic ganglia.

MATERIALS AND METHODS

SCG of three adult female rabbits and four adult female guinea pigs were studied. 5-Hydroxydopamine (5-OHDA) (100 mg/kg i.p.) was administered 2 hr before perfusion to mark the catecholamine-containing elements. Under Nembutal anesthesia, the animals were perfused with 2.5% glutaraldehyde and 1.0% paraformaldehyde in 0.1 M Sorensen's phosphate buffer. Ganglia were processed for light and electron microscopy (4). After locating SIF cells in semithin sections stained with toluidine blue, adjacent thin sections were cut and stained for study in the electron microscope. Using thin sections mounted on Formvar-coated 100 mesh copper grids, some whole sections and some individual

---

[1] The author makes no personal judgment regarding different proposed mechanisms for s-IPSP production.

grid squares were surveyed to locate (a) catecholaminergic synapses and (b) profiles of axons and dendrites as well as cell bodies containing the marker.

RESULTS

Granular vesicles in nerve processes and cell bodies were arbitrarily classified by size into three categories: (a) large granular vesicles (LGVs) approximately 2,200 Å in diameter; (b) intermediate granular vesicles (IGVs), about 800 Å in diameter and (c) small granular vesicles (SGVs), which contain tiny dense cores approximately 220 Å in diameter. Since they were usually found in the vicinity of SIF cells, profiles containing LGVs were tentatively identified as SIF cell processes.

Rabbit

SIF cells in the rabbit occurred singly as well as in clusters adjacent to blood vessels. LGVs between 1,800 and 3,000 Å in diameter were the most characteristic feature of the SIF cell cytoplasm. With few exceptions, nerve processes containing LGVs were found within 150 μm of SIF cell bodies. Relatively rare examples of LGV-containing processes were found beside blood vessels in other parts of the section. These may have originated from SIF cells out of the plane of the section. Presumptive SIF cell processes containing LGVs made synaptic contact with PGN dendrites (Fig. 1b,c; Fig. 2b) or dendritic spines (Fig. 2a). It is of interest that these SIF–PGN dendrite contacts were only encountered within a short distance of the SIF cell bodies.

Numerous other profiles containing only IGVs and/or SGVs were located in all parts of the section. The cells of origin of these processes were not traced to their sources and are therefore unknown since IGVs and SGVs were also present in PGN cell bodies and processes. No synaptic terminals containing only IGVs and SGVs were observed in areas distant from the SIF cells; hence we

uncovered no evidence to support the claim that rabbit PGNs give rise to feedback collaterals. In the rabbit SCG, terminals of cholinergic type containing agranular vesicles together with some granular vesicles devoid of the 5-OHDA marker were very numerous. Some of these terminals were seen to make synaptic contact with SIF cells and processes of PGNs.

Portions of the bodies of SIF cells as well as many of their processes were seen to be in close proximity to the intraganglionic blood vessels, and the frequent absence of Schwann cell sheath coupled with fenestrations in the vascular endothelial cells supported the notion that the SIF cells engage in vascular secretion. Even in the absence of any reliable evidence that adrenergic processes secrete into the extracellular space, it was of interest to find many catecholamine-containing processes with increased numbers of IGVs and SGVs along portions of the plasma membrane that were devoid of Schwann cell sheath.

Guinea Pig

As a rule, SIF cells in the guinea pig were arranged in clusters adjacent to blood vessels. As in the rabbit ganglion, the SIF cell cytoplasm was characterized by many LGVs approximately 2,000 Å in diameter and containing the 5-OHDA marker. The relative number of IGVs and SGVs was greater in the SIF cell processes than in the cell body. Presumed SIF cell processes (containing LGVs, IGVs, and SGVs) synapsed consistently with SIF cell bodies and SIF cell processes. Although it would be difficult to rule out the possibility that exceptions may exist, we did not find any SIF cell processes synapsing with PGNs in the guinea pig ganglion. Many sections were surveyed, and although complete continuity between the presynaptic SIF cell processes and their cells of origin was not achieved, there seems virtually no doubt that SIF cells are in direct associative synaptic contact with other SIF cells. Terminals of cholinergic type were much less

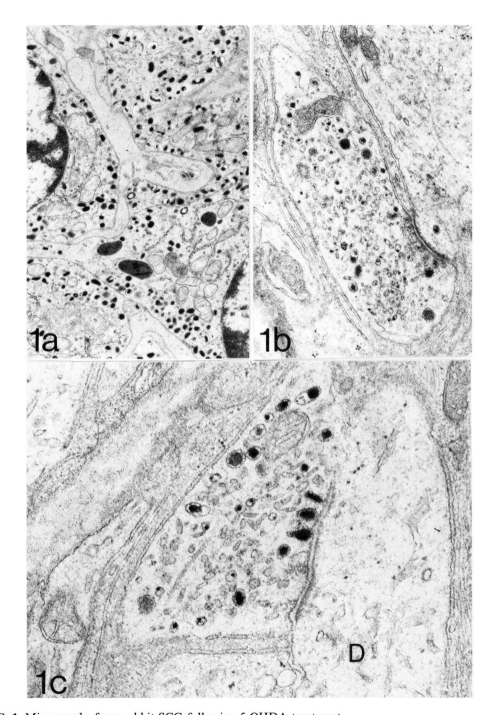

FIG. 1. Micrographs from rabbit SCG following 5-OHDA treatment.

    a: Field from a portion of a cluster of SIF cells. Cytoplasm contains LGVs which carry the electron-dense marker 5-OHDA. ×14,000.

    b: Catecholamine-containing terminal contains LGVs, IGVs, and SGVs. The presumed SIF cell efferent makes synaptic contact with a PGN dendrite. PGN dendrites frequently contain one or more IGVs. ×18,000.

    c: Presumed SIF cell terminal contains several LGVs, some of which are lined up against the presynaptic membrane. (D) Dendrite of PGN. ×48,000.

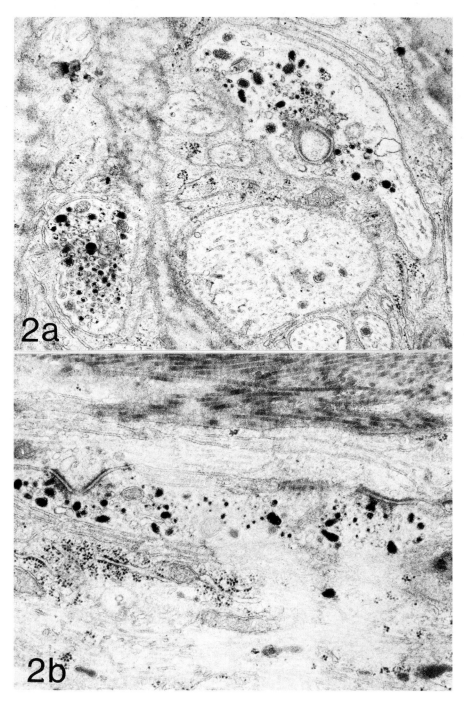

FIG. 2. Fields from SCG of a rabbit that received 5-OHDA treatment.
    a: Two SIF cell processes contain LGVs, IGVs, and SGVs. One process is in synaptic contact with a presumptive dendritic spine. ×31,500.
    b: A SIF cell terminal makes multiple synaptic contacts with a PGN dendrite. ×33,000.

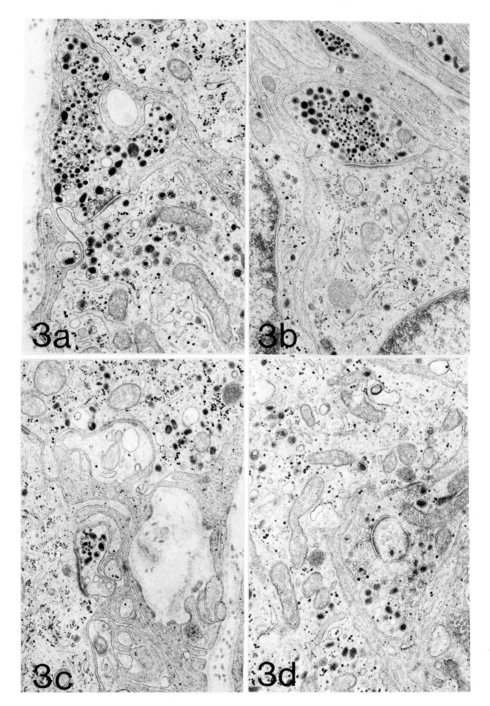

FIG. 3. Micrographs of guinea pig SCG. The animals were treated with the catecholamine marker analog 5-OHDA.

a: A SIF cell terminal, packed with granular vesicles, synapses with a large SIF cell process in the lower part of the field. Note that the LGVs in the terminal are similar to those in the large SIF cell process. ×27,000.

b: A SIF cell terminal forms an axosomatic connection with the cell body of a SIF cell. ×24,800.

c: A SIF cell terminal makes synaptic contact with an SIF cell soma (at left). ×22,500.

d: A SIF cell terminal synapses with a presumed SIF cell dendritic spine. ×21,800.

numerous in the guinea pig sections than in the rabbit.

In spite of the recognized inadequacy of nonquantitative surveys, it was evident that SIF cell processes were more numerous in the guinea pig section and that they often extended further from the SIF cell bodies. As with the rabbit ganglion, SIF cell processes were found in close relation to capillary walls, and the morphological features resembled those described in the rabbit ganglion. Again as in the rabbit, dense-cored vesicles were sometimes clustered against regions of the plasma membrane of unidentified neurites at sites where the plasma membrane was in direct continuity with the extracellular space. Some examples of associative synaptic connections between SIF cells are illustrated in Fig. 3a–d.

DISCUSSION

SIF cells receiving cholinergic inputs were identified in the rabbit. The SIF cells make efferent synaptic connections with PGNs. Dail and Evan (1) observed adrenergic terminals in the rabbit SCG and reached the probable conclusion that these were "endings of sympathetic ganglion cells," although they did not exclude the possibility that they were of SIF cell origin. Except for one terminal that contained a single LGV, the adrenergic terminals observed in Dail's study lacked the LGVs that are so characteristic of rabbit SIF cells. Although we are aware that our own observations do not exclude PGN feedback collaterals, we infer that the terminals seen in our study (Figs. 1 and 2) are indeed derived from SIF cells, a conclusion based on similar vesicle appearance and proximity to SIF cells. For practical purposes, this finding correlates well with electrophysiological (3) and biochemical (5) studies which indicate that the rabbit SIF cells are local circuit neurons (interneurons) that are involved in the production of the s-IPSP in the PGN via a cyclic AMP-mediated hyperpolarization of the PGN. On the other hand, it may be worth-

while to acknowledge that this study can report SIF cell connections on PGNs only in the vicinity of the SIF cells. We make no claim that *all* PGNs receive a direct synaptic SIF cell input.

A different SIF cell circuitry has been perceived in the guinea pig superior ganglion, and this must be considered in light of differences observed in the process of ganglionic transmission in this species. In particular, the s-IPSP has not been demonstrated in guinea pig superior ganglion (2); preganglionic physiological stimulation of guinea pig SCG fails to elevate cyclic AMP; and the neurotransmitter released by the SIF cells is norepinephrine rather than dopamine. The observation of associative connections between SIF cells (rather than SIF-PGN synaptic contact) is not an unexpected result since it is in accord with the apparent absence of an s-IPSP mechanism that is believed to include PGN inhibition by SIF cells. It is not yet possible to explain with certainty the associative SIF cell connections in the guinea pig, but it is possible to postulate an autoinhibitory mechanism that can suppress SIF cell firing.

Although we have not wholly explained the absence of an s-IPSP and associated mechanisms in the guinea pig SCG, it is not unreasonable to think that the SIF cells may, in another fashion, regulate the flow of signals from the PGNs, but the mechanism whereby this is accomplished is not yet clear.

ACKNOWLEDGMENTS

The author wishes to thank Paul Reimann, Evelyn Jew, Steve Simonson, and Wenan Wang for able assistance. Support for this work was provided by grants NS–11650 to T.H. Williams and HL–21914 to J. Jew.

REFERENCES

1. Dail, W.G., and Evan, A.P. (1978): Ultrastructure of adrenergic terminals and SIF cells in the superior cervical ganglion of the rabbit. *Brain Res.,* 148:469–477.

2. Dun, N., and Karczmar, A.G. (1977): The presynaptic site of action of norepinephrine in the superior cervical ganglion of guinea pig. *J. Pharmacol. Exp. Ther.,* 200:328-335.

3. Libet, B. (1970): Generation of slow inhibitory and excitatory postsynaptic potentials. *Fed. Proc.,* 29:1945-1956.

4. Williams, T.H., and Jew, J. (1975): An improved method for perfusion fixation of neural tissues for electron microscopy. *Tissue Cell,* 7:407-418.

5. Williams, T.H., Black, A.C., West, J.R., Sandquist, D., and Gluhbegovic, N. (1980): Biochemical aspects of SIF cell function in guinea pig and rabbit superior cervical ganglia. *This volume.*

6. Williams, T.H., Chiba, T., Black, A.C., Bhalla, R.C., and Jew, J. (1975): Species variation in SIF cells of superior cervical ganglia: are there two functional types? In: *SIF cells: Structure and Function of the Small, Intensely Fluorescent Sympathetic Cells,* edited by O. Eranko, pp. 143-162. Fogarty International Center Proceedings No. 30. Government Printing Office, Washington, D.C.

Histochemistry and Cell Biology of Autonomic Neurons, SIF Cells, and Paraneurons,
edited by O. Eränkö et al.
Raven Press, New York © 1980.

Biochemical Aspects of SIF Cell Function in Guinea Pig and Rabbit Superior Cervical Ganglia

Terence H. Williams, Asa C. Black, Jr., James R. West, Dean Sandquist, and Nedzad Gluhbegovic

Department of Anatomy, College of Medicine, University of Iowa, Iowa City, Iowa 52242

The claim is indeed well justified that small intensely fluorescent (SIF) or small granule-containing (SGC) cells are one of the basic cytological elements of sympathetic ganglia (1,4,15,17–20). Their presence in sympathetic ganglia of all mammals examined assures us that, since this element is unlikely to be present so often without purpose, it must influence transmission directly and/or elicit its effect(s) by neurohumoral transport. It must be admitted, however, that crucial functional data are as yet either unavailable or speculative for the reasons that (a) we are not yet equal to the task of isolating SIF cells functionally so that we can study their performance as an independent element, and (b) our known information about one species cannot be extrapolated to all others. In the rabbit superior cervical ganglion (SCG), the slow inhibitory postsynaptic potential (s-IPSP) is an uncontested finding, whereas the guinea pig SCG reportedly lacks the s-IPSP (2,3,7). We are studying the biochemical and morphological aspects of SIF cell function in these two species, focusing on questions relating to s-IPSP generation.

Experimental procedures used were as follows: SCG for fluorescent immunohistochemical localization of dopamine-β-hydroxylase (DBH) were processed as described by Wamsley et al. (15). The catecholamine assay described by Schmidt and Bhatnagar (14) was applied to SCG frozen in liquid nitrogen immediately after removal from the animal. Ganglia were assayed for cyclic AMP as described in Wamsley et al. (16) after (a) incubation with 50 μM dopamine (DA), L-isoproterenol (ISO), or L-norepinephrine (NE); or (b) supramaximal preganglionic physiological stimulation at 10 Hz.

WHICH CATECHOLAMINES SERVE AS THE SIF CELL TRANSMITTERS IN GUINEA PIG AND RABBIT SCG?

As demonstrated by immunocytochemistry in our laboratory (15), as well as by Elfvin et al. (4), SIF cells of the guinea pig SCG contain DBH (which converts DA to NE). Although we do not make the blanket assertion that each and every SIF cell in guinea pig SCG contains NE, this catecholamine appears to be the principal SIF cell transmitter in this species. There is sound evidence that the SIF cells of the rabbit SCG contain DA (9).

The ratio of DA to NE in an SCG can provide valuable evidence regarding the nature of the SIF cell catecholamine transmitter. A very small amount of DA in a principal ganglionic neuron (PGN) may be

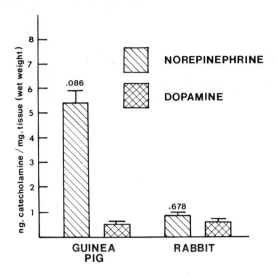

FIG. 1. Catecholamine levels in the SCG of rabbit and guinea pig. Numbers above the standard error bars for NE indicate the DA/NE ratio.

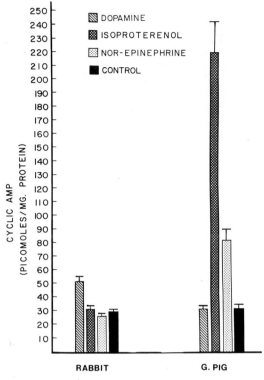

FIG. 2. Cyclic AMP generated by 12-min incubations with 50 μM concentrations of various catecholamines. Incubations were carried out at 37°C in the presence of 5 mM theophylline. Note that DA caused elevation of cyclic AMP in rabbit but not guinea pig, whereas the converse was true for NE and isoproterenol.

accounted for by its role as a precursor for NE synthesis in the PGNs. If the SIF cell transmitter is NE, the DA/NE ratio in the ganglion must be extremely low; whereas if DA is the SIF cell transmitter, the DA/NE ratio will be relatively high. The DA/NE ratio is about eight times higher in the rabbit SCG than in the guinea pig (Fig. 1), even though SIF cells are one order of magnitude more numerous in the guinea pig SCG than in the rabbit (on a per milligram of tissue basis). Therefore we feel that DA is virtually excluded as the major SIF cell transmitter in the guinea pig.

CYCLIC AMP RESPONSES COMPARED IN THE TWO SPECIES

Figure 2 demonstrates the cyclic AMP responses of rabbit and guinea pig SCG to pharmacological stimulation with 50 μM DA, ISO, and NE under conditions which produce maximal cyclic AMP accumulation. The rabbit SCG responds to DA, whereas the guinea pig SCG responds to NE and ISO. The rabbit SCG contains a DA receptor—adenylate cyclase complex—whereas the guinea pig SCG contains a β-adrenergic receptor, adenylate cyclase complex. Thus a

DA receptor-adenylate cyclase complex is not a component of the biochemical machinery of the guinea pig SCG.

If the NE in guinea pig SIF cells has a role in s-IPSP production, preganglionic physiological stimulation should lead to increased cyclic AMP synthesis; i.e., preganglionic physiological stimulation would cause NE release from the SIF cells, the NE then inducing more cyclic AMP synthesis in the PGNs. We studied the effect of supramaximal preganglionic physiological stimulation on cyclic AMP synthesis in both species. Stimulation of guinea pig SCG in the presence or absence of calcium failed to elevate cyclic AMP levels (Fig. 3), whereas the rabbit SCG demonstrated a marked increase, as observed by McAfee et al., (11). The increases seen in the rabbit SCG were calcium-dependent, which is consistent with current concepts about the role of calcium in neurotransmitter release.

Based in part on physiological data from the experiments of Libet, Greengard (5) postulated a mechanism for s-IPSP generation. This mechanism involves the presence of (a) a DA receptor-adenylate cyclase complex; (b) SIF cell-interneurons; (c) DA as the SIF cell transmitter; (d) the ability of preganglionic physiological stimulation to increase cyclic AMP levels in the PGN; and (e) the ability to generate an s-IPSP in the PGN. Since the rabbit SCG has elements (a), (c), (d), and (3), whereas the guinea pig SCG does not, it is natural to hypothesize that the rabbit SCG has SIF cell interneurons [element (b)] that synapse with PGNs and the guinea pig does not.

With respect to the β-adrenergic receptor complex in the guinea pig SCG, it should be noted that the bovine SCG reportedly contains a β-receptor complex located on nonneuronal elements (6), and the rabbit vagus nerve also contains such a complex which is clearly not involved in synaptic transmission (13). Thus the β-receptor complex seen in guinea pig SCG need not have any relationship to neural transmission (or to SIF cells, for that matter). It may be located on Schwann cells and/or blood vessel walls, perhaps controlling ion fluxes in the Schwann cells or blood vessel contractility.

Although there are 10 times as many SIF cells (per milligram of tissue) in the guinea pig SCG as in the rabbit (18,19), the guinea pig SCG reportedly lacks an s-IPSP when studied by intracellular recording techniques (2,3,7). It is reasonable to predict that SIF cells regulate transmission in the guinea pig SCG by one or more of the following mechanisms: (a) inhibition via a presynaptic α-adrenergic receptor mechanism, as Nishi demonstrated (12); (b) augmentation and persistence of the s-EPSP, as Libet et al. demonstrated in the rabbit SCG (8,10); (c) SIF cells in series, with the SIF cells affecting release of the transmitter from each other (Jew, *this volume*); (d) control of blood flow

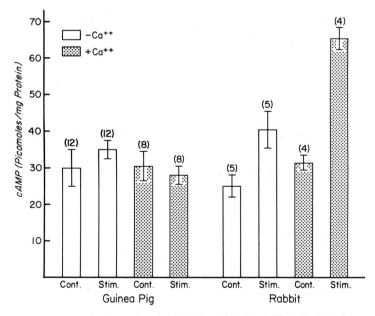

FIG. 3. Effects of supramaximal preganglionic physiological stimulation for 8 min on cyclic AMP levels in rabbit and guinea pig SCG. SCG were immersed at 37°C in Eagle's medium containing 5 mM theophylline with or without 2.2 mM CaCl$_2$. The preganglionic trunk was always maintained free of the medium. (From ref. 16.)

through the ganglion; (e) modulation of one PGN by another through recurrent collaterals or by some other means (dendrodendritic transmission?). Regulatory functions may be carried out by SIF cells acting via vascular channels (as paraneurons), as well as through local circuits (as interneurons) to PGNs or to other SIF cells.

CONCLUSIONS

1. The SIF cells of the rabbit SCG contain DA as transmitter, whereas those of the guinea pig SCG contain NE.

2. Whereas the rabbit SCG contains a DA receptor-adenylate cyclase complex, the guinea pig SCG contains a β-receptor-adenylate cyclase complex, which may have no direct effect on ganglionic transmission.

3. Preganglionic physiological stimulation elevates cyclic AMP levels in rabbit SCG but not in guinea pig SCG, the elevation in the rabbit being calcium-dependent.

We suggest that the regulatory mechanism(s) in which SIF cells function differs in the two species.

ACKNOWLEDGMENTS

This work was supported by NS-11650 to T.H.W. from the National Institutes of Health.

REFERENCES

1. Black, A.C., Jr., Jew, J.Y., West, J.R., Wamsley, J.K., Sandquist, D., and Williams, T.H. (1979): Catecholamines and cyclic nucleotides in the modulation of superior cervical ganglia. *Neurosci. Biobehav. Rev. (in press)*.

2. Dun, N., and Karczmar, A.G. (1976): Synaptic transmission and the effect of norepinephrine on the isolated superior cervical ganglion of the guinea pig. In: *Neuroscience Abstracts, Vol. II, Part 2: Society for Neuroscience, Sixth Annual Meeting*, p. 1001 (abstract 1446).

3. Dun, N., and Karczmar, A.G. (1977): The presynaptic site of action of norepinephrine in the superior cervical ganglion of guinea pig. *J. Pharmacol. Exp. Ther.*, 200:328-335.

4. Elfvin, L.G., Hökfelt, T., and Goldstein, M. (1975): Fluorescence microscopical, immunohistochemical, and ultrastructural studies on sympathetic ganglia of the guinea pig, with special reference to the SIF cells and their catecholamine content. *J. Ultrastruct. Res., 51:377-396.*

5. Greengard, P. (1978): *Cyclic Nucleotides, Phosphorylated Proteins, and Neuronal Function.* Raven Press, New York.

6. Kebabian, J.W., Bloom, F.E., Steiner, A.L., and Greengard, P. (1975): Neurotransmitters increase cyclic nucleotides in postganglionic neurons: Immunocytochemical demonstration. *Science,* 190:157-159.

7. Libet, B. (1970): Generation of slow inhibitory and excitatory postsynaptic potentials. *Fed. Proc.,* 29:1945-1956.

8. Libet, B. (1979): Which postsynaptic action of dopamine is mediated by cyclic AMP? *Life Sci.,* 24:1043-1048.

9. Libet, B., and Owman, Ch. (1974): Concomitant changes in formaldehyde-induced fluorescence of dopamine interneurones and in slow inhibitory postsynaptic potentials of the rabbit superior cervical ganglion, induced by stimulation of the preganglionic nerve or by a muscarinic agent. *J. Physiol. (Lond),* 237:635-662.

10. Libet, B., Kobayashi, H., and Tanaka, T. (1975): Synaptic coupling into the production and storage of a neuronal memory trace. *Nature,* 258:155-157.

11. McAfee, D.A., Schorderet, M., and Greengard, P. (1971): Adenosine 3',5'-monophosphate in nervous tissue: Increase associated with synaptic transmission. *Science,* 171:1156-1158.

12. Nishi, S. (1974): Ganglionic transmission. In: *The Peripheral Nervous System,* edited by J.I. Hubbard, pp. 225-255. Plenum Press, New York.

13. Roch, P., and Salamin, A. (1977): Effect of β-adrenergic drugs on adenosine 3',5'-monophosphate in rabbit vagus nerve. *J. Neurochem.,* 28:947-950.

14. Schmidt, R.H., and Bhatnagar, R.K. (1979): Regional development of norepinephrine, dopamine-β-hydroxylase, and tyrosine hydroxylase in the rat brain subsequent to neonatal treatment with subcutaneous 6-hydroxydopamine. *Brain Res. (in press)*.

15. Wamsley, J.K., Black, A.C., Jr., Redick, J.A., West, J.R., and Williams, T.H. (1978): SIF cells, cyclic AMP responses, and catecholamines of the guinea pig superior cervical ganglion. *Brain Res., 156:75-82.*

16. Wamsley, J.K., Black, A.C., Jr., West, J.R., and Williams, T.H. (1979): Cyclic AMP synthesis in guinea pig superior cervical ganglia: Response to pharmacological and preganglionic physiological stimulation. *Brain Res. (in press)*.

17. Williams, T.H. (1967): Electron microscopic evidence for an autonomic interneuron. *Nature,* 214:309-310.

18. Williams, T.H., Black, A.C., Jr., Chiba, T., and Jew, J. (1976): Interneurons/SIF cells in sympathetic ganglia of various mammals. In: *Chromaffin, Enterochromaffin, and Related Cells,* edited by R.E. Coupland and T. Fujita, pp. 95-116. Elsevier, Amsterdam.

19. Williams, T.H., Black, A.C., Jr., Chiba, T., and

Jew, J. (1977): Species differences in mammalian SIF cells. In: *Advances in Biochemical Psychopharmacology: Non-Striatal Dopaminergic Neurons,* edited by E. Costa and G.L. Gessa, 16:505–511. Raven Press, New York.

20. Williams, T.H., Chiba, T., Black, A.C., Jr., Bhalla, R.C., and Jew, J. (1976): Species variation in SIF cells of superior cervical ganglia: are there two functional types? In: *SIF cells: Structure and Function of the Small, Intensely Fluorescent Sympathetic Cells,* edited by O. Eränkö, pp. 143–162. Government Printing Office, Washington, D.C.

*Histochemistry and Cell Biology of
Autonomic Neurons, SIF Cells, and
Paraneurons,*
edited by O. Eränkö et al.
Raven Press, New York © 1980.

Neurons and Paraneurons in the Sympathetic Ganglia of Mammals

Tanemichi Chiba

Department of Anatomy, Saga Medical School, Saga 840-01, Japan

In addition to preganglionic cholinergic axons and principal ganglionic neurons, the existence of paraneurons and other input nerves have been revealed in the sympathetic ganglia of mammals. Small intensely fluorescent (SIF) cells are distributed widely in the sympathetic ganglia of most mammalian species (2,6,19,34–36) except for some rodents; they are small in population compared with the number of principal ganglionic neurons (2,6,35,36). Although Jacobowitz and Green (20) suggested that SIF cells have extensive branching processes in the ganglia, it is unlikely that SIF cells can influence most of the principal ganglionic neurons through synaptic junctions. Instead, most SIF cells in mammals seem to be neurosecretory cells (paraneurons). In the rat and the guinea pig, on the other hand, the SIF cell population is fairly large, and, at least in the rat, some SIF cells proved to be interneurons (7,24,32,33, 38). According to the results of serial section analysis, Kondo (22) claimed that some SIF cells in rat superior cervical ganglion (SCG) were sensory cells judged from their synaptic relationships.

A dense network of adrenergic nerves exists in the sympathetic ganglia of most mammalian species, and these fibers are thought to originate from the following four sources: (a) axon collaterals of principal ganglionic neurons; (b) incoming axons of the sympa-thetic postganglionic neurons; (c) dendritic processes of the principal ganglionic neurons; and (d) processes of SIF cells in the ganglia.

The existence of axon collaterals of the principal ganglionic neurons was suggested (13,37,39), but precise synaptic connections of axon collaterals in the ganglia have not been analyzed. The possibility of dendritic processes containing catecholamines in the sympathetic ganglia is shown in this chapter. In the substantia nigra, dendritic processes contained dopamine but their presynaptic nature has not been clearly demonstrated, although release of dopamine is verified biochemically (26). The processes of SIF cells are commonly observed in the ganglia. SIF cells contain characteristic large granules (more than 100 nm in diameter) in their cytoplasm and are usually identified without difficulty (4).

The axonal input from the other sympathetic ganglia was proposed to exist by Noon et al. (25). However, whether they terminate in, or simply pass through, the ganglia is not decided yet. Recently the existence of new types of nerve terminal as well as neuronal somata which showed polypeptide immunoreactivity have been found in the sympathetic ganglia. Substance P (17), VIP (18), and enkephalin (28) positive nerve terminals and enkephalin, VIP, and somatostatin (16) positive principal ganglionic neurons as well as enkephalin positive SIF cells have been observed.

Polypeptide immunoreactive neurons were distributed more densely in the abdominal ganglia than in other sympathetic ganglia. However, functional significance as well as synaptic organization of these polypeptide-containing neurons are not clear at present. Several types of granules were observed in neurons and SIF cells of the sympathetic ganglia (4), and the correlation between granule types and the transmitter (or hormone) contained need to be studied.

Some results of observations on adrenergic processes in the sympathetic ganglia of the guinea pig and human are described in the present chapter with special reference to their functional significance in modulation of synaptic transmission of the ganglia.

SYNAPTIC ORGANIZATION IN THE SCG OF THE GUINEA PIG

The unilateral SCG of the adult guinea pig was denervated almost completely and 5-hydroxydopamine (5-OHDA) was injected 2 days after denervation. Synaptic structures of the ganglia were examined in the operated as well as in the control side. In the control ganglia, small dense-cored vesicles were observed not only in the axonal varicosities but also in the dendrites (Fig. 1a). This probably corresponds to some of the fluorescent processes observed in the normal ganglia. Synaptic junctions occurred commonly between cholinergic axon varicosities and dendrites of principal ganglionic neurons. Synaptic junctions between adrenergic axon varicosities and undefined nerve processes (Fig. 1b) or SIF cells were also encountered. Synaptic junctions between SIF cells were often found. However, no synaptic junctions from SIF cells to principal ganglionic neurons have been confirmed in the SCG of the guinea pig (10,31). Norepinephrine was suggested as a neurotransmitter contained in SIF cells in this species (10).

In the denervated ganglia, synaptic junctions of any kind were few. Cholinergic axons were degenerated almost completely (Fig. 2a). Axon varicosities with small dense-cored vesicles were increased in number compared with the control side. Although synaptic-like close appositions were found between adrenergic axons and dendrites, typical synaptic differentiation was rarely encountered (Fig. 2b). SIF cells were found to be intact in the operated ganglia, but cholinergic synaptic input seemed to be eliminated from SIF cells. In the operated ganglia, only adrenergic elements of extraganglionic origin and cholinergic and other preganglionic axons were eliminated. As a result of axotomy of most principal ganglionic neurons, the concentration of catecholamine-containing vesicles were increased in adrenergic varicosities in axon collaterals and dendrites of the principal ganglionic neurons. At present, we have no conclusive criteria to differentiate between axons and dendrites in the sectional profiles. However, synaptic junctions were rarely encountered in the varicosities with small dense-cored vesicles, and it seemed likely that most varicosities with small dense-cored vesicles after denervation were dendrites (Fig. 2c).

SYNAPTIC ORGANIZATION IN THE HUMAN SYMPATHETIC GANGLIA

An extensive network of varicose fluorescent fibers was observed in lumbar sympathetic ganglia of the human (3). Some of these nerve fibers were located very close to principal ganglionic neurons. SIF cells, on the other hand, were small in number, formed small clusters, and had branching varicose processes (3,14,15). By electron microscopy, clustered SIF cells were found in most cases close to the small blood vessels, and processes were located in the perivascular space without any Schwann cell sheath (Fig. 3a). Thus most clustered SIF cells in human sympathetic ganglia seemed to be paraneurons which secrete transmitters into

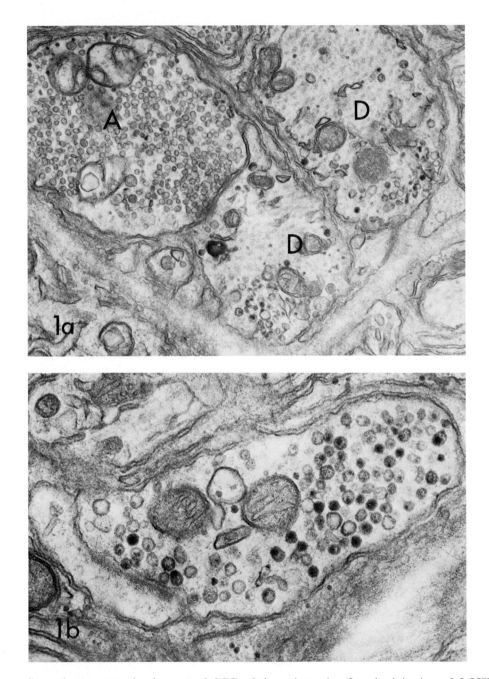

FIG. 1. Synaptic structures in the control SCG of the guinea pig after the injection of 5-OHDA.
a: A cholinergic axon (A) synapses on the dendrite (D) containing small dense-cored vesicles. ×30,000.
b: A presumed axon collateral with small dense-cored vesicles synapses on the spine. ×46,000.

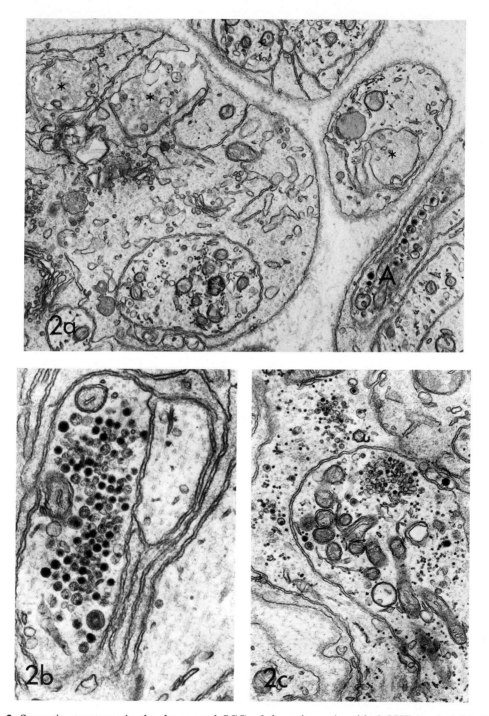

FIG. 2. Synaptic structures in the denervated SCG of the guinea pig with 5-OHDA administration.
  a: Most input nerve terminals were degenerated (*asterisk*). Survived dendrites and an adrenergic process (A) were also seen. ×17,000.
  b: A varicosity with small dense-cored vesicles in contact with a dendritic spine. ×34,000.
  c: Presumed dendrites with small dense-cored vesicles were increased in number. ×18,000.

extracellular space as well as blood vessels under the control of cholinergic preganglionic nerves (3,14,15). Axon varicosities with small dense-cored vesicles were occasionally encountered in human lumbar sympathetic ganglia after glutaraldehyde fixation, synapsing on presumed dendrites of principal ganglionic neurons (Fig. 3b). Sometimes synaptic vesicle-like structures were found in pre- and postsynaptic elements (Fig. 3c). However, the relationships of this kind were exceptional and should be studied more extensively.

MONOAMINE NEURONS IN PERIPHERAL VERSUS CENTRAL NERVOUS SYSTEM

The precise mechanism of function of monoamine neurons in the central nervous system (CNS) is still not clearly understood. The synaptic structure of monoamine nerve terminals was analyzed by several workers in order to reveal the functional significance of monoamines in the CNS. Descarries et al. (8,9) found less than 5% synaptic junctions in the monoamine axon varicosities of the brain cortex. However, synaptic junctions of monoamine nerve terminals were more frequent in the dorsal motor nucleus of vagus and nucleus of solitary tract in the medulla oblongata (5). Monoamine neurons in the peripheral nervous system, on the other hand, have no synaptic junctions with target cells. Monoamine-containing axon varicosities in the vas deferens contact the smooth muscle cytolemma within 20 nm (1,27). Monoamine nerve terminals in the arteriolar wall are located more than 100 nm away from smooth muscle cytolemma. The rather exceptional example of monoamine-containing cells in the peripheral nervous system which have synaptic junctions with the target cells are the SIF cells of a few mammalian species. Axon collaterals and dendrites containing monoamines in the sympathetic ganglia seem to have some synaptic junctions according to the present study in the guinea pig. Thus synaptic relationships between the monoamine nerve terminals and target cells are variable from specific synaptic junctions to nonspecific free nerve endings in the central and peripheral nervous systems. Nerve terminals with or without synaptic junctions seem to originate from a monoamine neuron in the CNS as well as in the sympathetic ganglia.

NEURONS AND PARANEURONS IN THE SYMPATHETIC GANGLIA

According to the paraneuron concept, neurons and peptide-hormone-producing cells are similar in that both are receptosecretory in function (12) and engage in reflex control of some biological activities. Paraneurons are cells which are located in non-neural tissues but which have neural nature. In contrast, neurosecretory cells are located in neural tissues but have hormone-secreting function. Neurosecretory cells and paraneurons are characterized by their content of synaptic vesicle-like granules as well as secretory granules in the same cells (4,12). It seems likely that there are several factors which convert neurons to paraneurons or vice versa. One of the factors which induce neurons in the sympathetic ganglia as well as in the adrenal medulla is nerve growth factor (23). On the other hand, one of the factors that induces paraneurons (SIF cells) in the sympathetic ganglia of the fetus is glucocorticoid (11). In this respect, it is of great interest that small granule-containing cells with processes have been detected in the adrenal medulla of the mouse (21) and the guinea pig (30).

In normal control sympathetic ganglia of guinea pigs and humans, enlarged processes containing a mixture of vesicles of variable size, tubules, and some dense bodies and mitochondria were observed consistently (Fig. 4a). Mitochondria-packed profiles were also encountered (Fig. 4b). As these processes increase in number after denervation, it is suggested that they represent degenerating as well as regrowing processes in the ganglia.

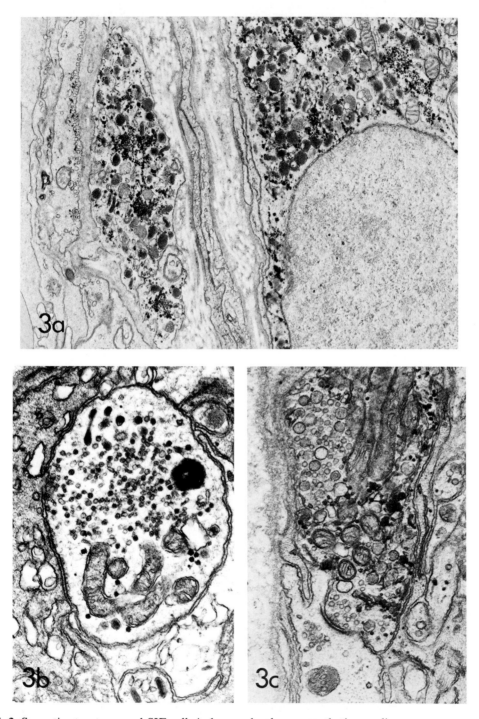

FIG. 3. Synaptic structures and SIF cells in human lumbar sympathetic ganglia.
    a: A SIF cell and a process of a SIF cell without Schwann cell sheath are located in the perivascular space. The capillary lumen is located at the left. ×14,000.
    b: An axon varicosity with small dense-cored vesicles synapses on the dendrite. ×30,000.
    c: Small clear vesicles are seen in pre- and postsynaptic processes. ×32,000.

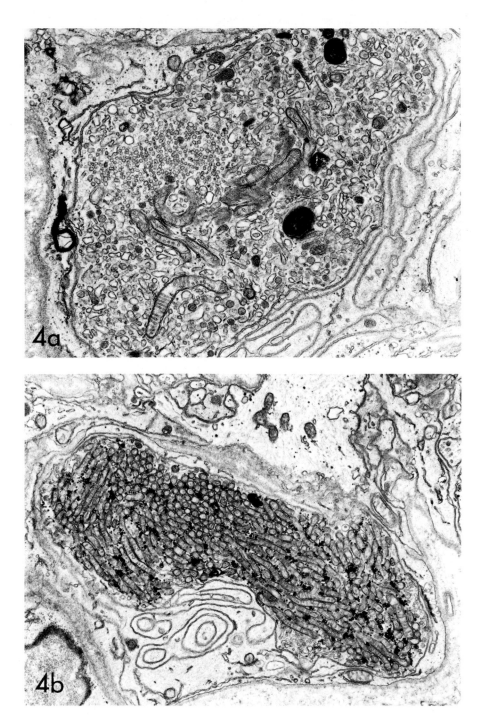

FIG. 4. Atypical processes found in normal human sympathetic ganglia. They probably represent the process of axonal and/or dendritic remodeling.
  a: An enlarged process containing synaptic vesicles, vesicles of variable configuration, dense bodies, and mitochondria. ×20,000.
  b: A process packed with numerous mitochondria, some glycogen particles, and vesicles.

Thus remodeling of the synaptic organization seemed to be a persistent event in the ganglia as well as in the CNS (29). Some dynamic aspects of the cell transformation as well as of the synaptic relationships are suggested to exist in the sympathetic ganglia of mammals.

REFERENCES

1. Chiba, T. (1973): Electron microscopic and histochemical studies on the synaptic vesicles in mouse vas deferens and atrium after 5-hydroxydopamine administration. *Anat. Rec.,* 176:35–48.
2. Chiba, T. (1977): Monoamine-containing paraneurons in the sympathetic ganglia of mammals. *Arch. Histol. Jpn. (Suppl.)* 40:163–176.
3. Chiba, T. (1978): Monoamine fluorescence and electron microscopic studies on small intensely fluorescent (granule-containing) cells in human sympathetic ganglia. *J. Comp. Neurol.,* 179:153–168.
4. Chiba, T. (1980): Synaptic relationships and granule morphology of SIF (GC) cells in sympathetic ganglia. *Biomed. Res. (Suppl.) (in press).*
5. Chiba, T. (1980): The synaptic structure of monoamine nerve terminals in the central nervous system. *Biomed. Res. (Suppl.) (in press).*
6. Chiba, T., and Williams, T.H. (1975): Histofluorescence characteristics and quantification of small intensely fluorescent (SIF) cells in sympathetic ganglia of several species. *Cell Tissue Res.,* 162:331–342.
7. Dail, W.G., Evan, A.P., Jr., and Eason, H.R. (1975): The major ganglion in the pelvic plexus of the male rat: A histochemical and ultrastructural study. *Cell Tissue Res.,* 159:49–62.
8. Descarries, L., Beaudet, A., and Watkins, K.C. (1975): Serotonin nerve terminals in adult rat neocortex. *Brain Res.,* 100:563–588.
9. Descarries, L., Watkins, K.C., and Lapierre, Y. (1977): Noradrenergic axon terminals in the cerebral cortex of rat III, topometric ultrastructural analysis. *Brain Res.,* 133:197–222.
10. Elfvin, L.G., Hökfelt, T., and Goldstein, M. (1975): Fluorescence microscopical, immunohistochemical and ultrastructural studies on sympathetic ganglia of the guinea pig, with special reference to the SIF cells and their catecholamine content. *J. Ultrastruct. Res.,* 51:377–398.
11. Eränkö, L., and Eränkö, O. (1972): Effect of hydrocortisone on histochemically demonstrable catecholamines in the sympathetic ganglia and extraadrenal chromaffin tissue of the rat. *Acta Physiol. Scand.,* 84:125–233.
12. Fujita, T. (1977): Concept of paraneurons. *Arch. Histol. Jpn. (Suppl.),* 40:1–12.
13. Grillo, M.A. (1965): Synaptic morphology in the superior cervical ganglion of the rat before and after preganglionic denervation. *J. Cell Biol.,* 27:136A.
14. Hervonen, A., and Kanerva, L. (1972): Catecholamine storing cells in human fetal superior cervical ganglion. *Acta Physiol. Scand.,* 84:538–542.
15. Hervonen, A., Alho, H., Helen, P., and Kanerva, L. (1979): Small intensely fluorescent cells of human sympathetic ganglia. *Neurosci. Lett.,* 12:97–101.
16. Hökfelt, T., Elfvin, L-G., Elde, R., Schultzberg, M., Goldstein, M., and Luft, R. (1977): Occurrence of somatostatin-like immunoreactivity in some peripheral sympathetic noradrenergic neurons. *Proc. Natl. Acad. Sci. USA,* 74:3587–3591.
17. Hökfelt, T., Elfvin, L-G., Schultzberg, M., Goldstein, M., and Nilsson, G. (1977): On the occurrence of substance P-containing fibers in sympathetic ganglia: Immunohistochemical evidence. *Brain Res.,* 132:29–41.
18. Hökfelt, T., Elfvin, L-G., Schultzberg, M., Said, S.I., Mutt, V., and Goldstein, M. (1977): Immunohistochemical evidence of vasoactive intestinal polypeptide-containing neurons and nerve fibers in sympathetic ganglia. *Neuroscience,* 2:885–896.
19. Jacobowitz, D. (1970): Catecholamine fluorescence studies of adrenergic neurons and chromaffin cells in sympathetic ganglia. *Fed. Proc.,* 29:1929–1944.
20. Jacobowitz, D., and Green, L.A. (1974): Histofluorescence study of chromaffin cells in dissociated cell cultures of chick embryo sympathetic ganglia. *J. Neurobiol.,* 5:65–83.
21. Kobayashi, S., and Coupland, R.E. (1977): Two populations of microvesicles in the SGC (small granule chromaffin) cells of the mouse adrenal medulla. *Arch. Histol. Jpn.,* 40:251–259.
22. Kondo, H. (1977): Innervation of SIF cells in the superior cervical and nodose ganglia: An ultrastructural study with serial sections. *Biol. Cell.,* 30:253–264.
23. Levi-Montalcini, R., and Calissano, P. (1979): The nerve growth factor. *Sci. Am.,* 240:44–53.
24. Matthews, M.R., and Raismann, G. (1969): The ultrastructure and somatic efferent synapses of small granule-containing cells in the superior cervical ganglion. *J. Anat.,* 105:255–282.
25. Noon, J.P., McAfee, D.A., and Roth, R.H. (1975): Norepinephrine release from nerve terminals within the rabbit superior cervical ganglion. *Naunyn Schmiedebergs Arch. Pharmacol.,* 291:139–162.
26. Reubi, J.C., and Sandri, C. (1979): Ultrastructural observations on intercellular contacts of nigral dendrites. *Neurosci. Lett.,* 13:183–188.
27. Richardson, K.C. (1962): The fine structure of autonomic nerve endings in smooth muscle of the rat vas deferens. *J. Anat.,* 96:427–442.
28. Schultzberg, M., Hökfelt, T., Terenius, L., Elfvin, L-G., Lundberg, J.M., Brandt, J., Elde, R.P., and Goldstein, M. (1979): Enkephalin immunoreactive nerve fibers and cell bodies in sympathetic ganglia of the guinea-pig and rat. *Neuroscience,* 4:249–270.
29. Sotelo, C., and Palay, S.L. (1971): Altered axons and axon terminals in the lateral vestibular nucleus of the rat. *Lab. Invest.,* 25:653–671.

30. Unsicker, K., Habwa-Fluh, O., and Zwarg, U. (1978): Different types of small granule-containing cells and neurons in the guinea-pig adrenal medulla. *Cell Tissue Res.,* 189:109–310.

31. Watanabe, H. (1971): Adrenergic nerve elements in the hypogastric ganglia of the guinea pig. *Am. J. Anat.,* 130:305–329.

32. Williams, T.H. (1967): Electron microscopic evidence for an autonomic interneuron. *Nature,* 214:309–310.

33. Williams, T.H., and Palay, S.L. (1969): Ultrastructure of the small neurons in the superior cervical ganglion. *Brain Res.,* 15:17–34.

34. Williams, T.H., Black, A.C., Jr., Chiba, T., and Jew, J. (1976): Interneuron/SIF cells in sympathetic ganglia of various mammals. In: *Chromaffin, Enterochromaffin and Related Cells,* edited by R.E. Coupland and T. Fujita, pp. 95–116. Elsevier, Amsterdam.

35. Williams, T.H., Black, A.C., Jr., Chiba, T., and Jew, J. (1976): Species differences in mammalian SIF cells. *Adv. Biochem. Psychopharmacol.,* 16:505–511.

36. Williams, T.H., Chiba, T., Black, A.C., Jr., Bhalla, R.C., and Jew, J. (1976): Species differences in structure and function of SIF cells in the superior cervical ganglion. In: *SIF cells,* edited by O. Eränkö, pp. 143–162. Government Printing Office, Washington, D.C.

37. Yamauchi, A. (1976): Ultrastructure of chromaffin-like interneurons in the autonomic ganglia. In: *Chromaffin, Enterochromaffin and Related Cells,* edited by R.E. Coupland and T. Fujita, pp. 117–130. Elsevier, Amsterdam.

38. Yokota, R. (1973): The granule-containing cell somata in the superior cervical ganglion of the rat, as studied by a serial sampling method for electron microscopy. *Z. Zellforsch.,* 141:331–345.

39. Yokota, R., and Yamauchi, A. (1974): Ultrastructure of the mouse superior cervical ganglion, with particular reference to the pre- and postganglionic elements covering the soma of its principal neurons. *Am. J. Anat.,* 140:281–298.

Histochemistry and Cell Biology of Autonomic Neurons, SIF Cells, and Paraneurons,
edited by O. Eränkö et al.
Raven Press, New York © 1980.

Effect of Aging on SIF Cells of the Rat

M. Partanen, A. Hervonen, and *R.M. Santer

*Department of Biomedical Sciences, University of Tampere, Tampere, Finland; and *Department of Anatomy, University College, Cardiff, Wales*

Although age-related changes have been studied in the central nervous system, little attention has been paid to the autonomic nervous system and even less to the system of SIF cells and paraganglia (6,7). The chromaffin system has been followed postnatally (15) and has been found to degenerate in the rat and in part (2) in man. No detectable changes in the fluorescence histochemistry (10) or ultrastructure (1,8) of SIF cells in man have been reported. Similarly, in the paracervical ganglion, no detectable color changes have been reported for SIF cells in newborn and young rats (11), and this is reflected in the ultrastructure of the catecholamine storage vesicles (12,14). However, changes with age in the fluorescence of SIF cells in rat superior cervical and celiac-superior mesenteric ganglia were previously reported (18).

The hypogastric ganglion was selected for study as it can be easily removed and is rich in SIF cells which occur singly or in large groups.

MATERIAL AND METHODS

Twelve newborn rats, 12 which were 2 to 3 months old, and 12 which were 28 to 31 months old were used. All were males.

Fluorescence Microscopy

The specimens were removed under light ether anesthesia and frozen immediately in liquid nitrogen before freeze-drying for 5 days under vacuum of 10^{-4} torr at $-40°C$. The specimens were warmed under vacuum and then exposed to paraformaldehyde vapor of 60% relative humidity at 80°C for 1 hr. The specimens were embedded in paraffin and sectioned serially at 10 μm. The microscope used was the Leitz MPV-2 microspectrophotometer equipped with an epi-illuminator, a band interference filter (Veril S-60) on the emission side, and a photomultiplier of S-20 type. The filters BG 3, KP 425, and TK 455 were used. The light source was an HBO 100 mercury lamp (Osram). The emission spectra were recorded by scanning the band interference filter (Veril S-60) with continuous registration. Kodak Tri-X (400 ASA) and Kodacolor (400 ASA) films were used.

Electron Microscopy

The young adult and the aged rats were perfused under light ether anesthesia via the abdominal aorta or the heart with glutaraldehyde or $KMnO_4$ fixatives. The tissue pieces from newborn rats were directly immersed in fixative. After the perfusion (15 min) the ganglia were immersed for a further 60 min in the same fixative. Glutaraldehyde 3% in 0.1 M phosphate buffer (pH 7.3) and concentrated $KMnO_4$ (3%) in 0.1 M phosphate buffer (pH 7.0) were used. Glutaraldehyde-fixed specimens were postfixed with 1% OsO_4. The $KMnO_4$-fixed specimens were block-stained with 1% uranyl acetate. The glutaraldehyde-OsO_4-fixed thin sections were grid-stained with lead citrate (17) and uranyl acetate (20). The specimens were studied with JEM 100 C at 60 kV.

RESULTS

Fluorescence Histochemistry

In the newborn and young adult rats, the SIF cells were distributed throughout the

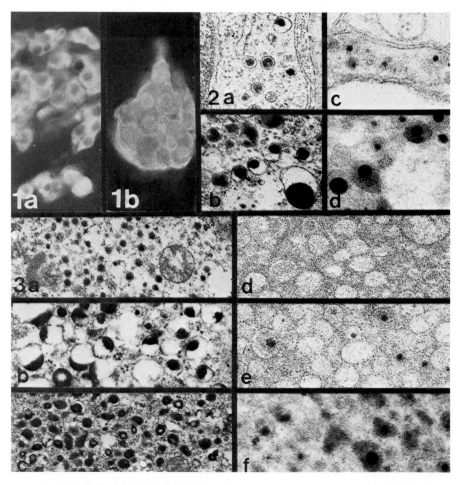

FIG. 1. The color of SIF cells in young adult rats **(a)** is greenish-yellow and in aged rats **(b)** greenish-yellow or yellowish-brown. (a) ×200. (b) ×500.

FIG. 2. Amine storage vesicles in two types of SGC cell after glutaraldehyde **(a, b)** and $KMnO_4$ **(c, d)** fixatives in young adult rats. ×30,000.

FIG. 3. Amine storage vesicles in three types of SGC cell after glutaraldehyde **(a–c)** and $KMnO_4$ **(d–f)** fixatives in aged rats. The classification into different classes between fixatives is not comparable. ×30,000.

whole ganglion separately or in groups. In adults the clusters were usually at the dorsal end of the ganglion. The capillarity of the clusters was usually richer in young adult rats than in the newborns. The SIF cells of the newborn and young adult rats emitted only greenish yellow fluorescence, with the emission maximum at 480 nm (Fig. 4). Based on the fluorescence microscopic observation, only one type of SIF cell was observed (Fig. 1a) in newborn and young adult rats.

In aged rats the SIF cells were also in vascularized clusters or separate. On the basis of the color of the fluorescence, two types of SIF cell could be distinguished in aged animals (Fig. 1b). The first type emitted greenish-yellow fluorescence and the second type yellowish-brown granular fluorescence. The SIF cells with yellowish-brown granular fluorescence usually had a greenish-yellow rim at the periphery of the cytoplasm (Fig. 1b). The emission maximum was at 480 nm in both cell types, but the yellowish-brown granular cells had a shoulder between 510 and 550 nm (Fig. 4). The intensity of the formaldehyde-induced fluorescence (FIF) was usually higher in the greenish-yellow SIF cells.

Electron Microscopy

After glutaraldehyde fixation two types of small granule-containing (SGC) cell could be distinguished in newborn and young adult rats. The first type contained vesicles of 50 to 150 nm (Fig. 2a), and the second type contained vesicles of 50 to 250 nm (Fig. 2b). The size and density of the core was variable in the second type, whereas in the first type the cores were more homogeneous in appearance. A similar distinction of cell types was clear after $KMnO_4$ fixation in newborn and young adult rats (Fig. 2c,d).

In aged rats three types of SGC cell could be distinguished after both fixatives, but the classification differs. After glutaraldehyde fixation the first type contained 50 to 150-nm round dense-cored vesicles (Fig. 3a). The second type contained 50 to 250-nm round vesicles with a dense core (Fig. 3b). The third type contained 50 × 250 nm elongated and variably sized dense-cored vesicles (Fig. 3c).

After $KMnO_4$ fixation the first type contained empty vesicles of 100 to 300 nm (Fig. 3d). The second type contained 100 to 300-nm vesicles with a small central core and a halo (Fig. 3e). The third type contained irregularly shaped vesicles (Fig. 3f). The size of these vesicles varied between 100 and 500 nm, and the vesicles were filled with electron-opaque material and had a very dense core (Fig. 3f).

DISCUSSION

This study revealed changes in the colors of fluorescence of SIF cells of the rat hypogastric ganglion with age and ultrastructural changes in amine-storing granules that were obvious especially after $KMnO_4$ fixation.

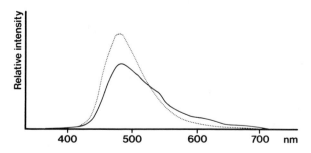

FIG. 4. The emission curves from the SIF cells of the aged rats. (....) From the greenish-yellow SIF cell. (—) From the yellowish-brown SIF cell.

Fluorescence Histochemistry

The changes in the fluorescence emission of SIF cells observed in the rat hypogastric ganglion with age are similar to observations on other sympathetic ganglia of the rat (18) in which yellow SIF cells appeared later in life than green SIF cells. A possible explanation of the presence of yellowish-brown SIF cells in the hypogastric ganglion of aged rats could be that a proportion of the greenish-yellow SIF cells seen in the younger age categories have greatly increased their amine content, with a corresponding shift in the fluorescence emission. This explanation is in accordance with the reports of a relatively constant number of SIF cells in sympathetic ganglia until the adult state (6,11). It should be noted that the emission maxima for the greenish-yellow and the yellowish-brown SIF cells was the same (480 nm) but that a shoulder was observed at 510 to 550 nm in the emission curve of the yellowish-brown SIF cells. This shoulder remains inexplicable.

Electron Microscopy

Many reports dealing with classification of SGC cells according to the storage granules are available (4,5,16,19). In the rat two kinds of SGC cell have been described in the hypogastric ganglion (4) in males and in the paracervical ganglion (14) in females. The first type contained 80 to 140-nm (14) or 90 to 120-nm (4) dense-cored vesicles and the second type contained 80 to 300-nm (14) or 120 to 250-nm (4) dense-cored vesicles. Lu et al. (16) described three types of SGC cell (50 to 150 nm; 100 to 300 nm; 50×170 nm elongated) in the superior cervical and celiac-superior-mesenteric ganglia of the rat. In the present study only two types of SGC cell were constantly found in the newborn and young adult rats. This result agrees with those of Kanerva and Teräväinen (14) and Dail et al. (4). In aged rats also a third type of 50×250 nm elongated vesicles was found in some SGC cells.

After $KMnO_4$ fixation, the SGC cells were seen by Kanerva and Hervonen (12,13) in newborn rats and by Hervonen et al. (9) in adult rats. In newborn rats only one type of SGC cell containing 80 to 200-nm dense-cored vesicles (12) or 90 to 150-nm dense-cored vesicles (6) has been described. However, the present study revealed two types of SGC cell in the hypogastric ganglion of the newborn rat, and the classification corresponds to that already described after glutaraldehyde fixation. In aged rats three types are seen according to the size, core, and shape of the storage granules; the classification clearly differs from that after glutaraldehyde fixation. The most prominent changes are the increase in the size of the vesicles and the irregularity in the shape of the third type. Vesicles of the third type are filled with dense precipitate.

The significance of the various types of storage granules is unclear. The second type with large vesicles after glutaraldehyde fixation resembles that of norepinephrine-containing cells in the adrenal medulla (3). The second type of SGC cell in young adult rats and the third type in aged rats after $KMnO_4$ fixation are proposed to correspond with each other, therefore probably containing norepinephrine. The SIF cells with yellowish-brown granular fluorescence may correspond with the SGC cells with large irregular dense-cored vesicles. No other age-dependent ultrastructural changes in the SGC cells can be correlated with changes in FIF. Thus the age-dependent changes are more evident in norepinephrine-containing SIF cells. This suspicion also supports the concept of the existence of different types of SIF cell. It is also concluded in this study that the cytoplasmic fluorescence and the ultrastructural morphology of the amine storage granules changes with age.

SUMMARY

The effect of aging on the SIF cells was studied with the formaldehyde-induced fluorescence (FIF) method and electron microscopy (EM). For EM the specimens were fixed with glutaraldehyde-OsO_4 and $KMnO_4$.

The emission spectra were recorded microspectrofluorimetrically.

Under fluorescence microscopy in newborn and young adult rats, only one type of SIF cell emitting greenish-yellow FIF could be found, whereas in aged rats another type of SIF cell emitting yellowish-brown, granular FIF was also found. The emission spectra of the greenish-yellow FIF of the newborn and young adults and the aged rats differed from that recorded from the yellowish-brown granular SIF cells.

According to the size of the dense-cored vesicles, two kinds of SGC cell could be distinguished in the hypogastric ganglion of the newborn and the young adult rats after glutaraldehyde fixation. The first type contained 50- to 150-nm vesicles, and the second type contained 50- to 250-nm vesicles. In aged rats a third type of SGC cell containing elongated 50 × 250 nm dense-cored vesicles was found.

After $KMnO_4$ fixation the classification was the same as after glutaraldehyde in newborn and young adult animals. In aged rats three types of SGC cell could be distinguished, according to the size, shape, and core of the storage granules; the first with 100- to 300-nm empty vesicles, the second with 100- to 300-nm vesicles with small cores, and the third with 100- to 500-nm irregular vesicles filled with electron-opaque material with a denser core.

REFERENCES

1. Chiba, T. (1978): Monoamine fluorescence and electron microscopic studies on small intensely fluorescent (granule-containing) cells in human sympathetic ganglia. *J. Comp. Neurol.*, 179:153-168.
2. Coupland, R.E. (1965): *The Natural History of the Chromaffin Cell.* Longmans, London.
3. Coupland, R.E., and Hopwood, D. (1977): The mechanism of differentiating staining reaction for adrenalin and noradrenalin storing granules in tissue fixed in glutaraldehyde. *J. Anat.*, 100:227-243.
4. Dail, W.G., Evan, A.P., and Eason, H.K. (1975): The major ganglion in the pelvic plexus of the male rat. *Cell Tissue Res.*, 159:49-62.
5. Elfvin, L.-G., Hökfelt, T., and Goldstein, M. (1975): Fluorescence microscopical, immunohistochemical and ultrastructural studies on sympathetic ganglia of the guinea-pig with special reference to the S.I.F. cells and their catecholamine content. *J. Ultrastruct. Res.*, 51:377-396.
6. Eränkö, L. (1972): Postnatal development of histochemically demonstrable catecholamines in the superior cervical ganglion of the rat. *Histochem. J.* 4:225-236.
7. Eränkö, O. (1976): *SIF Cells. Structure and Function of the Small, Intensely Fluorescent Sympathetic Cells.* Fogarty International Center Proceedings No. 30. DHEW publication No. (NIH) 76-942. Government Printing Office, Washington, D.C.
8. Hervonen, A., Alho, H., Helén, P., and Kanerva, L. (1979): Small intensely fluorescent cells of human sympathetic ganglia. *Neurosci. Lett.* 12:87-101.
9. Hervonen, A., Kanerva, L., and Teräväinen, H. (1972): The fine structure of the paracervical (Frankenhäuser) ganglion of the rat after permanganate fixation. *Acta Physiol. Scand.*, 85:506-510.
10. Hervonen, A., Vaalasti, A., Partanen, M., Kanerva, L., and Hervonen, H. (1978): Effects of ageing on the histochemically demonstrable catecholamines and acetylcholinesterase of human sympathetic ganglia. *J. Neurocytol.*, 7:11-23.
11. Kanerva, L. (1971): The postnatal development of monoamines and cholinesterases in the paracervical ganglion of the rat uterus. *Prog. Brain Res.*, 34:432-444.
12. Kanerva, L. (1972): Ultrastructure of sympathetic ganglion cells and granule-containing cells in the paracervical (Frankenhäuser) ganglion of the newborn rat. *Z. Zellforsch.*, 126:25-40.
13. Kanerva, L., and Hervonen, A. (1976): SIF cells, short adrenergic neurons and vacuolated nerve cells of the paracervical (Frankenhäuser) ganglion. In: *SIF Cells, Structure and Function of the Small, Intensely Fluorescent Sympathetic Cells,* edited by O. Eränkö, pp. 19-34. Fogarty International Center Proceedings No. 30. DHEW Publication No. (NIH) 76-942. Government Printing Office, Washington, D.C.
14. Kanerva, L., and Teräväinen, H. (1972): Electron microscopy of the paracervical (Frankenhäuser) ganglion of the adult rat. *Z. Zellforsch.*, 129:161-177.
15. Lempinen, M. (1964): Extra-adrenal chromaffin tissue of the rat and the effect of cortical hormones on it. *Acta Physiol. Scand. [Suppl. 231]*, 62:1-91.
16. Lu, K-S., Lever, J.D., Santer, K.M., and Presley, R. (1976): Small granulated cell types in rat superior cervical and coeliac-mesenteric ganglia. *Cell Tissue Res.*, 172:331-343.
17. Reynolds, E.S. (1963): The use of lead citrate at high pH as an electron-opaque stain in electron microscopy. *J. Cell Biol.*, 17:208-212.
18. Santer, R.M., Lu, K-S., Lever, J.D., and Presley, R. (1975): A study of the distribution of chromaffin-positive (CH+) and small intensely fluorescent (SIF) cells in sympathetic ganglia of the rat at various ages. *J. Anat.*, 119:589-599.
19. Watanabe, H. (1971): Adrenergic nerve elements in the hypogastric ganglion of the guinea pig. *Am. J. Anat.*, 130:305-330.
20. Watson, M.L. (1958): Staining of tissue sections for electron microscopy with heavy metals. *J. Biophys. Biochem. Cytol.*, 4:475-478.

Histochemistry and Cell Biology of Autonomic Neurons, SIF Cells, and Paraneurons,
edited by O. Eränkö et al.
Raven Press, New York © 1980.

Ultrastructure and Histochemistry of Human SIF Cells and Paraganglia

Pauli Helén, Hannu Alho, and Antti Hervonen

Department of Biomedical Sciences, University of Tampere, SF-33101 Tampere 10, Finland

Small intensely fluorescent (SIF) cells in adult human sympathetic ganglia have been reported by several authors (1,6–9,11,12). Ultrastructurally these cells have been shown to be identical with the small granule-containing cells (4). SIF cells have been extensively examined in several species (1–3,14) and in human fetal material (5). Because of their scanty existence in adult human ganglia, the ultrastructure was described only recently by Chiba (1), Kondo and Fujiwara (12), and Hervonen et al. (6,7).

MATERIAL AND METHODS

Human sympathetic ganglia—43 ganglia from 18 patients of various ages and both sexes—were obtained from sympathectomies performed because of various peripheral circulatory diseases. For electron microscopy, the thoracic and lumbar ganglia were immediately sliced in 0.9% saline and immersed in 3% glutaraldehyde buffered with phosphate at pH 7.3 for 1 to 3 hr and postfixed with 1% OsO_4. After embedding in Epon, the specimens were stained with uranyl acetate and lead citrate. For formaldehyde-induced fluorescence (FIF), pieces of the same ganglia were frozen in liquid nitrogen, freeze-dried, and exposed to paraformaldehyde vapor. The specimens were then embedded in paraffin under vacuum (10). A Leitz-MPV 2 microspectrofluorimeter was used for recording the fluorescence intensities and emission spectra.

RESULTS AND DISCUSSION

The SIF cells occurred as solitary cells (Fig. 1) or small clusters composed of 3 to 20 cells. Their intensive fluorescence among the principal neurons (filled with yellow or orange fluorescence) made them easy to detect. However, since the ratio of the SIF cells of the total number of nerve cells was usually under 0.1%, they were only occasionally seen. With electron microscopy they were found after extensive serial sectioning. In semithin sections, after toluidine blue staining they were detected as brownish, small granule-containing cells.

They were also randomly scattered among the connective tissue and in the nerve trunks between the ganglia. In these locations they can be referred to as paraganglia (6). However, electron and fluorescence microscopy showed them to have the same features as intraganglionic SIF cells.

Electron microscopically, the SIF cells were found to be surrounded almost entirely by satellite cells (Fig. 1). The rich network of synapses on SIF cells contained small agranular vesicles and larger granular vesicles, and could be classified as nonadrener-

paraganglia of adult man. *Am. J. Anat.,* 153:563–572.

9. Hervonen, A., Vaalasti, A., Partanen, M., Kanerva, L., and Vaalasti, T. (1976): The paraganglia, a persisting endocrine system in man. *Am. J. Anat.,* 146:207–210.

10. Hervonen, A., Vaalasti, A., Partanen, M., Kanerva, L., and Hervonen, H. (1978): Effects of ageing on the histochemically demonstrable catecholamines and acetylcholinesterase of human sympathetic ganglia. *J. Neurocytol.,* 7:11–23.

11. Hino, O., and Tsunekawa, K. (1978): The occur-

rence of small, intensely fluorescent (SIF) cells in human sympathetic ganglia. *Experientia,* 34:1359.

12. Kondo, H., and Fujiwara, S. (1979): Granule-containing cells in the human superior cervical ganglion. *Acta Anat. (Basel),* 103:192–199.

13. Libet, B. (1970): Generation of slow inhibitory and excitatory postsynaptic potentials. *Fed. Proc. Fed. Am. Soc. Exp. Biol.,* 29:1945–1956.

14. Williams, T.H., Black, A.C., Jr., Chiba, T., and Bhalla, R.C. (1975): Morphology and biochemistry of small, intensely fluorescent cell of sympathetic ganglia. *Nature,* 256:315–317.

Histochemistry and Cell Biology of Autonomic Neurons, SIF Cells, and Paraneurons,
edited by O. Eränkö et al.
Raven Press, New York © 1980.

Ultrastructural Study of the Frog Sympathetic Ganglia

Hiroshi Watanabe

Department of Anatomy, Tohoku University, School of Medicine, Sendai, 980, Japan

Previous morphological studies have shown that some of the mammalian sympathetic ganglion cells receive direct synaptic contacts from adrenergic elements which are probably derived from small granule-containing (GC) cells [small intensely fluorescent (SIF) cells] and/or axon collaterals of the principal sympathetic neurons (for review see refs. 2,8,19). These preganglionic adrenergic elements have been considered to modulate slow inhibitory postsynaptic potentials (s-IPSP) (for review see refs 6,7,9).

GC cells are also present in various amphibian sympathetic ganglia (3,4,11,17). However, no conclusive morphological evidence has yet been obtained of the synaptic contact between adrenergic presynaptic elements and principal sympathetic neurons. Thus the generation mechanism of the s-IPSP is a controversial subject in the amphibian sympathetic ganglia.

In this chapter fine structural features of the amphibian sympathetic ganglia are briefly summarized. The functional significance of the GC cells is also reviewed in relation to previous physiological studies on the synaptic transmission of the amphibian sympathetic ganglia.

PRINCIPAL SYMPATHETIC GANGLION CELLS

The amphibian sympathetic ganglion cells are unipolar and consist of two populations, large and small principal neurons (5,10,14, 15). These large (30 to 35 μm in diameter) and small (approximately 20 μm in diameter) cells comprise the fast B-conduction system (B-neurons) and the slow C-system (C-neurons), respectively (5). The two types of ganglion cell contain distinct postsynaptic specializations: highly flattened subsurface cisterns (Fig. 1) in the large B-neurons and postsynaptic bars (Fig. 2) in the small C-neurons (14,15). The subsurface cisterns appear in postsynaptic (Fig. 1) and extrasynaptic cytoplasm of the B-neurons. They are often associated with endoplasmic reticulum and/or mitochondria. The postsynaptic bars are situated in the cytoplasm subjacent to the postsynaptic membrane showing an asymmetrical density increase (Fig. 2). In addition to these specialized synapses, usual synapses without association of these postsynaptic specializations are present in the large and small ganglion cells.

It is generally accepted that only a single type of nerve terminal is present in various amphibian sympathetic ganglia (10,12,15,18). The small agranular vesicles apparent in the preganglionic terminals are consistent with the view that the transmitter released from the preganglionic terminals may be acetylcholine (ACh) (1,18). In spite of numerous morphological studies, there is no clear evidence for adrenergic or catecholamine-containing elements forming synapses with the

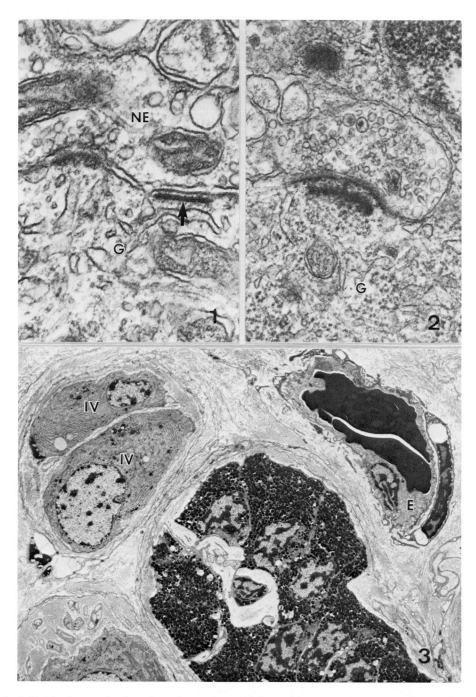

FIG. 1. A highly flattened subsurface cistern *(arrow)* lying subjacent to a postsynaptic membrane of a large ganglion cell (G). At the site of the subsurface cistern, the pre- and postsynaptic membranes are devoid of specializations or an aggregation of synaptic vesicles. (NE) Nerve ending. ×70,000.

FIG. 2. A postsynaptic bar apparent in a small ganglion cell. The bar is associated with a postsynaptic membrane specialization and an aggregation of synaptic vesicles. ×38,000.

FIG. 3. Low magnification of a micrograph showing a group of GC cells. An intercellular gap containing densely arranged collagenous fibers and thin processes from fibroblasts intervenes between the GC cells and a blood vessel. Type IV cells (IV) are also seen. (E) Endothelial cell. ×3,200.

large or small ganglion cells of various amphibian species (4,15,17,18).

GC CELLS

GC cells in the amphibian sympathetic ganglia may occur alone or, more frequently, in clusters of variable size. Some previous studies documented that the GC cells have no synapses in the toad *Bufo arenarum* (11) or in the bullfrog (17). Fujimoto (3), on the other hand, found synapses on the GC cells in the toad *Bufo vulgaris japonicus*. Uchizono and Ohsawa (13) also confirmed the presence of GC cells innervated by cholinergic nerves in the toad and frog sympathetic ganglia.

Four types of GC cell have been classified in the frog *(Limnodynastes dumerili)* sympathetic ganglia (4) based on the morphology of their granule components (Fig. 3). Types I and II GC cells resemble norepinephrine-storing and epinephrine-storing cells in the adrenal medulla, respectively. Type III cells contain small granular vesicles (100 to 300 nm in diameter) with a halo of variable width, but their other features are similar to the type I and II cells. Type IV cells are clearly different from the other three types with respect to the size (100 to 150 nm) and population of the granular vesicles (Fig. 3) (4,14). From the serial section study, the type IV cell is suggested to be a special type of sympathetic neuron rather than an interneuron or neurosecretory cell (14). The great majority of these four types of GC cell receive presynaptic nerve endings which are probably cholinergic in nature (4,14,15). There are also a number of GC cells without presynaptic innervation. Presumable gap junctions have been observed between the GC cells (14).

RELATIONSHIP OF GC CELLS TO BLOOD VESSELS

An intimate topographical relationship between GC cells and blood vessels has been noted in some amphibian sympathetic ganglia (3,4,17). Endocrine function in a manner similar to that of adrenomedullary cells has been also suggested in the GC cells. However, in contrast to the adrenomedullary cells, GC cells appear less frequently in close proximity to blood vessels (Fig. 3). Although some GC cells are located close to blood capillaries, the endothelial cells are of a continuous type (Figs. 3 and 4). Fenestrated capillaries, which are common in the adrenal medulla, have never been seen in the frog sympathetic ganglia (Figs. 3 and 4). It was confirmed by the serial section study that a certain number of solitary GC cells and some of large clumps of GC cells are tightly encapsulated by connective tissue without any topographical relationship to the blood vessels. It seems reasonable to assume that the GC cells release large amounts of catecholamines under the presumable cholinergic innervation. However, the structural features described above suggest that the secretion of catecholamines into the bloodstream may be of low efficiency, even if their major function is an endocrine secretion.

RELATIONSHIP OF GC CELLS TO THE PRINCIPAL GANGLION CELLS

Previous studies on the amphibian sympathetic ganglia, including serial section studies, revealed that none of the GC cells or their processes form synaptic contact with the ganglion cells (4,14,18). The majority of solitary and clustered GC cells are separated from the ganglion cells by a wide intercellular space containing many collagenous fibers and fibroblasts (Fig. 5). Some of the GC cells in the frog sympathetic ganglia are situated close to the ganglion cells. The closest distance between the GC cells and the ganglion cells is 0.5 μm. Even in such a case, thin processes of satellite cells completely cover the ganglion cells (Fig. 5). In addition, thin processes from perineurial cells usually separate the two cells (Fig. 5).

No one can deny the possibility that large amounts of catecholamines released from the GC cells may indirectly affect the synaptic transmission in the amphibian sympathetic

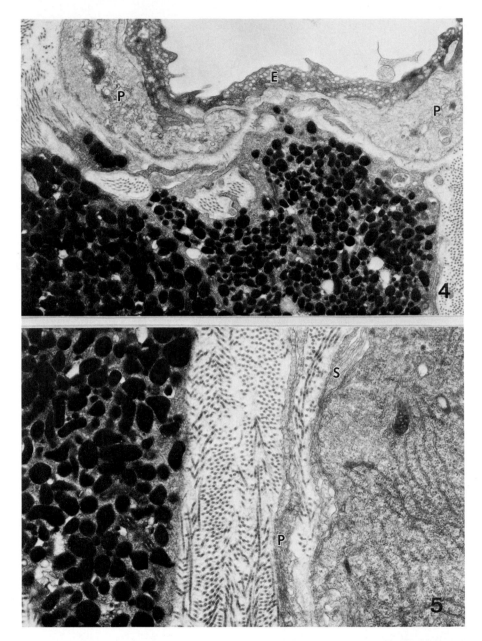

FIG. 4. GC cells lying in close apposition to a granular type of blood capillary. (E) Endothelial cell. (P) Pericytes. ×13,000.
FIG. 5. A principal ganglion cell lying opposite a GC cell. The ganglion cell is completely covered by thin processes of satellite cells (S). Thin processes (P) from perineurial cells and many collagenous fibers lie in the intracellular space. ×18,000.

ganglia. However, based on all morphological features so far reported, it seems more likely that several types of postsynaptic potentials including the s-IPSP are generated by the direct action of ACh, not by the action of catecholamines in the amphibian sympathetic ganglia (16,18). More detailed morphological data on the generation of s-IPSP are reported elsewhere (14,15).

ACKNOWLEDGMENTS

The author wishes to thank Professor Toshi Yuki Yamamoto for correcting the manuscript. This work was supported by grants 244014 and 237043 from the Ministry of Education, Science and Culture, Japan.

REFERENCES

1. Blackman, J.G., Ginsborg, B.L., and Ray, C. (1963): Synaptic transmission in the sympathetic ganglion of the frog. *J. Physiol. (Lond),* 167:355-373.
2. Chiba, T. (1977): Monoamine-containing paraneurons in the sympathetic ganglia of mammals. *Arch. Histol. Jpn. (Suppl.),* 40:163-176.
3. Fujimoto, S. (1967): Some observations on the fine structure of the sympathetic ganglion of the toad, *Bufo vulgaris japonicus. Arch. Histol. Jpn.,* 28:313-335.
4. Hill, C.E., Watanabe, H., and Burnstock, G. (1975): Distribution and morphology of amphibian extra-adrenal chromaffin tissue. *Cell Tissue Res.,* 160:371-387.
5. Honma, S. (1970): Functional differentiation in sB and sC neurons of toad sympathetic ganglia. *Jpn. J. Physiol.* 20:281-295.
6. Kuba, K., and Kotetsu, K. (1978): Synaptic events in sympathetic ganglia. *Prog. Neurobiol.,* 11:77-169.
7. Libet, B. (1970): Generation of slow inhibitory and excitatory postsynaptic potentials. *Fed. Proc.,* 29:1945-1956.
8. Matthews, M.R. (1974): Ultrastructure of ganglionic junctions. In: *The Peripheral Nervous System,* edited by J.I. Hubbard, pp. 111-150. Plenum Press, New York.
9. Nishi, S. (1974): Ganglionic transmission. In: *The Peripheral Nervous System,* edited by J.I. Hubbard, pp. 225-256. Plenum Press, New York.
10. Nishi, S., Soeda, H., and Koketsu, K. (1967): Release of acetylcholine from sympathetic preganglionic nerve terminals. *J. Neurophysiol.,* 30:114-134.
11. Piezzi, R.S., and Rodriguez Echandia, E.L. (1968): Studies on the pararenal ganglion of the toad *Bufo arenarum* Hensel. I. Its normal fine structure and histochemical characteristics. *Z. Zellforsch.,* 88:180-186.
12. Taxi, J. (1976): Morphology of the autonomic nervous system. In: *Frog Neurobiology,* edited by R. Llinás and W. Precht., pp. 93-150. Springer-Verlag, Berlin.
13. Uchizono, K., and Ohsawa, K. (1973): Morphophysiological consideration on synaptic transmission in the amphibian sympathetic ganglion. *Acta Physiol. Pol.,* 24:205-214.
14. Watanabe, H. (1977): Ultrastructure and function of the granule-containing cells in the anuran sympathetic ganglia. *Arch. Histol. Jpn., (Suppl.),* 40:177-186.
15. Watanabe, H., and Burnstock, G. (1978): Postsynaptic specializations at excitatory and inhibitory cholinergic synapses. *J. Neurocytol.,* 7:119-133.
16. Weight, F.F., and Padjen, A. (1973): Acetylcholine and slow synaptic inhibition in frog sympathetic ganglion cells. *Brain Res.,* 55:225-228.
17. Weight, F.F., and Weitsen, H.A. (1977): Identification of small intensely fluorescent (SIF) cells as chromaffin cells in bullfrog sympathetic ganglia. *Brain Res.,* 128:213-226.
18. Weitsen, H.A., and Weight, F.F. (1977): Synaptic innervation of sympathetic ganglion cells in the bullfrog. *Brain Res.,* 128:197-221.
19. Williams, T.H., Black, A.C., Jr., Chiba, T., and Jew, J. (1976): Interneurons/SIF cells in sympathetic ganglia of various mammals. In: *Chromaffin, Enterochromaffin and Related Cells,* edited by R.E. Coupland and T. Fujita, pp. 95-116. Elsevier, Amsterdam.

Histochemistry and Cell Biology of
Autonomic Neurons, SIF Cells, and
Paraneurons,
edited by O. Eränkö et al.
Raven Press, New York © 1980.

Small Intensely Fluorescent Cells and the Generation of Slow Postsynaptic Inhibition in Sympathetic Ganglia

Forrest F. Weight and Peter A. Smith*

Laboratory of Preclinical Studies, National Institute on Alcohol Abuse and Alcoholism, Rockville, Maryland 20852

The physiology and pharmacology of neurons and synapses in sympathetic ganglia have been investigated for a number of years. The simplicity of this neuronal system, together with its accessibility, and the possibility of investigating several types of synaptic and pharmacological interactions *in vitro*, have made the sympathetic ganglion an attractive model for studying neuronal and synaptic physiology and pharmacology. In view of this, it is somewhat surprising that at the present time scientists in this area of investigation do not agree on the identification of the neurotransmitters mediating certain synaptic potentials or on the mechanism of generation of these postsynaptic potentials (PSPs). Particularly controversial is the identification of the neurotransmitter that mediates the generation of the slow inhibitory postsynaptic potential (slow IPSP) and the mechanism of generation of this PSP. In this chapter we review the hypothesis that small intensely fluorescent (SIF) cells are inhibitory interneurons that mediate the genera-

tion of the slow IPSP. We then present the results of our studies on bullfrog sympathetic ganglia, which indicate that the slow IPSP is mediated by a direct postsynaptic action of the neurotransmitter acetylcholine that is released from the cholinergic preganglionic fibers which innervate these neurons; SIF cells are thus not involved in the generation of the slow IPSP.

THE SLOW IPSP AND SIF CELL HYPOTHESIS

The slow IPSP in sympathetic ganglia was first described in 1950 by Laporte and Lorente de No in turtle superior cervical ganglion (13). In 1952 Eccles studied this synaptic potential in rabbit superior cervical ganglion and found that the slow IPSP was potentiated by inhibitors of acetylcholinesterase. On the basis of these data, she suggested that the neurotransmitter mediating the slow IPSP was acetylcholine (5). Eccles and Libet (6) subsequently reported in 1961 that dibenamine reduced the amplitude of the P wave (a surface potential recording of the slow IPSP). Since at that time dibenamine was thought to be a specific adrenergic

---

*Present address: Department of Pharmacology and Therapeutics, McGill University, Montreal, P.Q. H3G 1Y6, Canada.

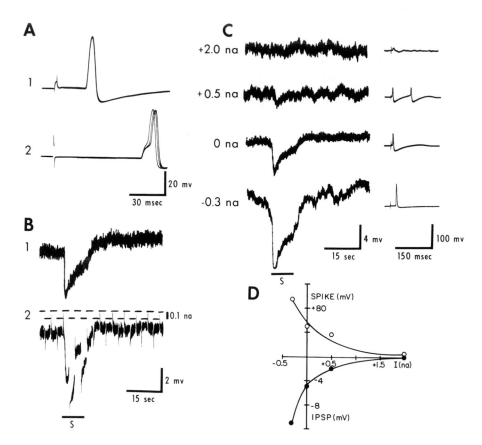

FIG. 4. Fast EPSP and slow IPSP recorded intracellularly in a C cell in the tenth paravertebral sympathetic ganglion of bullfrog. **A1:** Antidromic action potential. Antidromic conduction velocity of 0.24 meter/sec identifies the neuron as a C cell. **A2:** Fast EPSP and orthodromic action potential generated by stimulation of preganglionic C fibers in the eighth spinal nerve. **B1:** Slow IPSP generated in the same C cell, after nicotinic blockade, by stimulation of preganglionic C fibers in the eighth nerve at a frequency of 50 Hz. The period of stimulation is indicated by a line labeled S under the record in **B. B2:** *Upper record:* Hyperpolarizing constant current pulses of -0.1 nA. *Lower record:* Resistance change during slow IPSP. The bridge was balanced before stimulation such that current pulses produced no voltage deflection. Note that during the slow IPSP the current pulse produced a hyperpolarizing voltage deflection of 2 mV, indicating a resistance increase of 20 M ohms. Resting input resistance of this cell, determined from I-V curves, was 78 M ohms. **C:** Effect of membrane polarization by a steady d.c. current. *Left:* Amplitude of slow IPSP as a function of depolarizing ($+$) and hyperpolarizing ($-$) current. *Right:* Amplitude of antidromic spike recorded during the same polarizing current. **D:** Amplitude of slow IPSP *(filled circles)* and antidromic spike *(open circles)* in **C,** represented graphically as a function of the polarizing current. (From ref. 29.)

membrane that appears similar to the slow IPSP.

The slow IPSP has certain electrophysiological properties that differ from the properties of the common type of fast IPSP (27,28), i.e., an increase in membrane resistance during the slow IPSP (Fig. 4B) and an increase in slow IPSP amplitude by moderate hyper-

polarizing current and a decrease by depolarizing current (Fig. 4C) (29).

The electrophysiological properties of the membrane hyperpolarization produced by the iontophoretic administration of ACh were found to mimic the electrophysiological properties of the slow IPSP (29). In Fig. 5A the increase in membrane resistance during

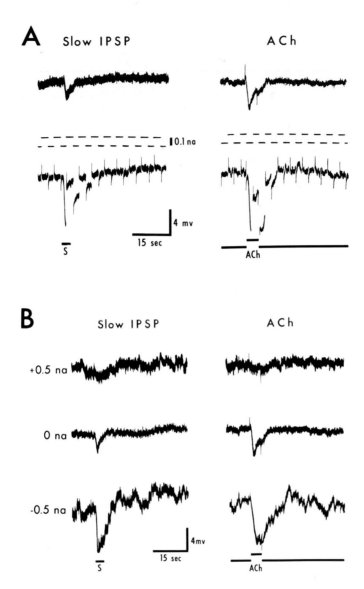

FIG. 5. Resistance change during, and effect of polarization on, slow IPSP and ACh hyperpolarization. Intracellular recording in a C cell. **A:** *Top left:* Slow IPSP generated by stimulation of preganglionic C fibers in the eighth spinal nerve at a frequency of 50 Hz for the period indicated by the line labeled S under the lower record. *Top right:* Hyperpolarization produced by the iontophoretic administration of ACh for the period indicated by the bottom line. *Bottom:* Hyperpolarizing constant current pulse of −0.1 nA during slow IPSP *(left)* and ACh hyperpolarization *(right)*. The bridge was balanced before stimulation such that current pulses produced no voltage deflection. Note that the hyperpolarizing current pulses produced hyperpolarizing voltage deflections during the slow IPSP and the ACh hyperpolarization, indicating that membrane resistance increased. **B:** *Middle:* Amplitude of control slow IPSP *(left)* and ACh hyperpolarization *(right)*. *Top:* Effect of depolarizing current of +0.5 nA on slow IPSP *(left)* and ACh hyperpolarization *(right)*. *Bottom:* Effect of hyperpolarizing current of −0.5 nA on slow IPSP *(left)* and ACh hyperpolarization *(right)*. (From ref. 30.)

the ACh hyperpolarization is compared to the increase in resistance during the slow IPSP in the same neuron. Similarly, in Fig. 5B the effect of membrane polarization on the ACh hyperpolarization is compared to the effect of polarization on the slow IPSP in the same neuron. Finally, it should be noted that the cholinergic hyperpolarization can also be elicited by the muscarinic agonist methacholine (MCh) (Figs. 6 and 7), and that the ACh and the MCh hyperpolarizations are antagonized by atropine.

CHOLINERGIC HYPERPOLARIZATION A DIRECT POSTSYNAPTIC ACTION

The fact that the iontophoretic administration of ACh can be quite localized suggests that the ACh hyperpolarization is due to a direct action of ACh on C cells. It does not exclude the possibility, however, that ACh might release another transmitter from synaptic terminals on C cells. This possibility was tested by using treatments known to prevent excitation–secretion coupling. If the ACh hyperpolarization results from ACh-induced release of a neurotransmitter or a neurohormone, inhibition of excitation–secretion coupling would prevent the release of the neurotransmitter or hormone, and the ACh hyperpolarization would be abolished. Figure 6 illustrates that cobalt (Co) blocks synaptic transmission to C cells in less than 15 min but does not affect the MCh hyperpolarization. In fact, the MCh hyperpolarization is still present after 45 min in Co, although such a long period of Co application has deleterious effects resulting in some nonspecific reduction in response amplitude.

Similarly, Fig. 7 illustrates that the MCh hyperpolarization can still be elicited despite the fact that synaptic transmission has been blocked by a calcium-free/high-magnesium Ringer's solution. Short periods of superfusion with calcium-free Ringer's solution (less than 30 min) totally block synaptic transmission and often reduce the hyperpolarizing re-

sponse to MCh (Fig. 7B). However, after longer periods of superfusion with Ca-free Ringer's, the MCh hyperpolarization recovers (Fig. 7c) although synaptic transmission remains blocked. Addition of magnesium (10 mM Mg) to Ca-free Ringer's usually promotes further recovery of the MCh hyperpolarization, although Mg acts synergistically with Ca removal to prevent excitation–secretion coupling. Despite the fact that prolonged superfusion of the ganglion with Ca-free/high-Mg is deleterious to cell function, a MCh hyperpolarization can still be elicited after more than 2 hr in Ca-free solution plus more than 1 hr in high-Mg Ringer's (Fig. 7f).

These experiments show that the inhibition of excitation–secretion coupling does not block the cholinergic hyperpolarization, as would occur if ACh or MCh was releasing a neurotransmitter or neurohormone (20). Thus, these experiments provide evidence that the cholinergic hyperpolarization is due to a direct postsynaptic action of ACh or MCh. The fact that the cholinergic hyperpolarization is elicited by MCh in the presence of nicotinic antagonists (*d*-tubocurarine or nicotine), together with the fact that the ACh and MCh hyperpolarizations are antagonized by atropine (2 μM), indicates that the cholinergic hyperpolarization is due to the direct activation of muscarinic postsynaptic receptors on sympathetic neurons.

SUMMARY AND CONCLUSIONS

We have reviewed our studies on the nature of the synaptic pathway and the identification of the neurotransmitter involved in the generation of the slow IPSP in sympathetic ganglia. We used the sympathetic ganglion of the bullfrog, since this preparation exhibits a number of experimental advantages. One is the fact that the slow IPSP may be recorded independently from the slow EPSP in an identifiable cell type (C cells) with a separate presynaptic innervation (the eighth nerve). The following points summa-

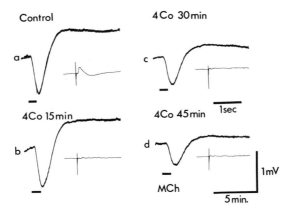

FIG. 6. Effect of cobalt on the MCh response. **a:** Control response to 1 mM MCh superfused for 60 sec *(bar)*. **Inset:** Fast EPSP elicited by stimulation of the eighth spinal nerve. **b:** Response to MCh after superfusion for 15 min in Ringer's solution containing 4 mM cobalt chloride (Co). **Inset:** Note the abolition of fast EPSP. **c** and **d:** Responses to MCh and fast EPSP after 30 and 45 min in 4 mM Co, respectively. Recorded from bullfrog sympathetic ganglion using the sucrose gap technique. The 5-min calibration refers to MCh response records; 1-sec calibration refers to fast EPSP records. Ringer solution contained 70 μM *d*-tubocurarine.

rize the data relevant to the nature of the synaptic pathway and the identification of the neurotransmitter mediating the slow IPSP. First, in histochemical and electron microscopic investigations, we were unable to find evidence that SIF cells in bullfrog sympathetic ganglia make synaptic connections with sympathetic neurons (32). The SIF cells in the ganglia appeared morphologically identical to chromaffin cells in the adrenal. In addition, SIF cells in the ganglia were located in close association to blood vessels. These observations suggest that the ganglionic SIF cells function to release their catecholamine content into the circulatory system as part of the extra-adrenal chromaffin system (4). A hormonal type of catecholamine release, however, would not explain the selective activation of the slow IPSP in C cells. There is evidence that catecholamines hyperpolarize B and C neurons (24). Thus a hormonal release of a catecholamine would hyperpolarize B and C cells; however, a hyperpolarization of B cells is not recorded after preganglionic C fiber stimulation (19,24,-29). Second, if the slow IPSP is mediated by a catecholamine released from SIF cells, an antagonist that blocks catecholamine actions

would be expected to block the slow IPSP. However, the alpha-adrenergic blocking agent DHE effectively antagonizes the catecholamine responses but does not reduce the slow IPSP (26). Third, C neurons are innervated by cholinergic preganglionic fibers in the eighth nerve, and repetitive stimulation of this input elicits the slow IPSP in type C neurons (12,24,27). Fourth, the iontophoretic administration of ACh directly to C cells produces a membrane hyperpolarization which has electrophysiological properties identical to those of the slow IPSP (30). Fifth, treatments which block excitation–secretion coupling do not block the cholinergic hyperpolarization, indicating that the cholinergic response is a direct postsynaptic action on sympathetic neurons (30). Sixth, the slow IPSP and the cholinergic hyperpolarization are antagonized by atropine, which indicates that both hyperpolarizing responses are due to the activation of muscarinic postsynaptic receptors.

On the basis of the electrophysiological, pharmacological, and morphological data presented above, we conclude that the slow IPSP in curarized and nicotinized ganglia (15) is a muscarinic postsynaptic response

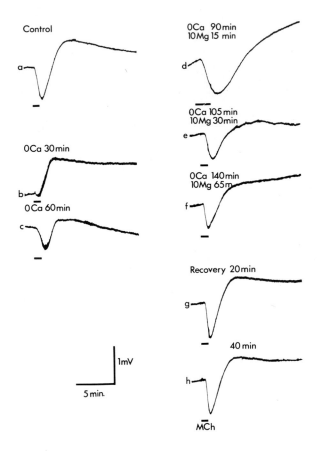

FIG. 7. Effect of Ca-free and high-Mg Ringer's solution on MCh response. **a:** Control response to 1 mM MCh superfused for 60 sec *(bar)* in normal Ringer's solution. **b:** MCh response after superfusion for 30 min in Ca-free (0 Ca) Ringer's solution. **c:** MCh response after 60 min in Ca-free Ringer's. **d:** After 90 min in Ca-Free Ringer and 15 min in Ca-free Ringer's containing 10 mM Mg. **e:** After 105 min in Ca-free Ringer's and 30 min in Ca-free 10 mM Mg Ringer's. **f:** After 140 min in Ca-free and 65 min in Ca-free 10 mM Mg Ringer's. **g** and **h:** Recovery recorded 20 and 40 min after return to normal Ringer's solution. Records were obtained from bullfrog sympathetic ganglion using the sucrose gap technique. Ringer's solution contained 70 μM *d*-tubocurarine. Ca-free Ringer's was prepared by omitting calicum chloride from the solution.

produced by the direct action of ACh. A corollary of this conclusion is that SIF cells are not interneurons which mediate the slow IPSP.

Figure 8 outlines our current understanding of the synaptic organization and transmitter identification in the bullfrog sympathetic ganglion. On the right of Fig. 8., the cholinergic preganglionic fibers from the eighth nerve (C fibers) are shown making monosynaptic connection to the smaller C

neuron. Cholinergic activation of nicotinic postsynaptic receptors elicits a fast EPSP, whereas cholinergic activation of muscarinic postsynaptic receptors generates the slow IPSP. On the left of the figure, the cholinergic preganglionic fibers that descend in the sympathetic chain (B fibers) are shown making synaptic contact with the larger B neuron. In these neurons, cholinergic activation of the nicotinic postsynaptic receptor also elicits a fast EPSP, but cholinergic activation

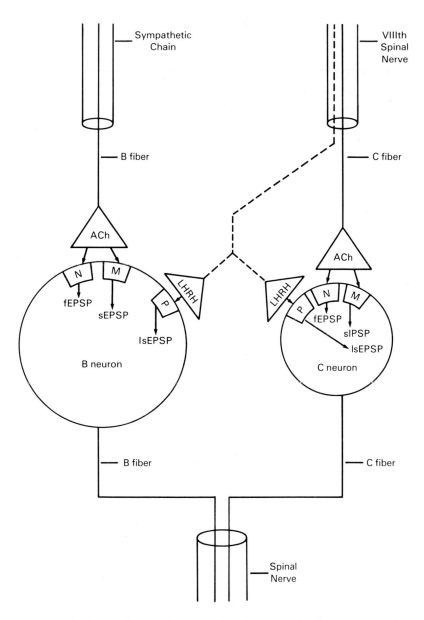

FIG. 8. Synaptic organization and transmitter identification in the bullfrog sympathetic ganglion. On the right, the smaller C neuron is innervated by cholinergic preganglionic axons in the eighth spinal nerve that have C fiber conduction velocity (C fiber). The neurotransmitter ACh is illustrated activating nicotinic postsynaptic receptors (N) to elicit the fast EPSP (fEPSP) and muscarinic postsynaptic receptors (M) to generate the slow IPSP (sIPSP). On the left, the larger B neuron is innervated by cholinergic preganglionic axons that descend in the sympathetic chain and have B fiber conduction velocity (B fiber). ACh activation of nicotinic postsynaptic receptors (N) on the B neuron also elicits a fast EPSP, but activation of muscarinic postsynaptic receptors (M) on the B neuron generates a slow EPSP (sEPSP). See ref. 34 for review. The dashed line indicates the pathway that generates the noncholinergic late-slow EPSP (lsEPSP). The transmitter mediating the late-slow EPSP was recently proposed to be similar to LHRH (10), so that this proposed transmitter is illustrated activating a peptide receptor (P).

of muscarinic postsynaptic receptors generates a slow EPSP (31,34). Also illustrated by a dashed line, but not discussed in this review, is the pathway that generates the noncholinergic late-slow EPSP (21), which was recently proposed to be mediated by a peptide similar to luteinizing hormone-releasing hormone (LHRH) (10).

Finally, it should be noted that our conclusion that the slow IPSP is mediated by the postsynaptic action of ACh is not unique to the sympathetic ganglion of the bullfrog. Recently Hartzell and co-workers (9) reported that the slow IPSP in parasympathetic neurons of the mudpuppy heart is also mediated by ACh. In mammalian ganglia, SIF cells have been studied morphologically predominantly in rat ganglia. However, most physiological and pharmacological studies on the slow IPSP have been conducted in the rabbit (14). In relation to this point, recent studies report that there are very few SIF cells in rabbit ganglia (2,3,38). In addition, efferent synapses from SIF cells to sympathetic neurons have been found only in rat sympathetic ganglia (2,3,38). In view of these data, it may not be possible to quantitatively account for the fact that the slow IPSP is reported to occur in the majority of rabbit sympathetic neurons (16). It should also be noted that in many respects the slow IPSP in mammalian and amphibian sympathetic ganglia are similar in their physiological and pharmacological properties (14). This suggests that the slow IPSP in mammalian ganglia may also be mediated by ACh released by the preganglionic fibers that innervate the sympathetic neurons. In view of this possibility, a critical re-evaluation of the synaptic pathway and the identification of the neurotransmitter mediating the slow IPSP in mammalian sympathetic ganglia is clearly indicated.

ACKNOWLEDGMENTS

We thank Ms. Catherine Straub for typing the manuscript.

REFERENCES

1. Busis, N.A., Weight, F.F., and Smith, P.A. (1978): Synaptic potentials in sympathetic ganglia: Are they mediated by cyclic nucleotides? *Science,* 200:1079–1081.
2. Chiba, T. (1977): Monoamine-containing paraneurons in the sympathetic ganglia of mammals. *Arch. Histol. Jpn. (Suppl.),* 40:163–176.
3. Chiba, T., and Williams, T.H. (1975): Histofluorescence characteristics and quantification of small intensely fluorescent (SIF) cells in sympathetic ganglia of several species. *Cell Tissue Res.,* 162:331–341.
4. Coupland, R.E. (1965): *The Natural History of the Chromaffin Cell.* Longmans, London.
5. Eccles, R.M. (1952): Responses of isolated curarized sympathetic ganglia. *J. Physiol. (Lond),* 117:196–217.
6. Eccles, R.M., and Libet, B. (1961): Origin and blockade of the synaptic responses of curarized sympathetic ganglia. *J. Physiol (Lond),* 157:484–503.
7. Eränko, O., editor (1976): *SIF Cells: Structure and Function of the Small, Intensely Fluorescent Sympathetic Cells.* Fogarty International Center Proceedings No. 30. Government Printing Office, Washington, D.C.
8. Feldberg, W., and Gaddum, J.H. (1934): The chemical transmitter at synapses in a sympathetic ganglion. *J. Physiol. (Lond),* 81:305–319.
9. Hartzell, H.C., Kuffler, S.W., Stickgold, R., and Yoshikami, D. (1977): Synaptic excitation and inhibition resulting from direct action of acetylcholine on two types of chemoreceptors on individual amphibian parasympathetic neurones. *J. Physiol. (Lond),* 271:817–846.
10. Jan, Y.N., Jan, L.Y., and Kuffler, S.W. (1979): A peptide as a possible transmitter in sympathetic ganglia of the frog. *Proc. Natl. Acad. Sci. USA,* 76:1501–1505.
11. Kobayashi, S., and Chiba, T., editors (1977): Paraneurons: New concepts on neuroendocrine relatives. *Arch. Histol. Jpn. (Suppl.),* 40.
12. Koketsu, K. (1969): Cholinergic synaptic potentials and the underlying ionic mechanisms. *Fed. Proc.,* 28:101–112.
13. Laporte, Y., and Lorente de No, R. (1950): Potential changes evoked in a curarized sympathetic ganglion by presynaptic volleys of impulses. *J. Cell. Comp. Physiol. (Suppl.),* 35:61–106.
14. Libet, B. (1970): Generation of slow inhibitory and excitatory postsynaptic potentials. *Fed. Proc.,* 29:1945–1956.
15. Libet, B., and Kobayashi, H. (1974): Adrenergic mediation of slow inhibitory postsynaptic potential in sympathetic ganglia of the frog. *J. Neurophysiol.,* 37:805–814.
16. Libet, B., and Tosaka, T. (1969): Slow inhibitory and excitatory postsynaptic responses in single cells of mammalian sympathetic ganglia. *J. Neurophysiol.,* 32:43–50.

17. Libet, B., and Tosaka, T. (1970): Dopamine as a synaptic transmitter and modulator in sympathetic ganglia: A different mode of synaptic transmission. *Proc. Natl. Acad. Sci. USA,* 67:667–673.

18. Libet, B., Chichibu, S., and Tosaka, T. (1968): Slow synaptic responses and excitability in sympathetic ganglia of the bullfrog. *J. Neurophysiol.,* 31:383–395.

19. Nishi, S., and Koketsu, K. (1968): Early and late afterdischarges of amphibian sympathetic ganglion cells. *J. Neurophysiol.,* 31:109–121.

20. Rubin, R.P. (1970): The role of calcium in the release of neurotransmitter substances and hormones. *Pharmacol. Rev.,* 22:389–428.

21. Schulman, J.A., and Weight, F.F. (1976): Synaptic transmission: Long-lasting potentiation by a postsynaptic mechanism. *Science,* 194:1437–1439.

22. Weight, F.F., and Smith, P.A. (1977): Role of electrogenic sodium pump in slow synaptic inhibition is re-evaluated. *Nature,* 267:68–70.

23. Smith, P.A., and Weight, F.F. (1980): IPSP reversal potential evidence for decreased sodium conductance with increased potassium conductance. *XXVIII International Congress Physiol. Sci. Abstracts, (in press).*

24. Tosaka, T., Chichibu, S., and Libet, B. (1968): Intracellular analysis of slow inhibitory and excitatory postsynaptic potentials in sympathetic ganglia of the frog. *J. Neurophysiol.,* 31:396–409.

25. Watanabe, H. (1977): Ultrastructure and function of the granule containing cells in the anuran sympathetic ganglia. *Arch. Histol. Jpn. (Suppl.),* 40:177–186.

26. Weight, F.F. (1973): Slow synaptic inhibition and adrenergic antagonism in sympathetic ganglion. In: *Society for Neuroscience Third Annual Meeting,* p. 312.

27. Weight, F.F. (1974): Synaptic potentials resulting from conductance decreases. In: *Synaptic Transmission and Neuronal Interaction,* edited by M.V.L. Bennett, pp. 141–152. Raven Press, New York.

28. Weight, F.F. (1974): Physiological mechanisms of synaptic modulation. In: *The Neurosciences: Third Study Program,* edited by F.O. Schmitt and F.G. Worden, pp. 929–941. MIT Press, Cambridge, Mass.

29. Weight, F.F., and Padjen, A. (1973): Slow synaptic inhibition: Evidence for synaptic inactivation of sodium conductance in sympathetic ganglion cells. *Brain Res.,* 55:219–224.

30. Weight, F.F., and Padjen, A. (1973): Acetylcholine and slow synaptic inhibition in frog sympathetic ganglion cells. *Brain Res.,* 55:225–228.

31. Weight, F.F., and Votava, J. (1970): Slow synaptic excitation in sympathetic ganglion cells: Evidence for synaptic inactivation of potassium conductance. *Science,* 170:755–758.

32. Weight, F.F., and Weitsen, H.A. (1977): Identification of small intensely fluorescent (SIF) cells as chromaffin cells in bullfrog sympathetic ganglia. *Brain Res.,* 128:213–226.

33. Weight, F.F., Petzold, G., and Greengard, P. (1974): Guanosine 3′,5′-monophosphate in sympathetic ganglia: Increase associated with synaptic transmission. *Science,* 186:942–944.

34. Weight, F.F., Schulman, J.A., Smith, P.A., and Busis, N.A. (1979): Long-lasting synaptic potentials and the modulation of synaptic transmission. *Fed. Proc.,* 38:2084–2094.

35. Weight, F.F., Smith, P.A., and Schulman, J.A. (1978): Post-synaptic potential generation appears independent of synaptic elevation of cyclic nucleotides in sympathetic neurons. *Brain Res.,* 158:197–202.

36. Weitsen, H.A., and Weight, F.F. (1973): Chromaffin cells in frog sympathetic ganglion: Morphology not consistent with role in generation of synaptic potentials. *Anat. Rec.,* 175:467.

37. Weitsen, H.A., and Weight, F.F. (1977): Synaptic innervation of sympathetic ganglion cells in the bullfrog. *Brain Res.,* 128:197–211.

38. Williams, T.H., Black, A.C., Chiba, T., and Jew, J.Y. (1977): Species differences in mammalian SIF cells. *Adv. Biochem. Psychopharmacol.,* 16:505–511.

Histochemistry and Experimental Cytology of Adrenal Chromaffin Cells and Glomus Cells

*Histochemistry and Cell Biology of
Autonomic Neurons, SIF Cells, and
Paraneurons,*
edited by O. Eränkö et al.
Raven Press, New York © 1980.

Influence of Drugs Affecting the Pituitary Adrenal Axis on Chromaffin Cells

R. E. Coupland and F. Bustami

Department of Human Morphology, University of Nottingham, Nottingham, England

During recent years work in our laboratory has been concerned with amine turnover in chromaffin cells (5). For this purpose the mouse is a convenient experimental animal. An incidental finding during this period was that, in addition to typical epinephrine-secreting (E) and norepinephrine-secreting (NE) cells, an extra category of small granule-containing chromaffin (SGC) cells also exists (6,7,13). These cells contain synaptic-type vesicles (STVs) in addition to small-diameter chromaffin granules (12). Dichromate-glutaraldehyde fixation showed that the STVs also contain catecholamines (Coupland and Bustami, unpublished). Although SGC cells exist in many species, ones containing STV have been reported only in mice.

Recent observations on the postnatal development of SGC cells showed that SGC cells appear to achieve their greatest complement of secretory granules including STVs at the time when the X-zone of the mouse shows maximal development (2). Thereafter the size of aggregates of STVs and the number of secretory granules decrease, and STVs are either few in number or absent in SGC cells of postpubertal animals 9 or more weeks old. However, the STVs reappear following postpubertal castration as a secondary X-zone develops, and their number decreases

following parenteral testosterone proprionate, which precipitates the disappearance of the X-zone. This effect may be due to either a direct effect of testosterone on the adrenal medulla and/or its effect on the hypothalamus–pituitary leading to a reduction in output of luteinizing hormone (LH) and/or degeneration of the X-zone.

Since an androgenic effect was involved in these changes, acting directly or indirectly through the pituitary, we decided to administer cyproterone acetate, a well accepted antiandrogen, to mice of different ages and in different states in order to observe its effect on SGC cells. However, events proved that the drug had a marked effect on other elements of the adrenal medulla, and the present work describes some of the changes observed with particular reference to E and NE cells.

MATERIALS AND METHODS

Prepubertal and 5- to 6-week-old CS1 strain mice were given cyproterone acetate (CPA) 30 mg/kg daily by mouth in two equally divided doses for 2 to 4 weeks. They were anesthetized with nembutal and then fixed by cardiac perfusion with 3% glutaraldehyde in 0.1 M cacodylate buffer pH 7.2. After 1 hr the adrenal glands were removed, sectioned, and postosmicated prior to dehydration and embedding in Araldite. Sections were cut on a Reichert OMU3 microtome, stained with alkaline lead citrate, and examined in a Philips 300 electron microscope. Some animals were also given the corticosteroid synthesis inhibitor metyrapone, 200 mg/kg daily in three equally divided doses (every 8 hr), either throughout the period or during the last 7 days of CPA administration. Normal untreated and control animals receiving the suspension vehicle *per os* were also examined.

RESULTS

Cyproterone Acetate

Essentially similar changes occurred in prepubertal and older animals, the rate of development depending on age. It was approximately twice as rapid in prepubertal animals as in those 5 to 6 weeks old.

Adrenal Cortex

CPA resulted in a marked degeneration of the inner zona fasciculata and zona reticularis. The X-zone also showed degenerative changes, but it must be remembered that this zone normally regresses during the period of drug administration (5 to 9 weeks of age). It did not appear to be specifically affected by CPA. The general appearance was that of a gland after hypophysectomy (Fig. 1).

Adrenal Medulla

NE cells.

A characteristic feature of NE cells, apart from highly electron-dense granules, is the presence of a paranuclear zone of cytoplasm with fine granular cytosol; this zone contains the Golgi apparatus, centrioles, and multivesicular bodies but is relatively free from secretory granules (3). In CPA-treated animals after 16 days this paranuclear zone enlarges

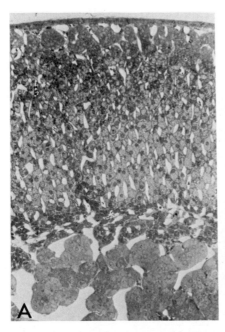
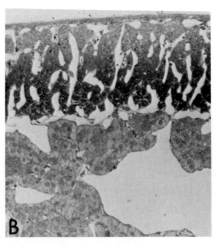

FIG. 1. Toluidine blue stain. **A:** Normal 9-week-old male mouse adrenal. **B:** CPA-treated (5 to 9 weeks) mouse. Note the marked atrophy of the inner zones of adrenal cortex. × 130.

in some cells. Mitochondria increase in length and possibly number, and form a prominent feature of a cell in which the region containing secretory granules becomes progressively reduced. At the same time the Golgi apparatus becomes more extensive, with a marked increase in tubular and especially vesicular components; microfibrils become more evident, endoplasmic reticulum increases in amount, and the cytoplasm takes on a neuronal-type appearance (Figs. 2 and 3) although the nuclear form does not usually appear to change. Ultimately, secretion granules either disappear or persist as isolated, small, highly dense elements (Fig. 4). These changes tend to affect groups of cells in the peripheral parts of the medulla, although not all of the peripheral medulla is affected and some peripheral regions as well as the central zone may be relatively unaffected.

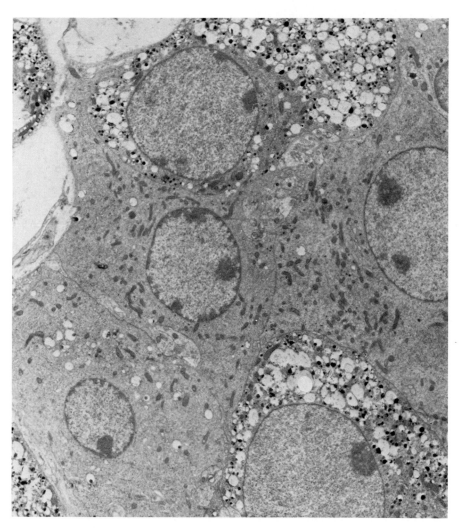

FIG. 2. Nine-week-old mouse. NE cells show varying degrees of cytoplasmic change after 4 weeks of CPA. ×5,800.

changes occur in both sexes but appear more marked in male animals.

Prepubertal animals show extensive changes and total transformation after as little as 16 to 21 days' CPA treatment; sexually mature animals show fully developed changes only after 3 to 4 weeks. A few affected (completely transformed) cells undergo cytoplasmic degeneration after 4 weeks of CPA as evidenced by cytoplasmic vacuolation and a disappearance of all organelles except for nucleus and lysosomes.

Cyproterone Acetate and Metyrapone

The administration of metyrapone for 4 weeks along with CPA largely prevents atrophy of the adrenal cortex. No fully transformed E or NE cells are evident, and only occasionally are minor changes suggestive of the earlier stage of change evident. A typical feature, however, is the appearance of typical electron-dense NE granules in some E cells, although many remain strictly E- or NE-secreting.

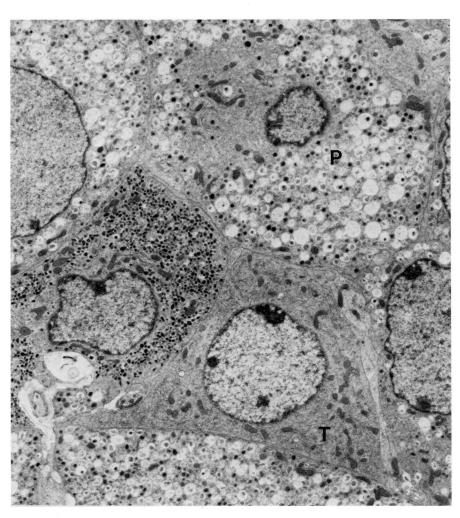

FIG. 5. Section through a mouse adrenal medulla after 4 weeks of CPA treatment showing partially (P) and completely (T) transformed E cells that lie adjacent to an SGC cell of normal appearance. ×6,800.

When metyrapone is given only during the last week of a 4-week period of CPA administration, the changes typical of CPA administration alone are reduced in degree, and partial rather than total cell transformation is observed in E and NE cells. SGC cells have not been observed to undergo transformation, and again STVs persist and may form medium-sized aggregates. Other elements best described as medium granule chromaffin (MGC) cells (a typical feature of the mouse adrenal medulla) also appear to be unaffected by the drugs. Counts reveal that throughout the experiments involving drug administration the number of SGC cells remains constant at 5 to 6%.

In animals treated with CPA alone, preganglionic nerve fibers often form a prominent feature in and between elements undergoing the characteristic changes, and nerve endings are very evident. Stereology would have to be applied to determine if this represents a significant increase in these elements.

DISCUSSION

It is now well established that chromaffin cells exist in all vertebrates and that, in general, in mature animals individual cells store either E or NE (for review see ref. 4). The mouse adrenal gland is unusual insofar as it has a characteristic X-zone in the cortex that is largely under the control of androgens, progesterone, and gonadotropin (9-11) and which regresses after puberty, quickly in the male and slowly in the female unless pregnancy supervenes. In addition, it has an adrenal medulla in which SGC cells appear to be more numerous than in other species and which contain synaptic-type vesicles in addition to typical secretion granules. Moreover, it is the only species in which in normal young adult animals occasional NE granules have been observed in what are otherwise E cells (Coupland, *unpublished*). In the mouse adrenal medulla, one often notes a marked variation in the range of granule sizes in E

and NE cells, a point made in respect to NE cells by Gorgas and Böck (7). This variation may be such as to justify reference to NE (typical), SGNE (smaller than average NE granules), E and SGE cells, as well as SGC cells and what appears to be a further distinct entity, a cell packed with medium-sized granules whose contents have moderate uniform electron density (MGC).

The changes involving E and NE cells consequent to CPA administration were unexpected but interesting. If, as seems likely, the contents of SGC cells vary with age and in parallel with changes in the X-zone—or in consequence of the secretions of the X-zone or of androgens (2)—it would be expected that CPA would ensure a persistence of STVs, and this it appears to do. However, the general effects of the drug are far-reaching, including action as a progestagen or an antiandrogen, inhibition of gonadotropin release, and suppression of ACTH output (14). As a result of the latter action, the adrenal cortex atrophies and takes on an appearance similar to that resulting from hypophysectomy.

It seems that the changes observed in E and NE cells following CPA may reflect (a) an action of the drug on the central nervous system resulting in adrenal medullary stimulation; (b) a decrease in ACTH output, acting directly and/or indirectly via the adrenal cortex; (c) a peripheral effect on androgen receptors; or (d) all combined. The fact that the changes were inhibited or reduced by the simultaneous administration of metyrapone suggests that ACTH production is important in these changes since adrenal atrophy is prevented by metyrapone, probably in consequence of the negative feedback of corticosteroid synthesis inhibition increasing ACTH production. Furthermore, another observation (Bustami and Coupland, *unpublished*) that the effects of CPA administration on E and NE cells can be prevented by simultaneous administration of dexamethasone indicates that adrenocorticosteroids are also important. In this respect it is interesting to note that the peripheral (juxtacortical) zone of the medulla is the region most affected by CPA.

The changes in E and NE cells following CPA administration involve an increase in specific cell organelles including Golgi apparatus, microfibrils, endoplasmic reticulum, and possibly mitochondria (which become elongated), as well as a gradual reduction in the number of secretory granules. Ultimately, secretion granules are not observed or are only few in number, small, and possess highly electron-dense contents characteristic of NE storage. Immediately after the disappearance of the typical secretion granules (E or NE), the cells possess many features typical of adrenal medullary sympathetic neurons. In many respects the cellular changes resemble those described recently by Aloe and Levi-Montalcini (1) following the administration of nerve growth factor to perinatal rats, whereas in other respects they resemble the changes described by Unsicker et al. (15) in the rat following high doses of vinblastine. There is no doubt that in some cells the cytoplasmic changes progress to cell degeneration. It is worth recording that similar changes have been observed in young and adult rats after CPA administration (Bustami and Coupland, *unpublished*).

These changes demonstrate the remarkable plasticity of what are often considered to be fully differentiated endocrine cells. Although quantitation is not yet complete, the general appearance of sections suggests that preganglionic nerve fibers and endings in the vicinity of the affected cells are more prominent than usual. This fact, together with signs of cell stimulation, suggests that the nervous system may be important in the changes. Experiments are in progress to test this possibility. Since it has been reported that nerve growth factor is produced by explants of mouse adrenal medulla (8), it is possible that some of the observed changes may be a consequence of local synthesis of this drug even though hormonal changes (ACTH and corticosteroids) almost certainly play a major role. Clearly further work is needed to explore the many possibilities.

ACKNOWLEDGMENTS

We thank Schering Chemicals Ltd. for supplies of cyproterone acetate, Ciba Laboratories for supplies of metyrapone, and Mrs. Annette Tomlinson and Mrs. Gail Edwards for invaluable technical and secretarial assistance.

REFERENCES

1. Aloe, L., and Levi-Montalcini, R. (1979): Nerve growth factor-induced transformation of immature chromaffin cells in vivo into sympathetic neurons: Effect of antiserum to nerve growth factor. *Proc. Natl. Acad. Sci. USA,* 76:1246–1250.
2. Bustami, F., and Coupland, R.E. (1979): Possible association between the presence of an X-zone in the mouse adrenal cortex and the stage of differentiation of small granular chromaffin (SGC) cells of the adrenal medulla. *J. Anat.,* 129:878p.
3. Coupland, R.E. (1972): The chromaffin system. *Handbuch. Exp. Pharmakol.,* 33:16–45.
4. Coupland, R.E. (1979): Catecholamines. In: *Hormones and Evolution,* edited by E.J.W. Barrington, pp. 309–340. Academic Press, London.
5. Coupland, R.E., and Kobayashi, S. (1976): Recent studies on the fixation of adrenaline and noradrenaline, and on amine synthesis and storage in chromaffin cells. In: *Chromaffin, Enterochromaffin and Related Cells,* edited by R.E. Coupland and T. Fujita, pp. 59–81. Elsevier, Amsterdam.
6. Coupland, R.E., Kobayashi, S., and Tomlinson, A. (1977): On the presence of small granule chromaffin (SGC) cells in the rodent adrenal medulla. *J. Anat.,* 124:488–489.
7. Gorgas, K., and Böck, P. (1976): Morphology and histochemistry of the adrenal medulla. I. Various cell types of primary catecholamine-storing cells in the mouse adrenal medulla. *Histochemistry,* 50:17–31.
8. Harper, G.P., Pearce, F.L., and Vernon, C.A. (1976): Production of nerve growth factor by the mouse adrenal medulla. *Nature,* 261:251–253.
9. Holmes, P.V., and Dickson, A.D. (1971): X-zone degeneration in the adrenal glands of adult and immature female mice. *J. Anat.,* 108:159–168.
10. Jones, I.C. (1949): The relationship of the mouse adrenal cortex to the pituitary. *Endocrinology,* 45:514–536.
11. Jones, I.C. (1952): The disappearance of the X-zone of the mouse adrenal cortex during first pregnancy. *Proc. R. Soc. Biol.,* 139:398–410.
12. Kobayashi, S., and Coupland, R.E. (1977): Two

populations of microvesicles in the SGC (small granule chromaffin) cells of the mouse adrenal medulla. *Arch. Histol. Jpn.,* 40:251-259.

13. Lassmann, H. (1974): Ein dritter Zelltyp im Nebennierenmark der Maus? *Verh. Anat. Ges.,* 68:641-644.

14. Neri, R.O., Monahan, M.D., Meyer, J.G., Afonso, B.A., and Tabachnick, I.A. (1967): Biological studies on an anti-androgen (SH714). *Eur. J. Pharmacol.,* 1:438-444.

15. Unsicker, K., Limmeroth-Erwert, B., Otten, U., Lindmar, R., Löffelholz, K., and Wolf, U. (1979): Short and long-term effects of vinblastine on the rat adrenal medulla. *Cell Tissue Res.,* 196:271-287.

Histochemistry and Cell Biology of Autonomic Neurons, SIF Cells, and Paraneurons,
edited by O. Eränkö et al.
Raven Press, New York © 1980.

Use of Antibodies to Norepinephrine and Epinephrine in Immunohistochemistry

A. A. J. Verhofstad, H. W. J. Steinbusch, *B. Penke, *J. Varga, and H. W. J. Joosten

*Department of Anatomy and Embryology, University of Nijmegen, 6500 HB Nijmegen, The Netherlands; and *Department of Medical Chemistry, University of Szeged, 6720 Szeged, Hungary*

Norepinephrine (NE) and epinephrine (E) play an important role in the nervous and endocrine systems. Their presence in several organs has been demonstrated by bioassay and biochemical methods. The first "attempts"[1] to study the cellular localization date back to the nineteenth century. In 1865 Henle (18) observed the brown coloration of the adrenal medulla after fixation in chromic acid or bichromate solutions. In fact, he described for the first time a chromaffin reaction. Since then, the chromaffin reaction has been modified several times (4,21), but because of its low sensitivity and specificity its use has been restricted to the adrenal medulla and related paraganglia. Other methods for the cellular localization have been described, e.g., the argentaffin reaction (6,13) and application of autoradiography after uptake of radioactive NE, E, or its precursor DOPA (19,33).

The most frequently applied technique is the histofluorescence reaction induced by formaldehyde or glyoxylic acid. Since the original descriptions by Eränkö (6,7), Falck and co-workers (8–10), and Axelsson et al.

(1; see also 24) many modifications have been published. It is now generally accepted that the induced fluorescence techniques are highly sensitive and can be used to visualize NE and E in the central as well as in the peripheral nervous system. However, they are rather unspecific; i.e., many closely related catecholamines [dopamine (DA), as well as NE and E], and indoleamines [e.g., serotonin (5-HT)] are demonstrated at the same time (2,22). Methods to differentiate between these related fluorophores are based on microspectrofluorimetry and/or pharmacological manipulations (2,22). However, these methods require complicated and expensive equipment and, moreover, can be used only in experimental conditions.

Immunohistochemistry using specific antibodies to catecholamine-synthesizing enzymes was introduced by Geffen et al. (11). They localized dopamine-β-hydroxylase [(DBH), an enzyme that marks the conversion of DA to NE] in the adrenal medulla and peripheral nerve fibers. Later, antibodies to other enzymes [e.g., phenylethanolamine N-methyltransferase (PNMT) marking the conversion of NE to E] were applied (12,20). These immunohistochemical methods offer possibilities to overcome the lack in specificity of the induced fluorescence techniques.

---

[1] We use quotation marks because it was not until the first decade of the twentieth century that norepinephrine and epinephrine were isolated and identified.

Furthermore, they enable investigations at an electron microscopic level as well (26). However, their applicability is limited for two reasons. First, the isolation of the synthesizing enzymes is complicated and time-consuming; and second, antibodies may show a low affinity to enzymes of other species (15). Furthermore, these antibodies do not provide direct information as to whether NE and E are actually synthesized at the site.

In this chapter we report another immunohistochemical approach using antibodies to NE and E.[2] These antibodies were raised with immunogens consisting of a carrier protein coupled with either NE or E. Thus we adopted the observations of Landsteiner (23) that small molecules, by themselves devoid of antigenicity, become immunogenic after coupling to an immunogenic carrier. Recently we presented an immunohistochemical procedure for serotonin using the same principle (31,32). Only a few reports have been published concerning the production of antibodies to NE and E (14,17,25,30,34). Furthermore, these antibodies were tested only by immunodiffusion, immunoelectrophoresis, and radioimmunoassay. No data about the applicability in immunohistochemistry could be found. We present here a brief description of some methodological aspects and some results. A detailed account will be published elsewhere.

MATERIALS AND METHODS

Preparation of Antibodies

Rabbits and sheep were injected with immunogens consisting of bovine serum albumin (BSA) coupled to NE or E. Formaldehyde was used as a coupling reagent. Each animal received 1 mg immunogen in 1 ml complete (first injection) or incomplete (booster injections) Freund's adjuvant. Se-

---

[2]The results were presented in part at the spring meeting of the Anatomical Society of Great Britain and Ireland, 1979 (abstract: *J. Anat.*) (*Lond*), 129:197, and at the 9th Congress of the Hungarian Society of Endocrinology and Metabolism, 1979 (abstract: *Acta Med. Acad. Sci. Hung.,* 26:69).

---

rum samples were tested by immunodiffusion according to Ouchterlony (28), immunoelectrophoresis, and immunofluorescence microscopy. It could be demonstrated that the raised antisera contained antibodies with a high affinity to the carrier protein. Since these antibodies caused unspecific staining of collagen-rich tissues (e.g., blood vessel wall), all sera to be used in immunofluorescence microscopy were absorbed with BSA (5 mg/ml).

Immunohistochemical Procedure

Adult male albino rats were anesthetized and perfused through the left ventricle with ice-cold (4°C) calcium-free Tyrode's, immediately followed by a fixative (4°C). Appropriate pieces of the brain, the superior cervical ganglion, stomach, small intestine, ductus deferens, and adrenal gland were dissected out and postfixed for 2 hr at 4°C followed by rinsing in 5% sucrose in 0.1 M phosphate buffer (pH 7.3) for 18 hr at 4°C. In some experiments unfixed tissues were used. Tissues were frozen on Dry Ice or isopentane cooled by liquid nitrogen and cut at 7 μm on a cryostat. Sections were stained according to the indirect immunofluorescence technique of Coons (3; see also 27). Triton X-100 (final concentration 0.1%) was added to all primary sera as well as to the fluorescein isothiocyanate (FITC) labeled conjugates (16). A Zeiss Universal microscope equipped for fluorescence with incident illumination was used for examination and photomicrography. Photographs were taken on Kodak Tri-X film.

Experiments

1. In order to determine which fixation conditions are required, either unfixed or perfusion-fixed tissues were used. Two of the most promising fixatives were tested further, i.e., 4% paraformaldehyde in 0.1 M sodium phosphate buffer pH 7.3, prepared according to Pease (29), and 4% paraformaldehyde-0.5% glutaraldehyde in the same buffer.

2. The antibody titer, as revealed in dilution experiments, was tested in the brain, small intestine, ductus deferens, and adrenal gland. In all experiments the dilution of the FITC conjugate was kept at 1:16. Incubation with the primary sera was performed at 4°C overnight.

3. The specificity of the antibodies was tested as follows: First, consecutive sections of the brain, stomach, small intestine, ductus deferens, and adrenal medulla were incubated with anti-NE, anti-E, or nonimmune serum. Second, consecutive sections of adrenal medulla fixed by 4% paraformaldehyde were stained with antibodies to bovine DBH, NE, bovine PNMT (12,20), or E. In a third type of experiment NE and E antisera were used in the brain, small intestine, ductus deferens, and adrenal medulla after absorption with NE, E, or 5-HT immunogen (1 and 2 mg/ml). The latter immunogen (31,32) was added to see if there was any cross reactivity to 5-HT.

RESULTS

1. No fluorescence was observed in unfixed tissues. In the adrenal medulla an intense cytoplasmic fluorescence was found with NE and E antisera after fixation in 4% paraformaldehyde. On the basis of the intensity of the fluorescence, two cell types could be distinguished (Table 1 and Fig. 1). Addition of 0.5% glutaraldehyde resulted in a less-brilliant fluorescence and a deterioration of the morphology of the cells. In the brain and the peripheral nervous tissue, only NE antisera gave positive staining. A bright fluorescence was observed if the paraformaldehyde–glutaraldehyde mixture had been used. Thus in the pons and the medulla oblongata, cell bodies known to contain NE (5) were easily seen (Fig. 2). The perikarya of the neurons in the superior cervical ganglion were negative. Fine fluorescing fibers with varicosities were seen in several parts of the brain (Fig. 2) and in the peripheral nervous system (Fig. 3). Nerve fibers in the ductus deferens were weakly stained after fixation in

TABLE 1. *Immunoreactivity[a] in the adrenal medulla of the adult rat[b]*

Serum and dilution	NE-storing cells	E-storing cells
Anti-DBH (1:32)	+	+ +
Anti-NA (1:500)	+ +[c]	+[c]
Anti-PNMT (1:300)	−	+
Anti-A (1:100)	−	+ +

[a]Immunoreactivity is expressed semiquantitatively: (+ +) strong; (+) moderate; (−) negative fluorescence.
[b]Paraformaldehyde 4% was used as a fixative.
[c]At higher dilutions the NE fluorescence in the NE- and E-storing cells changes to + / − .

4% paraformaldehyde alone. No fluorescence was detected in the other peripheral nervous tissues under these circumstances.

2. Dilution experiments revealed antibody titers of 1:1,000 (NE antibodies) and 1:100 (E antibodies), respectively.

3. After incubation with nonimmune serum, only a faint background fluorescence was observed. Comparing consecutive sections of the adrenal medulla incubated with antibodies to bovine DBH, NE, bovine PNMT, or E revealed that there is a good correlation between DBH and NE immunoreactivity in the adrenal medulla (Table 1). The same holds for PNMT and E immunoreactivity *(idem)*. The results of the absorption experiments using immunogen (2 mg/ml) are summarized in Table 2. The same results were achieved with immunogen 1 mg/ml. These experiments indicate that NE antibodies do not have a detectable degree of cross reactivity to E and 5-HT. The same holds for E antibodies with regard to NE and 5-HT.

DISCUSSION

NE and E by themselves are nonimmunogenic. However, after linkage to BSA, antibodies to NE and E could be raised. This principle, well known by the work of Landsteiner (23), was also adopted by those who reported on the preparation of NE and E antibodies (14,17,25,30,34). These authors characterized their antibodies by immunodif-

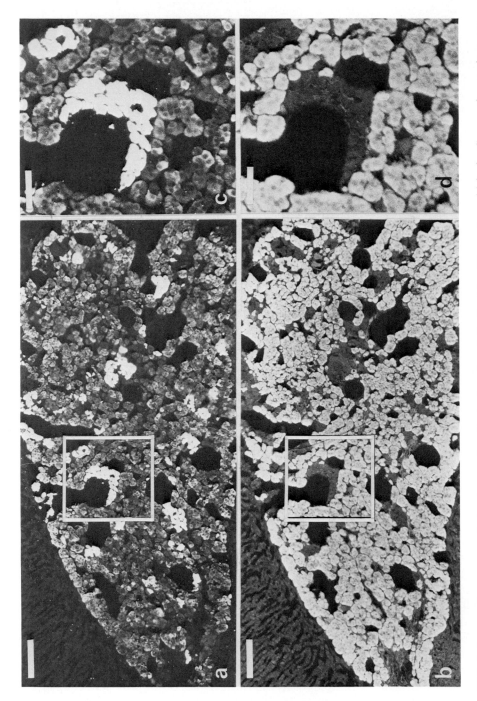

FIG. 1. Immunofluorescence photomicrographs of rat adrenal medulla fixed by 4% paraformaldehyde. Consecutive sections were stained with NE (**a**) and E (**b**) antiserum. NE-storing cells (NE-positive, E-negative) and E-storing cells (E-positive, moderately NE-positive) are demonstrated. (**c**) and **d**: Higher magnifications of areas indicated by the rectangle in **a** and **b**. Bars are 110 μm (**a**, **b**) and 44 μm (**c**, **d**).

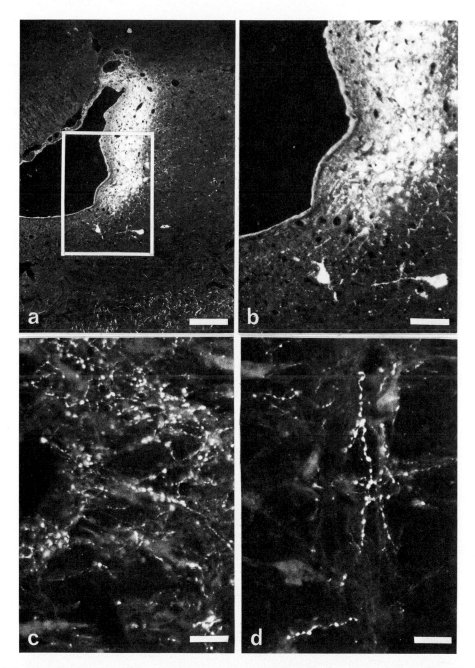

FIG. 2. Immunofluorescence photomicrographs of rat brain fixed by 4% paraformaldehyde-0.5% glutaraldehyde. Sections were stained with NE antiserum and demonstrate perikarya in the locus ceruleus area **(a)** and nerve fibers with varicosities in the nucleus cuneiformis **(c, d)**. **b:** Higher magnification of the area within the rectangle in **a.** Bars are 125 μm **(a)** and 230 μm **(b–d)**.

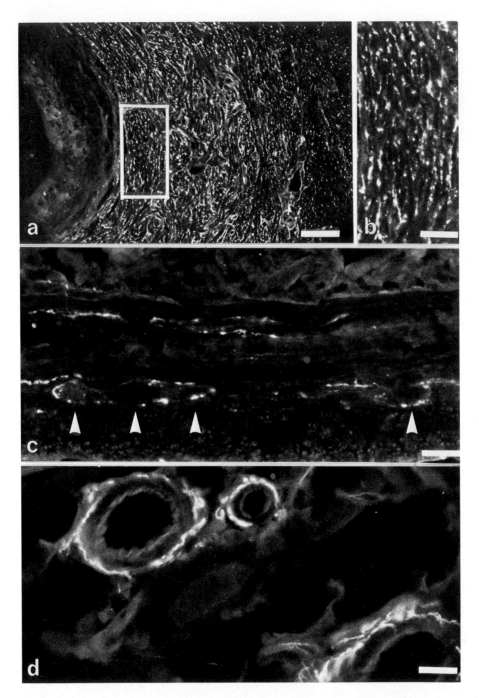

FIG. 3. Immunofluorescence photomicrographs of tissues fixed by 4% paraformaldehyde–0.5% glutaraldehyde. Sections are stained with NE antiserum. Nerve fibers with varicosities are in the muscular layer of the ductus deferens (**a**) and the small intestine (**c**). The nerve endings on the perikarya of Auerbach's plexus are indicated by arrowheads. **b:** Higher magnification of an area indicated by the rectangle in **a. d:** Small arteries in the submucosal layer of the stomach showing the perivascular nerve plexus. Bars are 80, 44, 960, and 600 μm for **a–d,** respectively.

TABLE 2. *Summary of the results of the absorption experiments*

Antiserum	BSA-NE	BSA-E	BSA-5-HT
NE	+	0	0
E[a]	0	+	0

Serum samples absorbed with immunogen 2 mg/ml were tested on adjacent sections. The decline in fluorescence intensity to negative or almost negative is expressed as + (absorption of antibodies) and 0 (no absorption).

[a] Tested only in the adrenal medulla.

fusion, immunoelectrophoresis, or radioimmunoassay. We demonstrated that after appropriate fixation an immunohistochemical staining of adrenal medullary cells, perikarya of NE-containing neurons in the brain, and nerve fibers in the central as well as the peripheral nervous system could be effected with the help of these antisera.

No fluorescence was observed in unfixed tissues. Thus fixation is of the utmost importance. However, different tissues appear to have different fixation requirements for optimal results. Fluorescence in the brain was observed only if a mixture of 4% paraformaldehyde and 0.5% glutaraldehyde was used. This mixture is also the best to use for peripheral nervous tissue. However, in the ductus deferens some fluorescence was seen using 4% paraformaldehyde alone. In contrast, in the adrenal medulla paraformaldehyde is the most favorable of the fixatives tested. We are aware that the fixation problem is not yet fully solved, as indicated by the absence of staining in the perikarya of the superior cervical ganglion, known to contain NE.

The matter of sensitivity of the presented immunohistochemical method also has not yet been fully explored. However, our first results look very promising. In this respect we would like to stress the high titer of both antibodies, which allows the detection of small concentrations of the antigens, i.e., NE and E. Furthermore, from our observations on the brain, peripheral nervous tissue (as far as the fibers are concerned), and adrenal me-

dulla, it may be concluded that the immunohistochemical method using antibodies to NE and E at least equals the sensitivity of the induced fluorescence techniques. At present there is no information on the sensitivity of the presented immunohistochemical method at the electron microscopic level.

Specificity is the most crucial point. We used three parameters (see also Materials and Methods). The value of each parameter alone may be questionable. However, the combined use of the presented data strongly supports our conclusion that NE antibodies are not demonstrating E or 5-HT in the tissues studied, at least under the conditions used (fixation, staining procedure, etc.). The same holds for E antibodies. Of course, cross reactivity to other substances (e.g., DA) cannot be excluded yet. Further characterization of the antisera by means of differential absorption and radioimmunoassay is in progress. A comparison with data of other authors (14,17,25,30,34) is hardly possible, since their data are rather limited or are based on other criteria (immunodiffusion, immunoelectrophoresis, and radioimmunoassay).

CONCLUSIONS

There is strong evidence in the rat that antibodies to NE and E can be used to demonstrate NE- and E-containing structures in the nervous as well as the endocrine systems. Preliminary observations show that this immunohistochemical method can also be applied in other species.

SUMMARY

Antibodies were raised in rabbits and sheep with antigens prepared by coupling NE or E to bovine serum albumin. Tissues were fixed with either 4% paraformaldehyde or 4% paraformaldehyde plus 0.5% glutaraldehyde in 0.1 M sodium phosphate buffer (pH 7.3). Cryostat sections were stained by the indirect immunofluorescence technique. Two parenchymal cell types, representing the

NE- and E-storing cells, were observed in the adrenal medulla. In the central nervous system, fibers and perikarya could be visualized by the antiserum to the NE antigen. In the peripheral nervous system, NE immunoreactive fibers were demonstrated. There was no detectable cross reactivity of NE antibodies with regard to E and 5-HT. Cross reactivity of E antibodies to NE and 5-HT equally could not be demonstrated.

ACKNOWLEDGMENTS

The antibodies to PNMT were kindly provided by Dr. M. Goldstein, Department of Psychiatry, New York University, Medical Center, New York. The help and advice of Mr. C.A. de Bruin and Mr. A.L.N. van Eupen in preparing the photographs is greatfully acknowledged. The authors wish to thank Dr. M. Weiss for critical reading and Miss Wanda de Haan for typing the manuscript.

REFERENCES

1. Axelsson, S., Björklund, A., Falck, B., and Lindvall, O. (1973): Glyoxylic acid condensation: A new fluorescence method for the histochemical demonstration of biogenic monoamines. *Acta Physiol. Scand.*, 87:57-62.
2. Björklund, A., Falck, B., and Lindvall, O. (1975): Microspectrofluorometric analysis of cellular monoamines after formaldehyde or glyoxylic acid condensation. In: *Methods in Brain Research*, edited by P.B. Bradley, pp. 249-294. Wiley, London.
3. Coons, A.H. (1958): Fluorescent antibody methods: In: *General Cytochemical Methods*, edited by J.F. Danielli, pp. 399-422. Academic Press, New York.
4. Coupland, R.E. (1965): *The Natural History of the Chromaffin Cell.* Longmans, London.
5. Dahlström, A., and Fuxe, K. (1964): Evidence for the existence of monoamine containing neurons in the central nervous system. I. Demonstration of monoamines in the cell bodies of the brain stem neuron. *Acta Physiol. Scand.* [*Suppl.*], 232:1-55.
6. Eränkö, O. (1952): On the histochemistry of the adrenal medulla of the rat, with special reference to acid phosphatase. *Acta Anat. (Basel)* [*Suppl.*], 17:1-60.
7. Eränkö, O. (1967): The practical histochemical demonstration of catecholamines by formaldehyde-

induced fluorescence. *J. R. Microsc. Soc.*, 87:259-276.
8. Falck, B. (1962): Observations on the possibilities of the cellular localization of monoamines by a fluorescence method. *Acta Physiol. Scand.* [*Suppl.* 197], 56.
9. Falck, B., and Owman, Ch. (1965): A detailed methodological description of a fluorescence method for the cellular localization of biogenic monoamines. *Acta Univ. Lundensis*, Section II, No. 7.
10. Falck, B., and Torp, A. (1961): A fluorescence method for histochemical demonstration of noradrenalin in the adrenal medulla. *Med. Exp.*, 5:429-432.
11. Geffen, L.B., Livett, B.G., and Rush, A. (1969): Immunohistochemical localization of protein components of catecholamine storage vesicles. *J. Physiol. (Lond)*, 204:593-605.
12. Goldstein, M., Fuxe, K., Hökfelt, T., and Joh, T.H. (1971): Immunohistochemical studies on phenylethanolamine-N-methyltransferase, DOPA-decarboxylase and dopamine-β-hydroxylase. *Experientia*, 27:951-952.
13. Gorgas, K., and Böck, P. (1976): Identification of chromaffin and enterochromaffin cells in semithin sections by means of argentaffin reaction. *Mikroskopie*, 32:57-63.
14. Grota, L.J., and Brown, G.M. (1976): Antibodies to catecholamines. *Endocrinology*, 98:615-622.
15. Grzanna, R., and Coyle, J.T. (1976): Rat adrenal dopamine-β-hydroxylase: Purification and immunologic characteristics. *J. Neurochem.*, 27:1091-1096.
16. Hartman, B.K. (1973): Immunofluorescence of dopamine-β-hydroxylase: Application of improved methodology to the localization of the peripheral and central noradrenergic nervous system. *J. Histochem. Cytochem.*, 21:312-332.
17. Hartman, B., and Spector, S. (1972): Application of immunology to catecholamine studies. In: *Methods in Investigative and Diagnostic Endocrinology, Vol. 1: The Thyroid and Biogenic Amines*, edited by J.E. Rall and I.J. Kopin, pp. 497-502. North-Holland, Amsterdam.
18. Henle, J. (1865): Über das Gewebe der Nebenniere und der Hypophyse. *Z. Rat. Med.*, 24:143-152.
19. Hökfelt, T., and Ljungdahl, Å. (1971): Uptake of [³H]noradrenaline and γ-[³H]aminobutyric acid in isolated tissues of rat: An autoradiographic and fluorescence microscopic study. *Prog. Brain Res.*, 34:87-102.
20. Hökfelt, T., Fuxe, K., Goldstein, M., and Joh, T.H. (1973): Immunohistochemical localization of three catecholamine synthesizing enzymes: Aspects on methodology. *Histochemie*, 33:231-254.
21. Hopwood, D. (1971): In: *Progress in Histochemistry and Cytochemistry, Vol. 3: The Histochemistry and Electron Histochemistry of Chromaffin Tissue*, pp. 1-66. Gustav Fischer Verlag, Stuttgart.
22. Jonsson, G. (1971): Quantitation of fluorescence of biogenic monoamines demonstrated with the formaldehyde fluorescence method. *Prog. Histochem. Cytochem.*, 2:229-331.

23. Landsteiner, K. (1945): *The Specificity of Serological Reactions.* Harvard University Press, Cambridge, Mass.

24. Lindvall, O., and Björklund, A. (1974): The glyoxylic acid fluorescence histochemical method: A detailed account of the methodology for the visualization of central catecholamine neurons. *Histochemistry,* 39:97–127.

25. Miwa, A., Yoshioka, M., Shirahata, A., and Tamura, Z. (1977): Preparation of specific antibodies to catecholamines and L-3,4-dihydroxyphenylalanine. I. Preparation of the conjugates. *Chem. Pharm. Bull.,* 25:1904–1910.

26. Nagatsu, I., and Kondo, Y. (1974): Immunoelectronmicroscopic localization of phenylethanolamine-N-methyltransferase in the bovine adrenal medulla. *Histochemistry,* 42:351–358.

27. Nairn, R.C. (1976): *Fluorescent Protein Tracing.* Churchill Livingstone, Edinburgh.

28. Ouchterlony, O. (1967): Immunodiffusion and immunoelectrophoresis. In: *Handbook of Experimental Immunology,* edited by D. Weir, pp. 665–706. Blackwell, Oxford.

29. Pease, D.C. (1962): Buffered formaldehyde as a killing agent and primary fixative for electron microscopy. *Anat. Rec.,* 142:342.

30. Spector, S. (1969): Application of immunology to neurochemistry. *Adv. Biochem. Psychopharmacol.,* 1:181–190.

31. Steinbusch, H.W.M., and Verhofstad, A.A.J. (1979): Immunofluorescent staining of serotonin in the central nervous system. *Adv. Pharmacol. Ther.,* 2:151–160.

32. Steinbusch, H.W.M., Verhofstad, A.A.J., and Joosten, H.W.J. (1978): Localization of serotonin in the central nervous system by immunohistochemistry: Description of a specific and sensitive technique and some applications. *Neuroscience,* 3:811–819.

33. Taxi, J. (1976): General principles of neurotransmitter detection: Problems and application to catecholamines. *J. Microsc. Biol. Cell.,* 27:243–248.

34. Went, I., and Kesztyus, L. (1939): Das Adrenalin; synthetische Herstellung eines Adrenalin Antigens: Aminoadrenalin und Adrenalin-Azoproteine. *Arch. Exp. Pathol. Pharmakol.,* 193:609–614.

Histochemistry and Cell Biology of Autonomic Neurons, SIF Cells, and Paraneurons,
edited by O. Eränkö et al.
Raven Press, New York © 1980.

Adrenal Chromaffin Cells in the Stressed Mouse

Shigeru Kobayashi and Yuriko Serizawa

Department of Anatomy, Niigata University School of Medicine, Niigata, 951 Japan

We reported previously that mouse adrenal medulla contains, in addition to epinephrine-storing (E) cells and norepinephrine-storing (NE) cells, a third type of chromaffin cell (3,4,7). This cell type is characterized by small secretory granules and hence is called a small granule chromaffin (SGC) cell (4).

One of the most peculiar cytological features of SGC cells is the constant existence of synaptic-type vesicles (6–8). These vesicles are dispersed throughout the cytoplasm, frequently forming aggregations, but show no tendency to gather in the cytoplasmic area where the SGC cells synapse with nerve terminals (6–8). Although the nature and function of the synaptic-type vesicles in the SGC cells remains mostly unknown (6,8), it seems worthwhile to examine whether the E and NE cells can also produce similar synaptic-type vesicles.

Restraint plus water-immersion stress has been employed by previous authors for producing so-called stress ulcers in the stomachs of various experimental animals (10). We thought it might be useful to apply similar procedures for the ultrastructural investigations on the secretory mechanisms of chromaffin cells because various types of stress produce a drastic depletion of catecholamines from the adrenal medulla (2). The purpose of this chapter is to describe some cytological changes of the adrenal chromaffin cells in the stressed mouse with particular reference to the synaptic-type vesicles in E and NE cells.

MATERIALS AND METHODS

Restraint Plus Water-Immersion Stress

Adult male mice of dd strain were fastened by the limbs with pins on a board in the supine position. The board was then made to stand vertically and immersed in water of about 20°C to the depth of the hind feet of the animals. As the controls, normal mice of the same strain were examined by the same methods without applying the stress.

Light Microscopy and Thin-Section Electron Microscopy

At 4 to 28 hr after the restraint plus water-immersion stress, mice were killed by cervical dislocation and were perfusion-fixed from the left ventricle of the heart with 2.5% glutaraldehyde in 0.1 M phosphate buffer of pH 7.4. Adrenal glands were postfixed with 1% osmium tetroxide. After dehydration using ethanol and propylene oxide, tissue blocks were embedded in Araldite. Semithin sections about 1.0 μm thick were stained with toluidine blue or were silver-impregnated using Battaglia's method (1); they were observed and photographed in the light microscope. Ultrathin sections were double-stained with

coated pits and vesicles were associated with the exocytotic invaginations of the cell membrane. Accumulations of synaptic-type vesicles were seen in the heavily degranulated chromaffin cells. Thus the results of this study indicate that the "exocytosis–vesiculation sequence" hypothesis holds true in the adrenal medulla of the stressed mouse.

In the present study it was shown that the occurrence of synaptic-type vesicles is not specific to the SGC cells, but E and NE cells also can produce them. However, this does not necessarily mean that the synaptic-type vesicles in the SGC cells are identical to those in the E and NE cells of the stressed mouse. In the latter, it is certain that the synaptic-type vesicles accumulate as a result of the increased incidence of exocytosis, whereas in the former exocytotic granule release could be not actively operating (3). Further studies are needed to clarify the functional significance of synaptic-type vesicles in the SGC cells of the mouse adrenal medulla.

SUMMARY

Mice were supinely restrained on a board by pinning the limbs. The board was then made to stand vertically and immersed in water to the height of the hind limbs of the animals. At 4 to 28 hr after this restraint plus water-immersion stress, a remarkable decrease in the number of secretory granules in the chromaffin cells was light microscopically demonstrated in silver-impregnated and toluidine-blue-stained semithin sections. Exocytotic invaginations containing pleomorphic granule materials were demonstrated by thin section and freeze-fracture replica electron microscopy. This stress-induced exocytotic degranulation of the chromaffin cells was accompanied by the occurrence of numerous small cytoplasmic vesicles resembling the synaptic-type vesicles of the small granule chromaffin cells.

ACKNOWLEDGMENTS

We would like to thank Professor T. Yamamoto and Dr. M. Fujita of the Department of Anatomy, Kyushu University Faculty of Medicine, Fukuoka, for the kind collaboration in the preparation of freeze-fracture replicas.

REFERENCES

1. Battaglia, G. (1969): Ultrastructural observations on the biogenic amines in the carotid and aortic-abdominal bodies of the human fetus. *Z. Zellforsch.,* 99:529–537.
2. Coupland, R.E. (1965): *The Natural History of the Chromaffin Cell.* Longmans, London.
3. Coupland, R. E., Kent, C., and Kobayashi, S. (1978): Amine turnover and the effects of insulin hypoglycaemia on small-granule chromaffin (SGC) cells of the mouse adrenal medulla. In: *Peripheral Neuroendocrine Interaction,* edited by R.E. Coupland and W.G. Forssmann, pp. 86-96. Springer, Berlin.
4. Coupland, R.E., Kobayashi, S., and Tomlinson, A. (1977): On the presence of small granule chromaffin (SGC) cells in the rodent adrenal medulla. *J. Anat.,* 123:488–489.
5. Douglas, W.W. (1973): How do neurones secrete peptides? Exocytosis and its consequences, including "synaptic vesicle" formation, in the hypothalamo-neurohypophyseal system. *Prog. Brain Res.,* 39:21–39.
6. Kobayashi, S. (1977): Adrenal medulla: Chromaffin cells as paraneurons. *Arch. Histol. Jpn. (Suppl.),* 40:61–79.
7. Kobayashi, S., and Coupland, R.E. (1977): Two populations of microvesicles in the SGC (small granule chromaffin) cells of the mouse adrenal medulla. *Arch. Histol. Jpn.,* 40:251–259.
8. Kobayashi, S., Serizawa, Y., Fujita, T., and Coupland, R.E. (1978): SGC (small granule chromaffin) cells in the mouse adrenal medulla: Light and electron microscopic identification using semi-thin and ultra-thin sections. *Endocrinol. Jpn.,* 25:467–476.
9. Nagasawa, J. (1977): Exocytosis: The common release mechanism of secretory granules in glandular cells, neurosecretory cells, neurons and paraneurons. *Arch. Histol. Jpn. (Suppl.),* 40:31–47.
10. Takagi, K., Kasuya, Y., and Watanabe, K. (1964): Studies on the drugs for peptic ulcer: A reliable method for producing stress ulcer in rats. *Chem. Pharmacol. Bull.,* 12:465–472.

Histochemistry and Cell Biology of Autonomic Neurons, SIF Cells, and Paraneurons,
edited by O. Eränkö et al.
Raven Press, New York © 1980.

Paraneurons and Paraganglia: Histological and Ultrastructural Comparisons Between Intraganglionic Paraneurons and Extra-Adrenal Paraganglion Cells

Joe A. Mascorro and Robert D. Yates

Department of Anatomy, Tulane University School of Medicine, New Orleans, Louisiana 70112

The term "paraneuron" was introduced recently to designate various cell types that are closely related to neurons (12,13). Paraneurons are not typical neurons, but they nevertheless possess similar characteristics and are said to be close relatives of neurons. More specifically, the chromaffin cells of the adrenal medulla are categorized as typical paraneurons because they contain and secrete catecholamines via exocytosis, receive a preganglionic sympathetic innervation, share a neural crest origin with sympathetic nerve cells, and are able to evoke action potentials (13). It is further hypothesized that paraneurons may be endocrine, sensory, or interneuronal in function (13).

The term "paraganglia," on the other hand, has been known for many years and designates cells that are similar to those of the adrenal medulla. This tissue type often was found in direct association with abdominal sympathetic ganglia and appropriately was termed the paraganglia (33). Many studies are currently available concerning the paraganglia. For example, it is now apparent that the term paraganglia is restrictive in accurately designating the expanse of extra-ad-

renal chromaffin tissue. Mapping studies, which utilize potassium dichromate to produce a gross chromaffin reaction, offer clear evidence that the paraganglia are, in actuality, a voluminous extra-adrenal chromaffin system extending throughout the retroperitoneum (26,28). Hence the paraganglia are more than simply paraganglionic in scope. The importance of this system is magnified by more recent studies which detail their presence in the retroperitoneal connective tissue spaces of man where they form a persisting endocrine system (15,16).

Several studies provided evidence that catecholamine-containing chromaffin-like cells also are located within sympathetic ganglia (3-6,22,30-32). In this intraganglionic environment, the chromaffin elements are connector neurons (interneurons) interposed between pre- and postganglionic elements where they assume an inhibitory role in ganglionic transmission. These interneurons are classified by their response to specific techniques. When visualized via electron microscopy their number of cytoplasmic granules is great, and thus they have been termed small granule-containing (SGC) cells. Fluorescence histochemistry results in such granules assuming a bright fluorescence, and hence the name small intensely fluorescent (SIF) cells. Abundant evidence has shown that interneurons possess the morphological characteristics necessary to affect ganglionic transmission. These intraganglion cells receive afferent synapses and, in turn, effect efferent connections by elongated processes emanating from their cell bodies. However, not all intraganglion chromaffin cells are interneuronal in function, as concluded by Heym and Williams (17). Clusters of SGC/SIF cells may represent neurosecretory rather than nervous elements. Furthermore, such cells may be classified as paraneurons (4).

· It is now believed that paraneurons possibly function as "interneurons," even though lacking synaptology or processes, by releasing catecholamines into the ganglion environment. Such a mechanism of action would imply a neurosecretory or endocrine action by the paraneurons (4,5).

The endocrine capacity of extra-adrenal chromaffin organs, collectively called the paraganglia, is well understood (9,14,15,28). Precise histological and ultrastructural comparisons have shown that the paraganglia are structurally similar to the cells of the adrenal medulla (9,26). In addition, various drug studies indicate that extra-adrenal chromaffin cells release catecholamines in response to sympathomimetic stimuli as do the adrenomedullary cells (23).

Preliminary observations in young New Zealand rabbits (27) indicate that intraganglionic chromaffin cells (paraneurons) as well as extra-adrenal chromaffin cells (paraganglia) possess similar morphology. The present study utilizes a different species and represents a more in-depth morphological comparison between paraneuronal and paraganglion cells. Indeed, if paraneuronal morphology closely resembles that of the paraganglia (a known neurosecretory system), then perhaps the idea that intraganglionic paraneurons can act by neurosecretion may be justified.

MATERIALS AND METHODS

Young kittens of either sex, weighing 200 to 400 g, were anesthetized with sodium pentobarbital (Nembutal, 40 mg/kg body weight, i.p.). The animals were sacrificed via intracardiac perfusion through the left ventricle with a solution of 3.125% glutaraldehyde buffered to pH 7.2 to 7.4 with 0.2 M sodium phosphate buffer. Following a 20 to 30-min perfusion, the retroperitoneal tissue block (RTB) from the level of the adrenal glands to the abdominal aortic bifurcation was removed *en toto* and immersed in fresh cold fixative for 2 hr. The RTBs subsequently were transferred to a solution of glutaraldehyde mixed with potassium dichromate (28). This combination produced a gross chromaffin reaction which vividly stained and thus localized the extra-adrenal chromaffin organs (paraganglia). An overnight phosphate buffer wash removed excess or nonreactive dichromate and prepared the RTBs for macrophotography. The localized paraganglia were dissected out of the RTBs and processed routinely for light and ultrastructural study.

Sections (0.5 μm) were prepared and stained with toluidine blue or acid fuchsin. These sections were utilized to locate favorable areas containing paraganglia and associated autonomic ganglia. Sections were searched further until granule-containing cells were located within the ganglia. The sections were then photographed with Plus-X panchromatic film utilizing a Zeiss photomicroscope II. Ultrathin sections of the desired blocks were prepared and stained sequentially with alcoholic uranyl acetate and lead citrate. Negatives were taken with a Siemens 101 electron microscope and developed routinely.

RESULTS

Distribution of Extra-adrenal Chromaffin Organs

Retroperitoneal tissue blocks stained with potassium dichromate displayed large chromaffin-reacting organs (paraganglia) at extra-adrenal sites (Fig. 1). The narrow and elongated organs occurred ventral and lateral to the abdominal aorta and displayed a brown coloration typical of the chromaffin reaction. The main extra-adrenal chromaffin organ was consistently present in the upper retroblock and attained lengths of 5 to 7 mm. The similarly stained adrenal medulla likewise exhibited the colorful dichromate (chromaffin) reaction, indicating the presence of catecholamines (Fig. 1). Minute bodies exhibiting chromaffinity were scattered throughout the retroperitoneum, particularly around the general area of origin of the inferior mesenteric artery. A continuation between intra- and extra-adrenal chromaffin tissue was not evident.

Histological Characteristics

Initial study with the light microscope provided evidence that sympathetic ganglia of the upper retroperitoneum were immediately opposed to the abdominal paraganglia, with only a connective tissue and vascular plane separating the two (Fig. 2). In survey sections examined at low magnifications, intraganglion chromaffin cells (paraneurons) were easily distinguished from the typical sympathetic neurons and connective tissue cells of the ganglion. The whole ganglion, in general, presented histological characteristics that allowed distinct differentiation from the adjoining paraganglion (Fig. 2).

The paraneurons most often appeared as clusters of varying size (Fig. 2). The cell groups were compactly arranged and abundantly vascularized. Occasionally the cells appeared as single or paired structures apposing blood vessels (Fig. 3). These features distinguished the intraganglionic chromaffin cells from their neural relatives, the sympathetic neurons. The nerve cells were noticeably larger, were always single rather than clustered, contained satellite cells, and showed nuclei often as large as the entire paraneuron.

The individual paraneuronal cells typically were small and rounded or polyhedral, and presented a central or eccentrically placed nucleus with clear chromatin and prominent nucleolus. The paraneuron cell bodies were distinguished by a marked basophilia not evident in the ganglion neurons (Fig. 4). Upon detailed examination by oil immersion optics, the basophilia was defined as individual dense vesicles scattered randomly throughout the cytoplasm (Figs. 3 and 4).

The paraganglia represented a homogeneous collection of epithelioid-type cells whose collective mass comprised an extra-adrenal chromaffin organ proper (Fig. 1). The chromaffin cells of the paraganglia occurred in cords or clusters and were surrounded by numerous thin-walled blood vessels that anastomosed freely throughout the gland (Fig. 5). The histological characteristics of individual paraneuronal and paraganglion cells were strikingly similar: small and polyhedral, prominent nuclei/nucleoli, and cytoplasmic vesicles subsequently identified as catecholamine granules via ultrastructural analysis. Mast cells occasionally were seen among the chromaffin cells, but typical neurons were never present.

Vascular elements and paraganglion cells were intimately associated, but distinct perivascular spaces intervened. Furthermore, connective tissue stroma radiated inward from a thick capsule which surrounded the paraganglion. These connective tissue strands caused paraganglion cells to be loosely arranged, whereas the intraganglion cell clusters appeared more compact. Except for this minor difference, however, the paraganglia and intraganglionic paraneurons were histologically identical (compare Figs. 4 and 5).

Ultrastructural Characteristics

The most prevalent observation at the electron microscope level was that intraganglion and extra-adrenal chromaffin cells were closely associated with blood vessels and

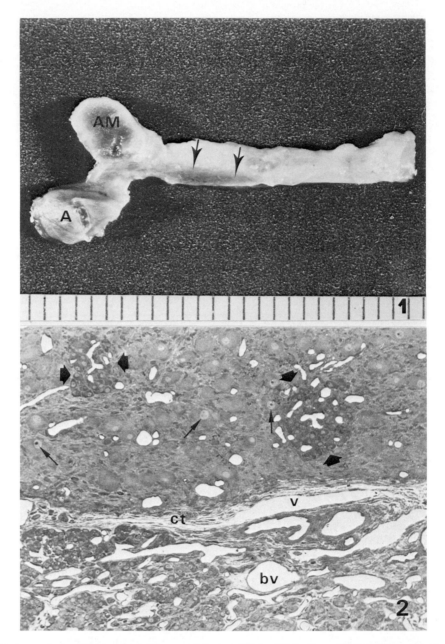

FIG. 1. Retroperitoneal tissue block from a young cat. Glutaraldehyde perfusion followed by immersion in glutaraldehyde/potassium dichromate. Large organ in the upper retroperitoneum represents the main abdominal paraganglion *(arrows)* and reacts to potassium dichromate as does the adrenal medulla (AM). (A) Right adrenal gland. Scale = millimeters.

FIG. 2. Low-power light micrograph illustrating an autonomic ganglion *(upper field)* adjacent to the abdominal paraganglion *(lower field)*. The two bodies are separated by a connective tissue (ct) and vascular (v) plane. Clusters of intraganglionic paraneurons *(arrowheads)* are distinguished easily from the surrounding neurons, which have pale cytoplasm and prominent nuclei/nucleoli *(arrows)*. Numerous blood vessels (bv) appear throughout the organs. ×293.

contained numerous cytoplasmic granules (Figs. 6 and 7). The granule complement in both cell types was distributed randomly and occupied large areas of the cytoplasm. The individual granules were round or oval, showed diameters ranging from 1,000 to 1,500 Å in cross section, and usually were surrounded by a membrane. The majority of cells possessed granules with opaque central cores following glutaraldehyde/osmium tetroxide fixation. However, a small proportion of paraganglion and paraneuronal cells contained granules whose central cores were reduced in density (Fig. 12). The cores of the lighter vesicles exhibited a homogeneous granularity which occupied most of the membrane-bound granule area. The darker granules, on the other hand, showed cores that occasionally were fragmented and often eccentrically placed or attached to the granule membrane (Fig. 12).

The cells displayed the usual complement of cytoplasmic organelles. However, the mitochondria, granular endoplasmic reticulum, and Golgi membranes were inconspicuous among the voluminous granules. Only the nuclei, with their multishapes and central or peripheral chromatin clumping, were obvious within the cells (Figs. 6 and 7).

Paraganglia and paraneurons exhibited an abundant blood supply. The main observation was that the cells apposed thin-walled blood vessels. The endothelium lining the vessels was continuous and showed varying pinocytotic activity. Pericapillary cell processes, collagen fibrils, and basal laminae occupied the subendothelial space. However, in areas where the above were absent, only a thin space separated the parenchymal and endothelial cells (Figs. 6 and 7).

As evidenced by the glutaraldehyde/potassium dichromate mapping technique, the abdominal paraganglia represented accumulations of chromaffin-positive, granule-containing cells. The organs were large, visible to the naked eye, and occupied large areas of the retroperitoneum (Fig. 1). By con-

trast, the intraganglionic chromaffin cells were microscopic structures which occurred singly, paired, or as clusters of varying size (Figs. 2–5, 10, 11).

Numerous ganglion components surrounded the paraneurons and precluded these cells (whether single, paired, or clustered) from making direct contact with postganglionic neurons (Figs. 10 and 11). The single cells were surrounded by many unmyelinated nerve fibers. The fibers, in turn, were enveloped by Schwann cells exhibiting large nuclei (Fig. 8). Prominent cytoplasmic processes originated from the somas of the single cells. The processes coursed through the ganglion but were not observed to synapse with ganglion elements. On the contrary, they always were separated from neuronal cell bodies by abundant connective tissue or nerve fibers (Figs. 8 and 9).

DISCUSSION

The complexity of peripheral catecholamine-containing chromaffin cells has become evident in recent years. Early histofluorescence reports by Eränkö and Härkönen (11) described monoamine-rich cells throughout the superior cervical ganglion of the rat. The cells also occurred as discrete organs adjacent to the ganglion. Similar cells were observed by Elfvin (10) in the inferior mesenteric ganglion of the rabbit. Using electron microscopy, this investigator observed many dense granules as well as cytoplasmic processes and designated these elements "granule-containing nerve cells." Since then, several studies of sympathetic ganglia revealed the small granule-containing cells and established their similarities with chromaffin cells of the adrenal medulla and sympathetic neurons. Williams and Palay (31) and Matthews and Raisman (29) best summarized the structure/function relationship of the cells by stating that they were connector neurons innervated by preganglionic fibers and situated between pre- and postganglionic elements.

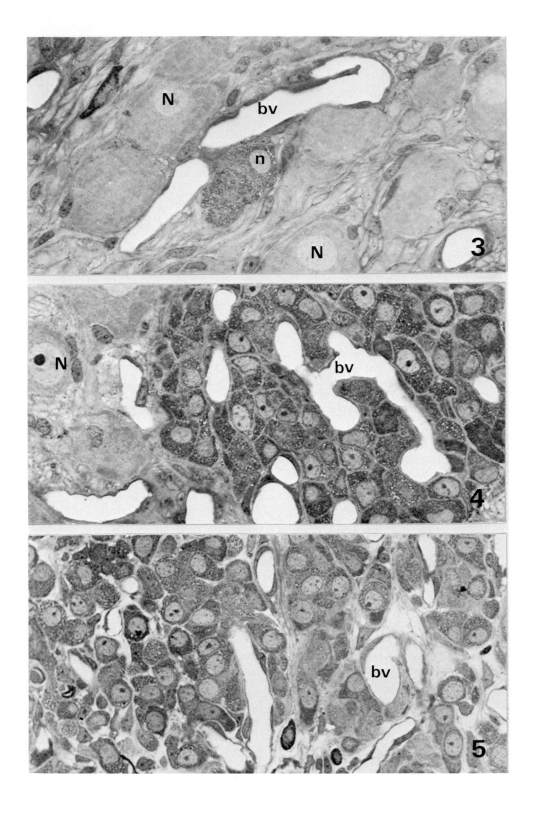

Furthermore, the cells formed synapses on the principal sympathetic neurons.

The paraneuron concept is new (12,13,20) and categorizes certain cells as "paraneurons" if they fulfill the following criteria: produce neurotransmitters and protein/polypeptide-like substances capable of hormone actions, originate in common with neurons (from neuroectoderm), and release neurosecretion-like vesicles at synaptic contacts. Since not all granule-containing cells within sympathetic ganglia possess the necessary synaptic connections to permit a direct interneuronal function, the paraneuron concept offers a viable explanation of how granule cells could affect postganglionic neurons and, ultimately, ganglionic transmission. Paraneurons secreting catecholamines into the ganglion environment theoretically could perform this function. This idea is rendered more attractive by the present work, which reports the morphological similarities shared by paraneurons and paraganglion cells known to function in an endocrine fashion.

The histological information presented here depicts abdominal sympathetic ganglia immediately juxtaposed to extra-adrenal chromaffin organs. In several fortuitous examples, intraganglionic clusters of chromaffin cells were within the same field of view as paraganglion cells. When such sections were examined with oil immersion optics, a pattern of histologic similarity became obvious. The intraganglionic and extra-adrenal chromaffin cells were interspersed among blood vessels with whom they shared a close relationship. As portrayed by ultrastructural examination, the separation often was minimal and included only basal laminae and collagen fibrils as barriers within the subendothelial space. This appears as an opportune arrangement for possible catecholamine diffusion into the vascularity and is in keeping with the idea that some interneurons are functionally related to blood vessels in an endocrine or chemoreceptive sense (22).

The present results do not offer novel information concerning the innervation of extra-adrenal chromaffin organs. Earlier studies utilizing staining techniques and silver impregnation reported that the paraganglia received a preganglionic sympathetic innervation like the adrenal medulla (18,21). Likewise, studies on chromaffin cells in the sympathetic trunk ganglia indicated that they also received a similar innervation (19). Ultrastructural documentation for these observations is largely lacking, and several investigators (9,14) were unable to locate nerve endings on abdominal paraganglion cells. Only few investigators have reported the occurrence of typical cholinergic-type synapses on these cells (24,25). The probability that paraganglion cells, at best, are sparsely innervated must be considered. It is interesting that the intraganglionic cells encountered during this study, whether single or in clusters, lacked innervation patterns. The single cells, conforming more to the type I interneuron of other studies (5,22,32), neither received afferent synapses nor made efferent contacts. In strictest definition, they could not function as interneurons via pre- or postganglionic contacts. On the other hand, clusters of the intraganglionic chromaffin cells resembled miniparaganglia and also lacked innervation. Only the cytoplasmic processes from the cell bodies of the single intraganglionic chromaffin cells suggested a possible nerve contact. However, such cytoplasmic

FIG. 3. A light micrograph illustrating intraganglionic chromaffin cells. This small cluster shows a nucleus (n) and its immediately neighboring blood vessels (bv). Several neurons (N) appear throughout the field. ×800.

FIGS. 4 and 5. Note the histological similarities between intraganglionic chromaffin cells (paraneurons, Fig. 4) and extra-adrenal chromaffin cells (paraganglia, Fig. 5). Both organs are richly vascularized (bv) and have cells and nuclei of similar size. All cells exhibit a vesicular, basophilic cytoplasm indicative of catecholamine granules when viewed ultrastructurally (see following figures). A neuron (N) is present in Fig. 4. ×800.

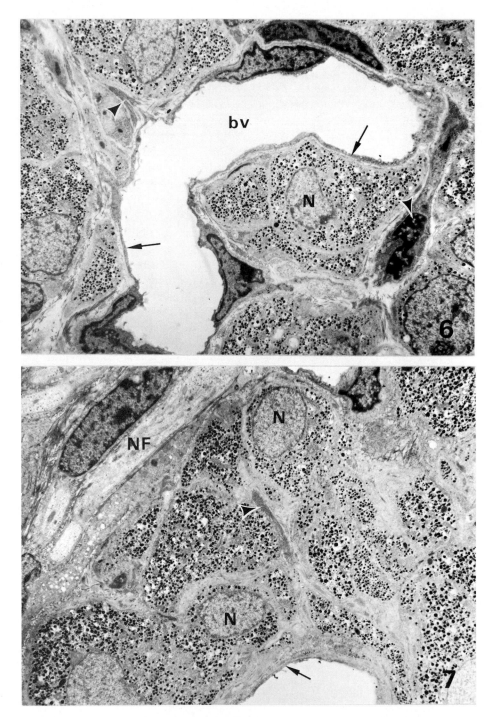

FIGS. 6 and 7. Low-power electron micrographs illustrating the fine structural similarities between paraganglion cells (Fig. 6) and intraganglionic paraneurons (Fig. 7). Individual cells in both organs contain prominent nuclei (N) and numerous cytoplasmic granules which correspond to the density observed with the light microscope (refer to Figs. 4 and 5). The parenchymal cells are in apposition to large blood vessels (bv), which are surrounded by thin and continuous endothelium *(arrows)*. Connective tissue cells and collagen fibers *(arrowheads)* interdigitate between cells. Unmyelinated nerve fibers are seen in paraneurons (NF in Fig. 7) but not in the paraganglia. ×4,800.

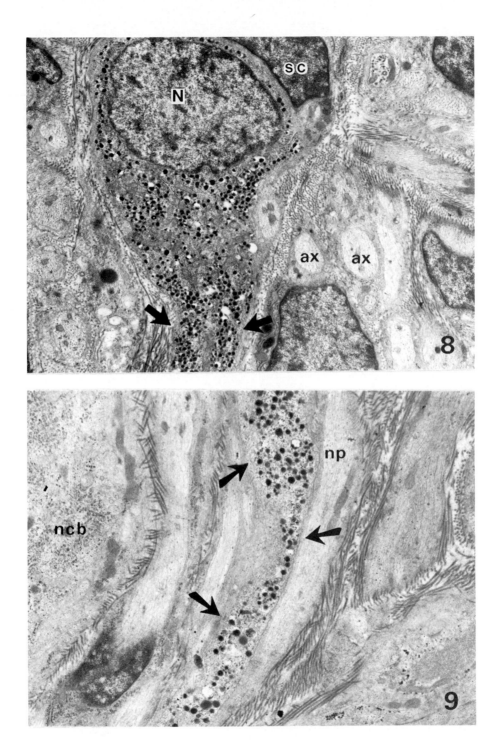

FIG. 8. A single paraneuron showing a large nucleus (N) and a satellite cell cap (sc). The cell body gives rise to an elongated process *(arrows)* which courses away from the paraneuron nucleus. Numerous unmyelinated axons (ax) and their Schwann cells surround the paraneuron whose cytoplasm is rich in catecholamine granules. ×10,000.

FIG. 9. A single paraneuronal process *(arrows)* presumably originating from a cell such as that seen in Fig. 8. The process contains catecholamine granules and is widely separated from a neuronal cell body (ncb). A nerve process (np) also courses through the field. ×11,000.

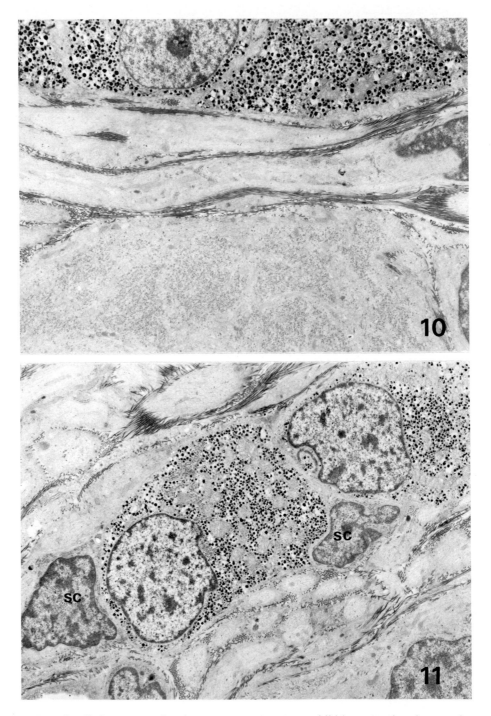

FIG. 10. Note the distinct separation between a paraneuron exhibiting cytoplasmic granules and a prominent nucleus *(upper field)*, and a neuronal cell body *(lower field)*. Collagen fibers and ganglion elements course between the two cells. ×4,560.

FIG. 11. An example of paraneurons occurring in pairs. Satellite cells (sc) appose the paraneurons. ×4,560.

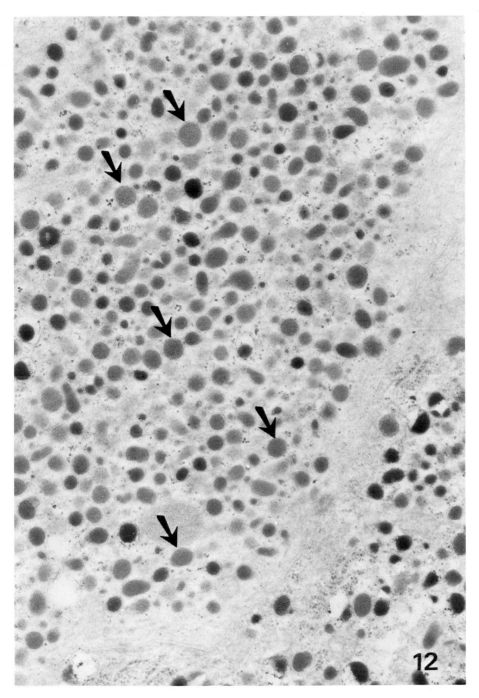

FIG. 12. Electron micrograph illustrating granules common to paraganglion and paraneuronal cells. Most cells contain granules of a very dense nature (see previous micrographs), but granules distinctly showing less central core density also occur *(arrows)*. This type of granule usually is round, membrane-bound, and exhibits an internal granularity. ×48,000.

prolongations never were observed in synapse with ganglion components, but this possibility cannot be precluded.

The granule content was voluminous in all of the intraganglionic and extra-adrenal cells studied. Although it was not within the scope of this morphological study to define the catecholamine nature of the granules, it should be mentioned that granule density varied following glutaraldehyde/osmium tetroxide fixation according to the method of Coupland and Hopwood (7,8). Researchers have applied the universally accepted concept that this fixation methodology differentiates epinephrine- from norepinephrine-storing granules. On this basis, granules with dense (norepinephrine) versus light (epinephrine) cores have been reported in many species (1,2,5,9,26). The majority of granules observed during this study corresponded to the dark-cored variety. However, aggregations of granules showing less-dense cores prevailed in certain cells of both organs.

A pioneer study by Williams and Palay (31) reported that interneurons in the rat sympathetic ganglion were few and therefore provided postganglionic connections for only a minority of ganglion cells. A later study by Chiba and Williams (5) reiterated that interneurons (SIF/SGC cells) in various species were sparse, and speculation has long existed concerning how a few interneurons could possibly modify transmission from large numbers of neurons. This study clearly illustrates intraganglionic clusters of chromaffin cells which are histologically and ultrastructurally identical to the individual extra-adrenal chromaffin cells that collectively comprise the endocrine abdominal paraganglia. These intraganglionic "miniparaganglia" may act as paraneuronal neurosecretory elements releasing catecholamines and, in some manner, affecting ganglion activity.

SUMMARY

Retroperitoneal tissue blocks containing extra-adrenal chromaffin organs and auto-nomic ganglia were used for morphological comparisons between intra- and extraganglionic chromaffin cells. The cells occasionally were found within the same tissue sections and thus presented an excellent opportunity for precise comparisons. Histologically, a distinct connective tissue and vascular plane intervened between the extra-adrenal chromaffin organs and sympathetic ganglia. The intraganglionic chromaffin cells occurred singly or in clusters and were surrounded by neurons and connective tissue. The largest cluster contained approximately 80 cells when cross sectioned, and numerous capillaries were interposed between the cellular elements. Paraganglia were many times larger than the intraganglionic cell clusters and likewise contained abundant vascularity. Numerous catecholamine granules filled the cytoplasm of the cells in both organs. The granules ranged from 1,000 to 1,500 Å in diameter and possessed dense cores. The paraganglion and intraganglionic cells were not innervated, but processes from the latter were encountered occasionally. Furthermore, synapses from the intraganglionic cells on postganglionic elements were not apparent. Considering the morphological similarities between the cell types, it is possible that the small intraganglionic cell clusters may represent a localized endocrine system secreting catecholamines into the ganglion environment.

REFERENCES

1. Benedeczky, I., and Smith, A.D. (1972): Ultrastructural studies on the adrenal medulla of golden hamster: Origin and fate of secretory granules. Z. Zellforsch. Mikrosk. Anat., 124:367–386.
2. Carmichael, S.W., and Blair, B.C. (1973): Normal ultrastructure of the day old dog adrenal medulla. J. Anat., 115:113–118.
3. Chiba, T. (1977): Monoamine containing paraneurons in the sympathetic ganglia of mammals. Arch. Histol. Jpn. (Suppl.), 40:163–176.
4. Chiba, T. (1978): Monoamine fluorescence and EM studies on SIF (granule-containing) cells in human sympathetic ganglia. J. Comp. Neurol., 179:153–168.

5. Chiba, T., and Williams, T.H. (1975): Histofluorescence characteristics and quantification of small intensely fluorescent (SIF) cells in sympathetic ganglia of several species. Cell Tissue Res., 162:331-341.

6. Chiba, T., Black, A.C., and Williams, T.H. (1977): Evidence for dopamine-storing interneurons and paraneurons in rhesus monkey sympathetic ganglia. J. Neurocytol., 6:441-453.

7. Coupland, R.E., and Hopwood, D. (1966): Mechanism of a histochemical reaction differentiating between adrenaline- and noradrenaline-storing cells in the electron microscope. Nature, 209:590-591.

8. Coupland, R.E., and Hopwood, D. (1966): The mechanism of the differential staining reaction for adrenaline- and noradrenaline-storing granules in tissues fixed in glutaraldehyde. J. Anat., 100:227-243.

9. Coupland, R.E., and Weakley, B.S. (1970): Electron microscopic observation on the adrenal medulla and extraadrenal chromaffin tissue of the postnatal rabbit. J. Anat., 106:213-231.

10. Elfvin, L-G. (1968): A new granule-containing nerve cell in the inferior mesenteric ganglion of the rabbit. J. Ultrastruct. Res., 22:37-44.

11. Eränkö, O., and Härkönen, M. (1965): Monoamine-containing small cells in the superior cervical ganglion of the rat and an organ composed of them. Acta Physiol. Scand., 63:511-512.

12. Fujita, T. (1976): The gastro-enteric endocrine cell and its paraneuronic nature. In: Chromaffin, Enterochromaffin and Related Cells, edited by R.E. Coupland and T. Fujita, pp. 191-208. Elsevier, Amsterdam.

13. Fujita, T. (1977): The concept of paraneurons. Arch. Histol. Jpn. (Suppl.), 40:1-12.

14. Hervonen, A. (1971): Development of catecholamine-storing cells in the human fetal paraganglia and adrenal medulla. Acta Physiol. Scand. [Suppl.], 368:1-94.

15. Hervonen, A., Vaalasti, A., Partanen, M., Kanerva, L., and Vaalasti, T. (1976): The paraganglia, a persisting endocrine system in man. Am. J. Anat., 146:207-210.

16. Hervonen, A., Partanen, S., Vaalasti, A., Partanen, M., Kanerva, L., and Alho, H. (1978): The distribution and endocrine nature of the abdominal paraganglia of adult man. Am. J. Anat., 153:563-572.

17. Heym, C., and Williams, T.H. (1979): Evidence for autonomic paraneurons in sympathetic ganglia of a shrew (Tupaia glis). J. Anat., 129:151-164.

18. Hollinshead, W.H. (1937): The innervation of abdominal chromaffin tissue. J. Comp. Neurol., 67:133-143.

19. Karpenko, V.P. (1964): The innervation of the chromaffin tissue of the sympathetic ganglia of the truncus sympatheticus. Tr. Volgogradskogo Med. Inst., 15:128-130.

20. Kobayashi, S. (1977): Adrenal medulla: Chromaffin cells as paraneurons. Arch. Histol. Jpn. (Suppl.), 40:61-79.

21. Kofman, U. (1935): Zur innervation des Paraganglion aorticum abdominale bei einigen Saugetieren. Z. Anat., 105:305-315.

22. Lu, K-S., Lever, J.D., Santer, R.M., and Presley, R. (1976): Small granulated cell types in rat superior cervical and coeliac-mesenteric ganglia. Cell Tissue Res., 172:331-343.

23. Mascorro, J.A., and Yates, R.D. (1971): Ultrastructural studies of the effects of reserpine on mouse abdominal sympathetic paraganglia. Anat. Rec., 170:269-280.

24. Mascorro, J.A., and Yates, R.D. (1972): Ultrastructural studies on paraganglia innervation. Arch. Mex. Anat., 39:27-28.

25. Mascorro, J.A., and Yates, R.D. (1974): Innervation of abdominal paraganglia: An ultrastructural study. J. Morphol., 142:153-164.

26. Mascorro, J.A., and Yates, R.D. (1977): The anatomical distribution and morphology of extraadrenal chromaffin tissue (abdominal paraganglia) in the dog. Tissue Cell, 9:447-460.

27. Mascorro, J.A., and Yates, R.D. (1979): Morphological similarities between intraganglion paraneurons and extraadrenal paraganglion cells in young New Zealand rabbits. J. Cell Biol., 83:139a.

28. Mascorro, J.A., Yates, R.D., and Chen, I-Li. (1976): A glutaraldehyde/potassium dichromate tracing method for the localization and preservation of abdominal extraadrenal chromaffin tissues. Stain Technol., 50:391-396.

29. Matthews, M.R., and Raisman, G. (1969): The ultrastructure and somatic efferent synapses of small granule-containing cells in the rat superior cervical ganglion. J. Anat., 105:255-282.

30. Siegrist, G., Dolivo, M., Dunant, Y., Foroglou-Kerameus, C., de Ribaupierre, Fr., and Rouiller, C.H. (1968): Ultrastructure and function of the chromaffin cells in the superior cervical ganglion of the rat. J. Ultrastruct. Res., 25:381-407.

31. Williams, T.H., and Palay, S.L. (1969): Ultrastructure of the small neurons in the superior cervical ganglion. Brain Res., 15:17-34.

32. Williams, T.H., Black, A.C., Chiba, T., and Bhalla, R.C. (1975): Morphology and biochemistry of small, intensely fluorescent cells of sympathetic ganglia. Nature, 256:315-317.

33. Zuckerkandl, E. (1901): Uber Nebenorgane des Sympaticus in Retroperitonealraum des Menschen. Anat. Anz., 15:97-107.

Histochemistry and Cell Biology of Autonomic Neurons, SIF Cells, and Paraneurons,
edited by O. Eränkö et al.
Raven Press, New York © 1980.

Microspectrofluorimetric Quantitation of Catecholamine Fluorescence in Rat SIF Cells and Paraganglia: Effect of Hypoxia

Hannu Alho, Teuvo Takala, and Antti Hervonen

Department of Biomedical Sciences, University of Tampere, SF-33101 Tampere 10, Finland

The paraganglia and the small intensely fluorescent (SIF) cells contain large amounts of catecholamines (4,6,8). The physiological role of SIF cells and paraganglia was widely discussed recently (3,6–8,15), but the question is still unresolved. Endocrine function for the paraganglia was suggested by several authors (4,8,11,12). On the other hand, the paraganglia of human fetus react clearly to strong hypoxia (9), and Brundin (2) showed that newborn rabbit paraganglia also release considerable amounts of catecholamines (CAs) under hypoxia. Hervonen and Partanen (10) did not find any changes in newborn rat paraganglia after severe hypoxia. However, paraganglia and SIF cells are particularly resistant to CA depletion by pharmacological agents (16).

Microspectrofluorimetric quantitation of biogenic monoamines gives linear results only up to certain amine concentration (13,14,17). However, Schipper (18) showed with a microspectrofluorimetric scanning procedure that the monoamine stores do not lose the linearity at low concentrations. The present study was undertaken to elucidate the functional role of paraganglia and SIF cells.

MATERIAL AND METHODS

Six litters of newborn rats (Sprague-Dawley), each containing 7 to 13 rats, were used. The animals were divided into three groups: (a) controls; (b) those exposed to 14 min of hypoxia; and (c) those exposed to 18 min of hypoxia. One newborn rat at a time was placed into a plexiglass chamber through which pure N_2 was flowing at a rate of 8 liters or more per minute. The animals were kept in these conditions 14 or 18 min, after which the animals were immediately sacrificed and the adrenal glands, abdominal paraganglia, and superior cervical ganglia removed. Adult rats were also used, and the exposure time under hypoxia was 2×2 min with 5-min intervals. The specimens were immersed quickly in liquid nitrogen, and a standardized formaldehyde-induced fluorescence (FIF) method was used (5). For the fluorescence intensity measurements or when the emission spectra only were to be recorded, the sections were placed on ordinary microscope slides and mounted under a coverslip with xylene.

Microfluorimetry

A modified Leitz MPV 2 microspectrofluorimeter was used. The excitation light source for the fluorescence intensity and emission measurements was the stabilized HBO 100 mercury lamp (Osram). The specimens were excited with the 405 nm HG peak obtained through the filter set D (containing BG 3 and KP 425 as excitation filters, a TK 455 chromatic beam splitter, and a K 460 barrier filter) of the Ploemopak II epi-illuminator. The second barrier filter was the Veril B 60 band interference filter. For the intensity measurements this filter was set at 485 nm (transmittance 50%, band with 20 nm). Each section was scanned by choosing the starting point at random. A total of 100 recordings were made from each section, and four longitudinal sections with 100 μm intervals from each specimen were measured. The values of these four sections were summarized and the results expressed as a mean histogram showing the fluorescence intensity groups on the x-axis and the percentage distribution of the values on the y-axis. The histogram describing the summarized fluorescence intensity values grouped as explained above is referred to here as the intensity profile. More details of microfluorimetric quantitation method used in this chapter can be found elsewhere (1).

RESULTS AND DISCUSSION

Observation by naked eye only did not reveal any difference between hypoxic and control specimens when comparing the fluorescence intensity with that of the abdominal paraganglia as seen in Figs. 1 through 3. Hervonen and Partanen (10) estimated the intensity visually, which might be the reason they found hypoxia to have no effect. By microspectrofluorimetry there was a statistically significant decrease of CA content in abdominal paraganglia of newborn rats after 18 min of hypoxia (X^2 test, $p < 0.05$). The mean intensity profile histograms of abdominal paraganglia and abdominal sympathetic ganglion cells of newborn rats are shown in Fig. 4. We did not observe any effect of hy-

poxia in SIF cells of adult rats. The symmetrical distribution curve (Fig. 4) from abdominal paraganglia might indicate that even the highest cellular concentrations can be registered. The concentration-dependent quenching has been demonstrated by several authors (13,14,17) in nonbiological models. It is possible that the biological monoamine stores behave in a different manner and do not lose the linearity so soon (18). The reliability of the intensity profiles of the most intense fluorescent cells is open to question, but the distribution of the lower intensities in the profile evidently describes the actual CA concentration within the cells.

Brundin (2) showed that newborn rabbit paraganglia release considerable amounts of norepinephrine in response to hypoxia. He suggested that the paraganglia are of importance in fetal hypoxia. The considerable decrease of the fluorescence intensity evidently indicates depletion of CA from the paraganglia of newborn rat. The mechanism for the release of CA from the fetal cells is obviously different from that in the adult. The discharge of CA from the newborn paraganglia is probably caused by direct stimulation of paraganglionic cells by the decreased oxygen concentration. No functional nerve endings can be demonstrated in the fetal paraganglia. In conclusion, the effect of hypoxia on fetal paraganglia is direct and is not mediated by the preganglionic innervation.

Electron microscopic studies on the changes of the fine structure of the CA-storing cells induced by the hypoxia are under preparation.

SUMMARY

Microfluorimetric studies were carried out on the FIF of CAs in SIF cells and abdominal paraganglia. Newborn and adult rats were placed in an atmosphere of pure nitrogen. The newborn rats survived 18 min (mean) in pure nitrogen, whereas the adults survived only 3 min (mean). The FIF inten-

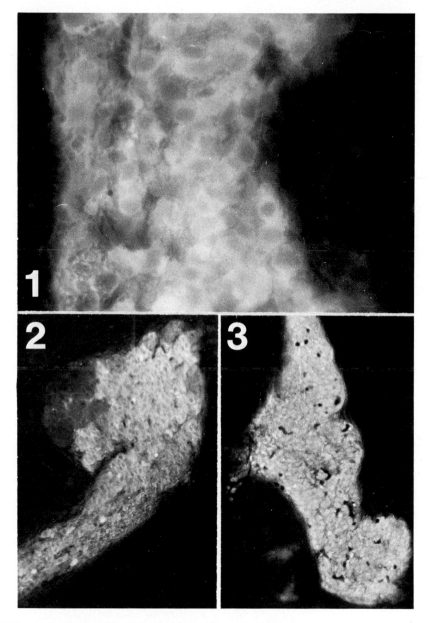

FIG. 1. Abdominal paraganglia of newborn rat. Control. ×450.
FIG. 2. Abdominal paraganglia of newborn rat. Survival time under hypoxia 16 min. ×180.
FIG. 3. Abdominal paraganglia of newborn rat. Survival time under hypoxia 18 min. ×180.

sity profiles of SIF cells of adult rats were similar to those of the controls. The FIF intensity profiles of abdominal paraganglia of newborn rats differs statistically from controls after 18 min of hypoxia. It is concluded that the paraganglia of newborn rats might release significant amounts of CAs under hypoxic conditions.

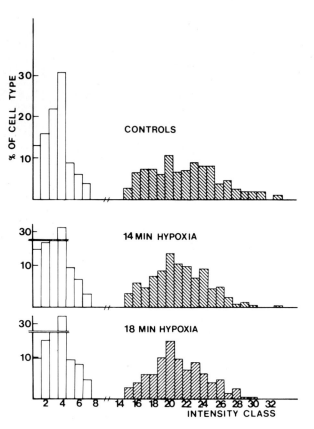

FIG. 4. Mean intensity profile histograms of abdominal paraganglia and abdominal ganglion cells, four scans per ganglion. On the ordinate is the frequency percentage of each fluorescence intensity class (abscissa). After 18 min of hypoxia there was a significant decrease in FIF intensity (x^2 test, $p < 0.05$); after 14 min of hypoxia there was no change (x^2 test, $p > 0.05$). Open bars, neurons of abdominal sympathetic ganglion; stippled bars, cells of abdominal paraganglion.

REFERENCES

1. Alho, H., and Hervonen, A. (1980): Microspectro-fluorimetric quantitation of catecholamine fluorescence in rat sympathetic ganglia. I. Estimation of the fluorescence profiles. *Histochem. J. (in press).*
2. Brundin, T. (1966): Studies on the preaortal paraganglia of newborn rabbits. *Acta Physiol. Scand.* [*Suppl.*], 70:290.
3. Burnstock, G., and Costa, M. (1975): *Adrenergic Neurons.* Chapman and Hall, London.
4. Coupland, R.E. (1965): *The Natural History of the Chromaffin Cell.* Longmans, London.
5. Eränkö, O. (1967): The practical histochemical demonstration of catecholamines by formaldehyde induced fluorescence. *J. R. Microsc. Soc.,* 87:259-276.
6. Eränkö, O., editor (1976): *SIF Cells. Structure and Function of the Small, Intensely Fluorescent Cells.* DHEW Publication No. (NIH) 76-942. Government Printing Office, Washington, D.C.

7. Eränkö, O., and Eränkö, L. (1971): Small intensely fluorescent granule-containing cells in the sympathetic ganglion of the rat. *Prog. Brain Res.,* 34:39-51.
8. Hervonen, A. (1971): Development of catecholamine storing cells in human fetal paraganglia and adrenal medulla. *Acta Physiol. Scand.* [*Suppl.*], 368:1-94.
9. Hervonen, A., and Korkala, O. (1972): The effect of hypoxia on the catecholamine content of human fetal abdominal paraganglia and adrenal medulla. *Acta Obstet. Gynecol. Scand.,* 51:17-24.
10. Hervonen, A., and Partanen, S. (1972): Histochemically demonstrable catecholamines of the newborn rat paraganglia after heavy hypoxia. *Experientia,* 28:1208-1209.
11. Hervonen, A., Partanen, S., Vaalasti, A., Partanen, M., Kanerva, L., and Alho, H. (1978): The distribution and endocrine nature of the abdominal paraganglia of adult man. *Am. J. Anat.,* 153:563-572.

12. Hervonen, A., Vaalasti, A., Partanen, M., Kanerva, L., and Vaalasti, T. (1976): The paraganglia, a persisting endocrine system in man. *Am. J. Anat.,* 146:207–210.

13. Jonsson, G. (1969): Microfluorimetric studies on the formaldehyde induced fluorescence of noradrenaline in adrenergic nerves of rat iris. *J. Histochem. Cytochem.,* 17:714–723.

14. Jonsson, G. (1971): Quantitation of fluorescence of biogenic monoamines demonstrated with the formaldehyde fluorescence method. *Prog. Histochem. Cytochem.,* 2:299–334.

15. Matthews, M.R., and Raisman, G. (1969): The ultrastructure and somatic efferent synapses of small,

granule-containing cells in the superior cervical ganglion. *J. Anat.,* 105:255–282.

16. van Orden, L.S., Burke, J., Geyer, M., and Lodoen, F. (1970): Differentiation of norepinephrine storage compartments in peripheral adrenergic nerves. *J. Pharmacol. Exp. Ther.,* 174:56.

17. Ritzén, M. (1967): Cytochemical identification and quantitation of biogenic monoamines—a microspectrofluorimetric and autoradiographic study. M.D. thesis, Stockholm.

18. Schipper, J. (1979): A scanning microfluorimetric study on noradrenergic neurotransmission. M.D. thesis, Amsterdam.

Histochemistry and Cell Biology of
Autonomic Neurons, SIF Cells, and
Paraneurons,
edited by O. Eränkö et al.
Raven Press, New York © 1980.

Comparative Studies on the Effects Elicited by Pre- and Postnatal Injections of Anti-NGF, Guanethidine, and 6-Hydroxydopamine in Chromaffin and Ganglion Cells of the Adrenal Medulla and Carotid Body in Infant Rats

Luigi Aloe and Rita Levi-Montalcini

Laboratory of Cell Biology, C.N.R., Rome, Italy

Previous studies have shown that injection of antibodies to the nerve growth factor (anti-NGF) (9), guanethidine (3,7), or 6-hydroxydopamine (6-OHDA) (2) in neonatal rodents produces precocious and irreversible lesions in the immature sympathetic nerve cells of para- and prevertebral ganglia, leading to destruction of 90 to 96% of noradrenergic neurons of these ganglia. No degenerative effects were apparent in the chromaffin cells of the adrenal gland and in the glomus cells of the carotid body of neonatal mice and rats submitted immediately after birth to injections of anti-NGF, guanethidine, or 6-OHDA.

We recently reported that injections of a specific antiserum to NGF in 16- to 17-day-old rat fetuses, resumed after birth and pursued for the first two postnatal weeks, result in a massive destruction of immature chromaffin cells in the adrenal gland (1). Discontinuation of the treatment is followed by progressive replacement of the damaged cells, probably by proliferation of residual immature chromaffin cells.

The object of this study is to compare the effects produced by anti-NGF, guanethidine, or 6-OHDA on chromaffin and ganglion cells of the adrenal medulla and on the glomus and ganglion cells of the carotid body of infant rats injected during pre- and postnatal life with one of these three sympathectomizing agents. The results reported below are based mainly on light and fluorescence microscopic studies of specimens dissected out from 1- to 2-week-old rats injected on the 16th and 19th days of fetal life and every other day after birth.

ADRENAL MEDULLA

Control

The adrenal medulla of an untreated 14-day-old rat contains 20,000 to 25,000 chromaffin cells and about 120 ganglionic cells whose function has not yet been elucidated. At this stage the two types of chromaffin cell, norepinephrine (NE) and epinephrine (E) storing cells, are well recognizable. The nature of these two cell types can be ascertained either with the formaldehyde-induced fluorescence (FIF) technique in view of the fact that E-storing chromaffin cells exhibit a yellow fluorescence whereas NE-storing cells emit a yellow-green light, or ultrastructurally because of the presence of electron-dense granules that are different in shape and dimension in their cell bodies (1,5,6). The presence of ganglion cells is easily detectable with common histological techniques, e.g., toluidine blue or hematoxylin and eosin.

Anti-NGF

The injection of NGF antibodies into rat fetuses results in the death of over 40% of chromaffin cells and widespread alterations of the majority of the surviving cells in 14-day-old pups (Table 1). Fluorescence microscopic studies showed that cell loss occurs more frequently in clusters (Fig. 1B) and that the cells which appear to be most affected by this treatment are the NE-storing cells. Preliminary ultrastructural studies revealed that the early stage of cell damage consists of disorganization of cytoplasmic organelles, followed by nuclear chromatin disaggregation and subsequent cell death. The ganglion cells, which in the control adrenal medulla range in number between 115 and 125, are reduced to an average of 26 in the anti-NGF injected rats. However, following prolonged treatment even these residual cells in light microscopic preparations show marked signs of hypotrophy or they disappear altogether (Fig. 2B).

Guanethidine

Guanethidine treatment produces structural alterations in the chromaffin cells of the adrenal medulla similar to those produced by anti-NGF injections, but the damage is somewhat less pronounced. Figure 1C shows fluorescence microscopy of the adrenal medulla of a guanethidine-treated rat. Also in this experimental group, as in the previous one, large areas devoid of chromaffin cells are apparent, but there is no pronounced reduction in the number of yellow-green fluorescent cells. The number of chromaffin and ganglion cells in the adrenal medulla of guanethidine-treated rats is markedly reduced (Fig. 2C and Table 1), and the remaining cells show signs of deterioration. Ultrastructural studies are in progress to determine if the toxic action produced by guanethidine injections on these cells is preceded by disruption of mitochondria, as was reported for the sympathetic nerve cells of newborn (3,7) and adult (8) injected rats.

6-Hydroxydopamine

As is shown in Table 1 and Figs. 1D and 2D, the number and structure of the adrenal chromaffin cells after 6-OHDA treatment are not apparently altered. The formaldehyde-induced catecholamine fluorescence of the chromaffin cells was occasionally less intense than that of controls, whereas in the toulidine blue preparations it was similar to that

TABLE 1.   *Number of cells in the adrenal medulla of 14-day-old rats treated pre- and postnatally[a]*

Treatment	Chromaffin cells[b] (No.)	Ganglion cells	
		No.	Ave. diameter
Saline	23,900 ± 1,200	120 ± 14	27
AS-NGF	13,550 ± 940	26 ± 5	13
Guanethidine	15,400 ± 1,280	32 ± 7	13
6-Hydroxydopamine	22,650 ± 1,660	43 ± 8	14

[a] Animals were injected on the 16th and 19th days of fetal life and every other day after birth.
[b] Standard error of the mean; at least four cases each.

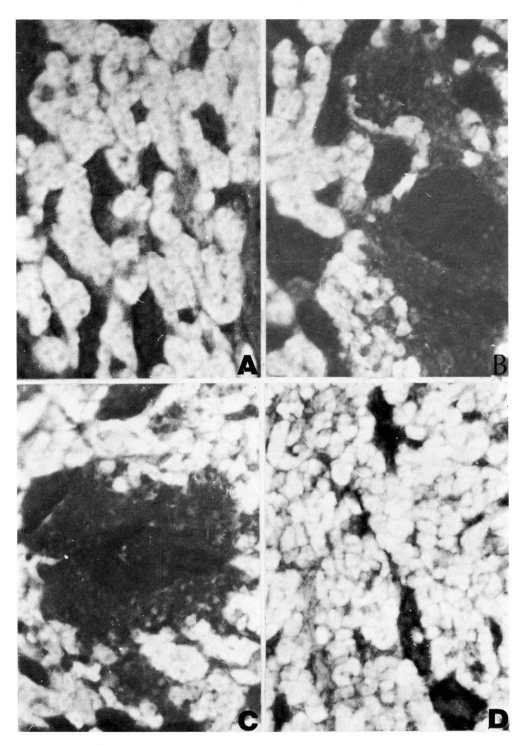

FIG. 1. Formaldehyde-induced catecholamine fluorescence of the adrenal medulla of a 10-day-old rat injected pre- and postnatally with vehicle solution (**A**), anti-NGF (**B**), guanethidine (**C**), and 6-OHDA (**D**). Large areas devoid of chromaffin cells are noticeable in **B** and in **C**, but not in **D**. ×288.

H
N

Pe

De

par
the
upt
Col
pho
witl
it se
ules
sitic
chro
kno
hav
ing
drol
that
sine
regu
caro
for
(15).
of pe
as a
ules

Th
izatic
cleot
of tw
micr
(10)
quina

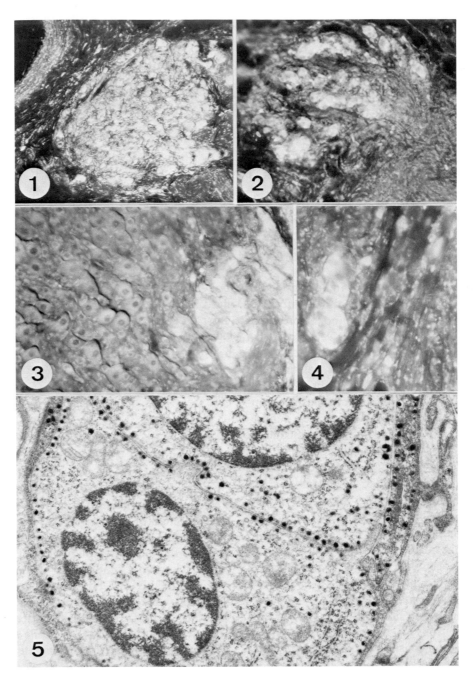

FIG. 1. Quinacrine fluorescence of rat carotid body chief cells 3 days after a single injection of quinacrine (200 mg/kg i.p.). ×140.

FIG. 2. Quinacrine fluorescence of guinea pig carotid body chief cells (conditions as in Fig. 1). Small groups of strongly fluorescent cells are embedded into connective tissue that gives a faint background fluorescence. ×220.

FIG. 3. Quinacrine fluorescence of mouse carotid body and superior cervical ganglion 1 day after a single injection of quinacrine (120 mg/kg i.p.). Background fluorescence is more prominent under these conditions. The bright fluorescence of carotid body parenchyma by far exceeds the also specific labeling of sympathetic perikarya. ×140.

up less quinacrine than do type I cells of guinea pigs or rats. Quinacrine fluorescence seems to outline the perikarya of type I cells or to be concentrated toward the surface of the cell body. In rats, as in guinea pigs, the entire cytoplasm of chief cells is brightly fluorescent. Granular quinacrine fluorescence was never observed.

Perikarya in the superior cervical ganglion are also seen to fluoresce after quinacrine administration. Nuclei and nucleoli of sympathetic neurons remain unlabeled under the conditions studied. Perikarya show faint, homogeneous fluorescence, with numerous tiny but brightly fluorescent grains. The dimensions of these grains allow them to be interpreted as lysosomes.

Single or grouped SIF cells (Fig. 4) are also strongly labeled with quinacrine. Their fluorescence does not differ from that of carotid body chief cells. As a topographical peculiarity in mice, the carotid body is firmly attached to the superior cervical ganglion (Fig. 3), and groups of chief cells are often seen to be intermingled with nerve fascicles coming from or running to the ganglion. These cells may be interpreted as transitional forms between SIF cells and carotid body tissue. SIF cells display homogeneous cytoplasmic fluorescence.

Mast cells in all three species under investigation are most intensively labeled with quinacrine. Only the specific granules are fluorescent.

The uranaffin reaction significantly enhances the electron density of chromatin, ribosomes, and specific storage granules in carotid body chief cells. This is convincingly demonstrated in unstained sections (Fig. 6). Staining the sections with lead citrate enables one to discern cell membranes and membrane-bound cellular compartments (Fig. 5).

Tissue preservation is fairly good. Matrix material in specific granules is separated from the limiting membrane by a clear halo. The membranes and the cores display enhanced electron opacity. No substructure can be discerned in the cores. Specific granules in mouse carotid body chief cells are preferentially located near the cell membrane (Fig. 5). Only a few granules are distributed throughout the cytoplasm.

DISCUSSION

Quinacrine has been shown to bind to adrenomedullary chromaffin granules and to peptidergic nerves in sufficient concentrations to be visualized in the fluorescence microscope (10). This is also the case for mouse, rat, and guinea pig carotid body chief cells. Quinacrine fluorescence in adrenomedullary cells and purinergic nerves is attributed to high levels of adenine nucleotides within these structures. The same is suggested for carotid body type I cells. Fixation with uranium ions has been proposed as a cytochemical method to visualize adenine nucleotides (12). A positive uranaffin reaction of the specific granules is observed in carotid body chief cells. This lends support to the assumption that adenine nucleotides are constituents of the matrix of these organelles. In the case of mouse carotid body chief cells, specific granules are known to be located preferentially near the plasmalemma (5), and quinacrine fluorescence is restricted to the periphery of the perikaryon. This may be taken as evidence that quinacrine fluorescence marks specific granules of chief cells, although no granularity of quinacrine fluorescence can be discerned. Obviously the granules are too small (in relation to the section thickness) to be identified in the fluorescence microscope.

◀
FIG. 4. Quinacrine fluorescence of rat SIF cells (conditions as in Fig. 3). SIF cells display bright fluorescence, showing no differences to carotid body chief cells. ×350.
FIG. 5. Mouse carotid body chief cells, uranaffin reaction, thin section stained with lead citrate. Fine structural details can be easily discerned. Nuclear chromatin, specific granules, and ribosomes appear highly electron-dense. ×9,000.

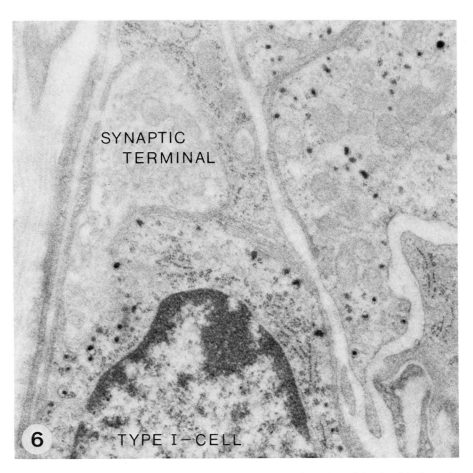

FIG. 6. Mouse carotid body tissue, uranaffin reaction, unstained section. Selective binding of uranium ions to specific granules, ribosomes, and nuclear chromatin is evident. Mitochondria and synaptic vesicles within axon terminals remain unstained. ×15,000.

The observation of quinacrine labeling of carotid body type I cells, as the uranaffin reaction of their storage granules, points to the close relationship between adrenomedullary chromaffin cells and paraganglionic cells. Additionally, it opens new aspects about the function of carotid body chief cells. There is no doubt that released (=extracellular) adenine nucleotides are rapidly degraded to adenosine by ektoenzymes (14). Adenosine is known to influence vascular smooth muscle cells, causing vasodilation in muscle, brain, intestine, and adipose tissue (2,9) but vasoconstriction in the kidney (13). It has been suggested to be involved in the local control of blood flow in response to the oxygen needs of the respective tissues. Adenosine may be directly released (muscle tissue) (1), or adenine nucleotides may be secreted and rapidly broken down to adenosine by ektoenzymes (proposed for the carotid body). Adenosine has been found to inhibit the release of norepinephrine from electrically stimulated blood vessels, as well as from nervously stimulated vas deferens, kidney, and adipose tissue (9). It can be assumed that adenosine evinces the same effect on adrenergic terminals in the carotid body. Besides local flow regulation and inhibition of adrenergic terminals, adenosine may also influence the secretory activity of chief cells and chemore-ceptive processes.

ACKNOWLEDGMENT

This study was supported by the "Deut-sche Forschungsgemeinschaft," grant Bo 525-3.

REFERENCES

1. Baer, H.P., and Drummond, G.I. (1968): Catabolism of adenine nucleotides by the isolated perfused rat heart. *Proc. Soc. Exp. Biol. Med.,* 127:33-36.
2. Berne, R.M. (1963): Cardiac nucleotides in hypoxia: Possible role in regulation of coronary blood flow. *Am. J. Physiol.,* 204:317-322.
3. Berne, R.M., Rubio, R., and Curnish, R.R. (1974): Release of adenosine from ischaemic brain: Effect on cerebral vascular resistance and incorporation into cerebral adenine nucleotides. *Circ. Res.,* 35:262-271.
4. Böck, P. (1980): Noradrenaline and acidic protein(s) in four types of cat carotid body chief cells. *Arch. Histol. Jpn. (in press).*
5. Böck, P., and Gorgas, K. (1976): Catecholamines and granule content of carotid body type I-cells. In: *Chromaffin, Enterochromaffin and Related Cells,* edited by R.E. Coupland and T. Fujita, pp. 355-374. Elsevier, Amsterdam.
6. Capella, C., and Solcia, E. (1971): Optical and electron microscopical study of cytoplasmic granules in human carotid body, carotid body tumours and glomus jugulare tumours. *Virchows Arch. [Zellpathol],* 7:37-53.
7. Fujita, T. (1976): The gastro-enteric endocrine cell and its paraneuronic nature. In: *Chromaffin, Enterochromaffin and Related Cells,* edited by R.E. Coupland and T. Fujita, pp. 191-208. Elsevier, Amsterdam.
8. Fujita, T. (1977): Concept of paraneurons. *Arch. Histol. Jpn. (Suppl.),* 40:1-12.
9. Hedqvist, P., and Fredholm, B.B. (1976): Effects of adenosine on adrenergic neurotransmission; prejunctional inhibition and postjunctional enhancement. *Naunyn Schmiedebergs Arch. Exp. Path. Pharmakol.,* 293:217-223.
10. Olson, L., Alund, M., and Norberg, K-A. (1976): Fluorescence-microscopical demonstration of a population of gastro-intestinal nerve fibres with a selective affinity for quinacrine. *Cell Tissue Res.,* 171:407-423.
11. Pearse, A.G.E. (1969): The cytochemistry and ultrastructure of polypeptide hormone-producing cells of the APUD series and the embryologic, physiologic and pathologic implications of the concept. *J. Histochem. Cytochem.,* 17:303-313.
12. Richards, J.G., and DaPrada, M. (1977): Uranaffin reaction: A new cytochemical technique for the localization of adenine nucleotides in organelles storing biogenic amines. *J. Histochem. Cytochem.,* 25:1322-1336.
13. Scott, J.B., Daugherty, R.M., Dabney, J.M., and Haddy, F.J. (1965): Role of chemical factors in regulation of flow through kidney, hindlimb and heart. *Am. J. Physiol.,* 208:813-824.
14. Trams, E.G., and Lauter, C.J. (1974): On the sidedness of plasma membrane enzymes. *Biochem. Biophys. Acta,* 345:180-197.
15. Winkler, H. (1976): The composition of adrenal chromaffin granules: An assessment of controversial results. *Neuroscience,* 1:65-80.

*Histochemistry and Cell Biology of
Autonomic Neurons, SIF Cells, and
Paraneurons,*
edited by O. Eränkö et al.
Raven Press, New York © 1980.

Some Observations on the Glomus (Aorticopulmonary) Bodies of the Developing Rabbit Heart

Raymond E. Papka

Department of Anatomy, University of Kentucky, School of Medicine, Lexington, Kentucky 40536

Recent investigations of the autonomic nervous elements associated with the developing rabbit heart (21–24) revealed the presence of numerous small, vascularized, glomus-like clusters of cells at the base of the heart. Some of these glomera are located between the aorta and pulmonary artery and thus correspond to the aorticopulmonary glomera. The principal cells in these clusters have intense catecholamine fluorescence, contain osmiophilic granules of the monoamine-storing type, are innervated, and probably have a receptor–secretory function (21). Such features are shared by other cells, e.g., small intensely fluorescent (SIF) cells, carotid body chief cells, and adrenal medullary cells. These features would qualify these cells as paraneurons according to the concept of Fujita (9).

The aorticopulmonary glomera have many structural features in common with the carotid body and are thought also to play a role in chemoreflexes (3,5,13). In continuing studies on the autonomic nervous elements associated with the developing heart, it is essential that these glomera be examined to assess their potential role in chemoreflexive and cardiovascular functions. In this study, aorticopulmonary glomera were studied by light, electron, and fluorescence microscopy in New Zealand white rabbits ranging in age from 20 days' gestation (31 days is term) to 35 days' postnatal.

GENERAL MORPHOLOGY AND CATECHOLAMINE HISTOCHEMISTRY

At 20 days' gestation, groups of intensely fluorescent cells are present in the connective tissue between the aorta and pulmonary artery (21) (Fig. 1a). With light microscopy various-sized groups of small dark-staining cells are evident in the same location (Fig. 1b). These cellular aggregates are often associated with nerve fibers, and often one or more capillaries are present within the cell group or nearby (Fig. 1b). By day 24 of gestation (and more so in older fetuses), several capillaries are evident within the cellular aggregates and many of the cells have a close association with the vasculature (21) (Fig. 1c). The catecholamine fluorescence of these cells does not appear to change in intensity, color, or distribution with increasing animal age. The color of the fluorescence is yellow-green which generally indicates the presence of a primary catecholamine. The nature of

glomus cells are grouped near the presynaptic membrane density of the glomus cell. The postsynaptic element is either a nerve ending or a small branch of a nerve fiber. These synapses are similar to those described for the carotid body (3,15–17,26). This particular type of interaction is seen first in 26-day fetuses but is more often encountered in postnatal animals. Finally, structures which resemble reciprocal synapses as seen in the carotid body (2,15–17,26) are rarely encountered in the developing aorticopulmonary glomera (Fig. 3g). It is possible—because of sampling limitations with electron microscopy, the small size of the synaptic junctions, and extensive surface area of plasma membranes—that many reciprocal synapses are not seen. Also many of the single synapses may be one part of reciprocal synapses. In the rat carotid body (15), 20% of nerve endings are presynaptic to glomus cells, 35% are postsynaptic to glomus cells, 5% are reciprocal synapses, and 40% do not show morphological evidence of synaptic contacts.

From the information available about the aorticopulmonary glomera, it is not possible to say whether most of the nerve endings are strictly presynaptic to glomus cells, strictly postsynaptic to glomus cells, or form reciprocal synapses. Consequently, it is difficult to know if these collections of cells are secretory paraneurons, chemoreceptors, or a mixture of both. However, there are many morphological similarities in the carotid body and aorticopulmonary glomera. The carotid body is generally accepted to be a chemoreceptor; likewise, the aorticopulmonary glomera also may be chemoreceptors. In line with this problematical function, several investigators (3,5,13) have shown in the dog that there are small bodies located between the aorta and pulmonary artery which receive systemic blood from a branch of the left coronary artery and are associated with the vagus nerve. Stimulation of these bodies induces a chemoreflex that results in hypertension (5,13).

ACKNOWLEDGMENT

Supported in part by DHEW grant 1R01 HL 22226-01 HED from the National Heart, Lung and Blood Institute.

REFERENCES

1. Ballard, K., and Jones, J. (1972): Demonstration of choline acetyltransferase activity in the carotid body of the cat. *J. Physiol. (Lond)*, 227:87–94.
2. Butler, P., and Osborne, M. (1975): The effect of cervical vagotomy (decentralization) on the ultrastructure of the carotid body of the duck, Anas platyrhynchos. *Cell Tissue Res.*, 163:491–502.
3. Coleridge, H., Coleridge, J., and Howe, A. (1970): Thoracic chemoreceptors in the dog: A histological and electrophysiological study of the location, innervation and blood supply of the aortic bodies. *Circ. Res.*, 26:235–247.
4. Dearnaley, D., Fillenz, M., and Woods, R. (1968): The identification of dopamine in the rabbits carotid body. *Proc. R. Soc. Lond. [Biol]*, 170:195–203.
5. Eckstein, R., Shintani, F., Rowen, H., Shimomura, K., and Ohya, N. (1971): Identification of left coronary blood supply of aortic bodies in anesthetized dogs. *J. Appl. Physiol.*, 30:488–492.
6. Eyzaguirre, C., and Zapata, P. (1968): The release of acetylcholine from carotid body tissues: Further study of the effects of acetylcholine and cholinergic blocking agents on the chemosensory discharge. *J. Physiol. (Lond)*, 195:589–607.
7. Eyzaguirre, C., Koyano, H., and Taylor, J. (1965): Presence of acetylcholine and transmitter release from carotid body chemoreceptors. *J. Physiol. (Lond)*, 178:463–476.
8. Fidone, S., Weintraub, S., Stavinoha, W., Stirling, C., and Jones, L. (1977): Endogenous acetylcholine levels in cat carotid body and the autoradiographic localization of a high affinity component of choline uptake. In: *Chemoreception in the Carotid Body*, edited by H. Acker, S. Fidone, D. Pallot, C. Eyzaguirre, D. Lubbers, and R. Torrance, pp. 106–113. Springer, Berlin.
9. Fujita, T. (1976): The gastro-enteric endocrine cell and its paraneuronic nature. In: *Chromaffin, Enterochromaffin and Related Cells*, edited by R. Coupland and T. Fujita, pp. 191–208. Elsevier, Amsterdam.
10. Furshpan, E., MacLeish, P., O'Lague, P., and Potter, D. (1976): Chemical transmission between rat sympathetic neurons and cardiac myocytes developing in microcultures: Evidence for cholinergic, adrenergic, and dual-function neurons. *Proc. Natl. Acad. Sci. USA*, 73:4225–4229.
11. Greene, L., and Rein, G. (1977): Synthesis, storage and release of acetylcholine by a noradrenergic pheochromocytoma cell line. *Nature*, 268:349–351.
12. Jacobowitz, D., and Koelle, G. (1965): Histochemical correlations of acetylcholinesterase and cate-

cholamines in postganglionic autonomic nerves of the cat, rabbit and guinea pig. *J. Pharmacol. Exp. Ther.,* 148:225–237.

13. James, T., Isobe, J., and Urthaler, F. (1975): Analysis of components in a cardiogenic hypertensive chemoreflex. *Circulation,* 52:179–192.

14. Korkala, O., and Waris, T. (1977): The acetylcholinesterase reaction and catecholamine fluorescence in the glomus cells of rat carotid body. *Experientia,* 33:1363–1364.

15. McDonald, D. (1976): Structure and function of reciprocal synapses interconnecting glomus cells and sensory nerve terminals in the rat carotid body. In: *Chromaffin, Enterochromaffin and Related Cells,* edited by R. Coupland and T. Fujita, pp. 375–394. Elsevier, Amsterdam.

16. McDonald, D., and Mitchell, R. (1975): The innervation of glomus cells, ganglion cells and blood vessels in the rat carotid body: A quantitative ultrastructural analysis. *J. Neurocytol.,* 4:177–230.

17. Morgan, M., Pack, R., and Howe, A. (1975): Nerve endings in rat carotid body. *Cell Tissue Res.,* 157:225–272.

18. Nachmansohn, D. (1959): *Chemical and Molecular Basis of Nerve Activity.* Academic Press, New York.

19. Nishi, K., and Stensaas, L. (1974): The ultrastructure and source of nerve endings in the carotid body. *Cell Tissue Res.,* 154:303–319.

20. Palkama, A. (1967): Demonstration of adrenomedullary catecholamines and cholinesterases at electron microscopic level in the same tissue section. *Ann. Med. Exp. Fenn.,* 45:295–306.

21. Papka, R. (1974): A study of catecholamine-containing cells in the hearts of fetal and postnatal rabbits by fluorescence and electron microscopy. *Cell Tissue Res.,* 154:471–484.

22. Papka, R. (1975): Localization of acetylcholinesterase in granule-containing cells of glomus-like bodies in pre- and postnatal rabbits by electron microscopy. *Cell Tissue Res.,* 162:185–194.

23. Papka, R. (1976): Studies of cardiac ganglia in pre- and postnatal rabbits. *Cell Tissue Res.,* 175:17–35.

24. Papka, R. (1978): Development of innervation to the atrial myocardium of the rabbit. *Cell Tissue Res.,* 194:219–236.

25. Tranzer, J., and Richards, J. (1976): Ultrastructural cytochemistry of biogenic amines in nervous tissue; Methodologic improvements. *J. Histochem. Cytochem.,* 24:1178–1193.

26. Verna, A. (1973): Terminaisons nerveuses afferentes et efferentes dans le glomus carotidien du lapin. *J. Microsc. (Paris),* 16:299–308.

Granular Vesicles and Experimental Cell Biology of Sympathetic Neurons and Paraneurons

Histochemistry and Cell Biology of Autonomic Neurons, SIF Cells, and Paraneurons,
edited by O. Eränkö et al.
Raven Press, New York © 1980.

Chromaffin Granules: A New Look at an Old Discovery

H. Blaschko

University Department of Pharmacology, Oxford OX 1 3 QT, England

The contribution of the Oxford laboratory to the study of the intracellular localization of the catecholamines did not arise as the result of a planned investigation but as the outcome of an enquiry that had been designed to answer an entirely different question. We had begun to speculate on the question of the biosynthesis of epinephrine many years earlier. This had started with the proposal that the enzyme L-DOPA decarboxylase, discovered by Holtz (10) in the guinea pig kidney, was one of the catalysts on the metabolic pathway that led to epinephrine (2). The first experimental support for this suggestion came in 1951 when my colleague Langemann discovered large amounts of this enzyme in the bovine adrenal medulla.

This discovery did not lead at once to an acceptance of the proposed biosynthetic pathway. At the Paris International Biochemical Congress, von Euler (7) expressed doubts about the postulated role of dopamine as a precursor of epinephrine. However, at Oxford it seemed to us worthwhile to follow the idea further and to find out if we could obtain evidence of the conversion of dopamine to norepinephrine. So when in 1952 Dr. A. D. Welch came to Oxford, we decided to tackle this question by experiment. This work was never completed because at the beginning we made an observa-tion that seemed to open up a line of study that was more exciting than the search for an enzyme we believed would surely be found sooner or later.

For our experiments we used homogenates of bovine adrenal medulla, the material already used by Langemann; we wanted to add dopamine to this homogenate to find out if norepinephrine had been formed. If the methods then available at Oxford had been less primitive, our subsequent observations would never have been made. However, radioactively labeled precursors were then not at our disposal, and so we had to devise a method by which we could rapidly eliminate the enormous amounts of norepinephrine and epinephrine initially present in the homogenates.

Our method worked very well when we tested it in model experiments, using aqueous solutions of epinephrine in lieu of the homogenates. We suspended a dialysis bag containing such a solution in a large beaker filled with a well-aerated suspension of an acetone-dried powder of mushrooms that had been thoroughly washed to remove all water-soluble constituents. Our expectations were fulfilled: The phenolase present in the powder rapidly destroyed the epinephrine. The dark red color of adrenochrome began to appear immediately in the beaker, and after a short time bioassay showed that all the pressor activity in the dialysis bag had disappeared, indicating that all the amine had been destroyed.

We thereupon did an experiment in which the aqueous epinephrine solution was replaced by an adrenal medullary homogenate. Everything went well at the start: The red adrenochrome color appeared rapidly in the beaker, and there was an initial loss of pressor activity in the dialysis bag. However, things then went differently; after about one-third of the pressor activity had been lost, the remaining activity was retained and disappeared only very slowly, over the course of hours.

What could be the reason for this behavior? We suspected that in the homogenate the pressor amine was not freely dissolved, but that it was being held back by some diffusion barrier within the dialysis bag.

To obtain some information about the localization of the amine in our homogenates, we decided on a more careful analysis. We prepared bovine adrenal medullary homogenates in isotonic sucrose and subjected them to high-speed centrifugation at low temperature. At the end of the centrifugation we recovered between one-fourth and one-third of the pressor activity in the supernatant; the sediment had retained the major part of the activity. When we resuspended the sediment in isotonic sucrose and centrifuged it again, the pressor activity was retained in the sediment (4).

This is how we set out to discover dopamine-β-hydroxylase and found instead the chromaffin granules. Welch saw the need for radioactive precursors, and in his laboratory at Yale we subsequently demonstrated the formation of norepinephrine first from L-DOPA and then from dopamine (5,9). The enzyme dopamine-β-hydroxylase was isolated soon afterward (12), and has since been studied by a great multitude of observers.

The isolation and characterization of the chromaffin granules, carried out simultaneously at Oxford and by the late Hillarp at Lund, was soon followed by the recognition

that there were many analogies between the storage of the catecholamines in the chromaffin cells and in the adrenergic neurons. In fact, the cytological studies followed a course parallel to the research on biosynthesis: Langemann's work was followed by the finding of L-DOPA decarboxylase in adrenergic neurons (13). Similarly, the study of the chromaffin granules led to that of the amine-containing vesicles of adrenergic neurons, first by von Euler and Hillarp (8) and subsequently by many others.

Between the amine-carrying organelles of the chromaffin cells and those of the adrenergic neurons, there exist many similarities as well as some differences. Among the similarities is the presence in both of one of the enzymes of the biosynthetic pathway. This enzyme is dopamine-β-hydroxylase. In the chromaffin granules this enzyme is known to occur in the granule membrane and the soluble contents. In the two locations the enzyme is catalytically active and immunochemically indistinguishable, but there is a recent report (1) that the enzyme in the membrane carries an amphiphilic tail which is absent in the soluble enzyme; the amphiphilic form can be converted into the soluble form by limited proteolysis. These findings are of interest also in regard to what is known of the enzyme in the large dense-cored nerve granules: According to Kirksey and co-workers (11), much of the enzyme is not membrane-bound but is present in a rather insoluble complex in the matrix. It is presumably that part of the enzyme that is available for exocytotic release. It will be interesting to determine if this enzyme also possesses an amphiphilic moiety in the vesicle and if the latter is lost upon release into the extracellular space.

The differences between the two types are obvious. Some differences in chemical composition are in fact determined by the difference in size, which affects the ratio of the membranal material to the matrix. The differences I am concerned with here are related to the functional differences between endocrine and neuronal organelles. The first of these differences is determined by a cytological feature, i.e., the distance between the site of protein synthesis and the site of release of the chemical messenger. In the neuron these two sites may be far apart, whereas in the chromaffin cell they are rather close to each other. Although in the neuron much of the amine synthesis appears to occur close to the endings, it seems likely that it is this fact, the intraneuronal distance, that is related to the fact that in the neuron the phenomenon of uptake is well developed.

The second difference is in the size of the amine-carrying organelles, which may be related to different functional requirements. The adrenal medullary hormones exert their function far away from the point of release; an extreme dilution takes place when they are discharged into the bloodstream. A large secretory organelle may ensure the liberation of a relatively large quantity of messenger at one and the same time. On the other hand, the situation at a conventional nerve ending is different: Here the point of release is close to the receptors in the effector tissue. The amount of amine required to elicit a response is small and may be contained in a smaller organelle.

It is in this context that the small, intensely fluorescent (SIF) cells of sympathetic ganglia are of particular interest (6). We are not interested here whether these cells are neuronal or endocrine; this is a distinction that has lost much interest in view of recent observations by cell biologists. Also, there is the possibility that some variety exists in this group of cells (3). There are three features that seem to be of relevance in regard to what has been said before: (a) These cells (and their processes) have organelles that resemble the chromaffin granules of the adrenal medulla in size and appearance. (b) They have processes that are brief, with their endings located in the same ganglia as the cells themselves. (c) Some of these cells have endings close to blood vessels within the ganglia, suggesting that the amine released may act at some distance from the site of release.

The account given here is necessarily condensed and incomplete. For instance, no mention has been made of the difference between large and small dense-cored vesicles of adrenergic nerves. Also, the chromaffin cells present in the carotid body and related structures have not been discussed. It will be interesting to see if these can be fitted into the picture that was tentatively developed here.

REFERENCES

1. Bjerrum, O.J., Helle, K.B., and Bock, E. (1979): Immunochemically identical hydrophilic and amphiphilic forms of the bovine adrenal medullary dopamine β-hydroxylase. *Biochem. J.,* 181:231–237.
2. Blaschko, H. (1939): The specific action of L-DOPA decarboxylase. *J. Physiol. (Lond),* 96:50P.
3. Blaschko, H. (1977): Biochemical aspects of transmitter formation and storage: comments on relationship. In: *The Synapse,* edited by G.A. Cottrell and P.N.R. Usherwood, pp. 102–116. Blackie, Glasgow.
4. Blaschko, H., and Welch, A.D. (1953): Localization of adrenaline in cytoplasmic particles of the bovine adrenal medulla. *Arch. Exp. Pathol. Pharmakol.,* 191:17–22.
5. Demis, D.J., Blaschko, H., and Welch, A.D. (1955): The conversation of dihydroxyphenylalanine-2-C14 (DOPA) to norepinephrine by bovine adrenal medullary homogenates. *J. Pharmacol. Exp. Ther.,* 113:14–15.
6. Eränkö, O., and Eränkö, L. (1971): Small, intensely fluorescent, granule-containing cells in the sympathetic ganglion of the rat. *Prog. Brain Res.,* 34:39–51.
7. Euler, U.S. von (1952): Noradrenaline. In: *Symposium sur les Hormones Protéiques et Derivées des Protéines,* pp. 39–55. No. 4 IIIème Congrès International de Biochimie. Sedes, Paris.
8. Euler, U.S. von, and Hillarp, N.Å. (1956): Evidence for the presence of noradrenaline in submicroscopic structures of adrenergic axons. *Nature,* 177:44–45.
9. Hagen, P. (1956): Biosynthesis of norepinephrine from 3,4-dihydroxyphenylethylamine (dopamine). *J. Pharmacol. Exp. Ther.,* 116:26.
10. Holtz, P. (1938): Fermentativer Abbau von L-Dioxyphenylalanin (Dopa) durch Niere. *Arch. Exp. Path. Pharmakol.,* 191:87–118.
11. Kirksey, D.F., Klein, R.L., Baggett, J.McC., and Gasparis, M.S. (1978): Evidence that most of the dopamine β-hydroxylase is not membrane-bound in purified large dense cored noradrenergic vesicles. *Neuroscience,* 3:71–81.
12. Kirshner, N. (1959): Biosynthesis of adrenaline and noradrenaline. *Pharmacol. Rev.,* 11:350–360.
13. Schümann, H.J. (1960): Formation of adrenergic transmitters. In: *Ciba Foundation Symposium on Adrenergic Mechanisms,* edited by J.R. Vane, G.E.W. Wolstenholme, and M. O'Connor, pp. 6–16. Churchill, London.

Histochemistry and Cell Biology of Autonomic Neurons, SIF Cells, and Paraneurons,
edited by O. Eränkö et al.
Raven Press, New York © 1980.

Compartmentation of Synaptic Vesicles in Autonomic Neurons: Morphological Correlates of Neurotransmitter Pools in Monoaminergic Synapses; the Vesicles as Dual Quantal Elements in Transmitter Storage and Release

Amanda Pellegrino de Iraldi

Instituto de Biología Celular, Facultad de Medicina, Universidad Nacional de Buenos Aires, Buenos Aires, Argentina

The idea that neurotransmitters could be stored in subcellular organelles emerged with the discovery of synaptic vesicles in electron microscopic studies (7) and their subsequent isolation by cell fractionation procedures (9). The finding by von Euler and Hillarp (11) that norepinephrine is present in a particulated fraction of bovine splenic nerves stimulated electron microscopists to find these subcellular particles. In nerve endings and varicosities from the peripheral autonomic nervous system (3,8) and in the central nervous system (16,30), two types of vesicles were identified: a predominant population of clear, granular vesicles of about 40 to 60 nm, and a minor population of dense-cored vesicles of about 80 to 90 nm. Both types of vesicles were also found in the perikaryon (10) and the large ones in the nerve trunks (14). Histochemical, pharmacological, and biochemical similarities and differences between the two types of vesicles have been reported.

From the first description (8) the granular vesicles were characterized by an electron-dense core and a clear space between the core and the surrounding membrane. It was soon demonstrated by histochemical and pharmacological methods, at the electron microscopic level, that the dense granules contain biogenic amines (2). The clear space was little or not at all considered. According to some authors, it could be an artifact produced by the fixation techniques.

We noted that the clear space is an important morphological component of the granular vesicles, that current fixations with osmium tetroxide stain only the core (Fig. 1A), that dichromate reacts only with the core in the histochemical technique of Wood (38) (Fig. 1B), and that the mixture of zinc iodide and osmium tetroxide (ZIO) stains the core and the clear space differentially (26) (Fig. 1C). With these facts in mind, we advanced the hypothesis (27) that two main compartments could be distinguished in the granular vesicles of the autonomic nerves: (a) a central compartment, the core, a storage site for biogenic amines; and (b) an outer compartment, the clear space, which we called the matrix. This hypothesis was supported by pharmacological studies combined with histochemistry, electron microscopy, and biochemistry, showing that tyramine affects each compartment differently (28).

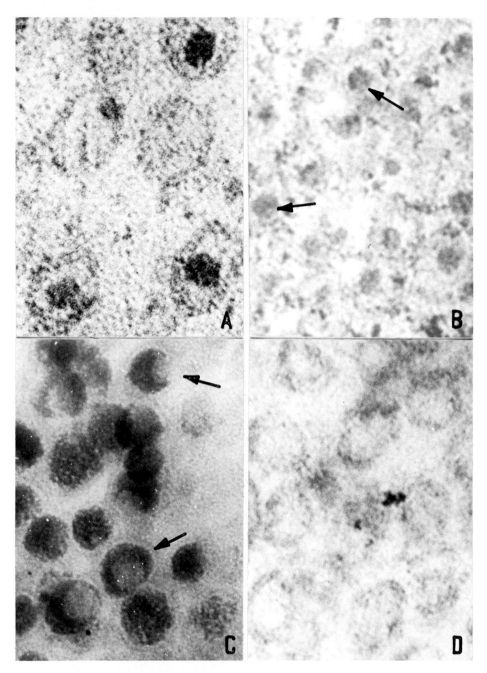

FIG. 1. Small synaptic vesicles from rat pineal nerves. **A**: osmium tetroxide fixation. **B**: Wood histochemical technique for catecholamines and indoleamines; staining on the grids with uranyl acetate and lead citrate. *Arrows* point to reactive cores. **C**: ZIO staining of the matrix; *arrows* point to unreactive cores. **D**: ZIO reaction after treatment with N-ethylmaleimide, which blocks -SH groups. ×360,000.

In this chapter I will consider the characteristics and significance of both compartments as revealed by morphological and histochemical methods, without treatment or after administration of drugs or electrical stimulation.

ANATOMY OF GRANULAR VESICLES

In small vesicles (Figs. 1 and 2) after osmium tetroxide fixation, we observed that the size of both compartments is variable and inversely related; i.e., the smaller the core, the larger is the outer compartment and vice versa (29). The core may even disappear, giving place to a small clear vesicle exclusively formed by the outer compartment. It can reach, through intermediate stages, a maximum diameter of about one-half of the mean diameter of the vesicle. In this case, assuming that the vesicle and the core are spherical and that the vesicle diameter is 60 nm, the core diameter 30 nm, and the thickness of the vesicle membrane about 6 nm, the maximum volume of the core would be 14,139 nm^3. In this case the matrix would reach its minimum volume, 43,886 nm^3. This is three times the maximum volume of the core. Considering that the maximum volume of the matrix could be 58,061 nm^3, the variation for this compartment is about 25% of the maximum volume, whereas the variation of the core size is about 100%. Thus the matrix is a constant component of the small granular vesicle, and the core is a transient one.

Interchange of materials between the matrix and the cytosol occurs through an area at constant or slightly changing size, whereas the interchange between the matrix and the core takes place through an area changing with the size of the core. The surface area of the vesicle is about 11,340 nm^2, and the maximum surface area of the core is about 2,835 nm^2.

The matrix does not completely surround the core, which is in contact with the vesicle membrane or linked to it by a pedicle, the width of which is about twice the thickness of the vesicle membrane. Morphological evidence suggests (Fig. 3, C,D, *arrows*) that the core is formed by invagination of the vesicle membrane. This configuration implies the existence of a special locus in the vesicle membrane, a kind of "door," which could be related to the influx or efflux of the neurotransmitter. Furthermore, it suggests that the volume of the vesicle must change inversely with the size of the core. Previous measurements (28) have not revealed significant changes, although electron micrographs of monoaminergic vesicles after tyramine depletion suggest that this could be the case.

The maximum size of the cores and their presence in small vesicles vary in different autonomic nerves, being scarce and almost lacking in some places [e.g., small arteries (Fig. 3A)] but abundant and well developed in others [e.g., the vas deferens (Fig. 3B) and the pineal gland].

In large granular vesicles the matrix may be conspicuous in some vesicles and almost absent in others (Fig. 3B, 1 and 2, respectively). This suggests that different kinds of vesicles could be included in this category.

HISTOCHEMISTRY OF BIOGENIC AMINES AT THE ELECTRON MICROSCOPIC LEVEL

The localization of biogenic amines in the core has been demonstrated with various se-

FI(
B:
tio
C,

FIG. 2. Various stages of both compartments in small synaptic vesicles.

material. We observed that in small vesicles this type of osmiophilia appears after glutaraldehyde-osmium tetroxide fixation (Fig. 4C); but when osmium tetroxide is employed alone, the osmiophilia is restricted to the core (Fig. 4B), as in the untreated controls (Fig. 4A). In most of the large vesicles, intense osmiophilia is present in both cases. Furthermore, the osmiophilic material produced by glutaraldehyde is observable only after a short fixation (30 to 60 min), whereas after prolonged fixation it seems to be washed out (31). We obtained similar results with the Wood histochemical technique (38) in 5-OHDA-treated animals (23). These observations strongly suggest that monoamines are located in both compartments, and that they are tightly bound in the core and loosely bound in the matrix.

ELECTRICAL STIMULATION AND VESICLE COMPARTMENTS

It is well known that electrical stimulation of adrenergic nerves induces release of biogenic amines from the nerve endings, and that synthesis is increased at the same time (6). Thus small quantitative changes may be produced in the total content of a given organ. The effect of electrical stimulation on vesicle compartments has not been previously investigated.

We stimulated the afferent trunks of both superior cervical ganglia in adult rats under chloral hydrate anesthesia (25). Stimulation was performed by applying square pulses (1 msec, 25 volts) through platinum electrodes at a frequency of 25/sec for 30 min. Stimulation was applied without previous treatment or after 5-OHDA administration. Rats injected with 5-OHDA (4×20 mg/kg i.p.) over a period of 48 hr were stimulated 4 hr after the last injection. In other cases they received an additional dose immediately before the administration of the anesthetic. Pineal glands of injected and stimulated animals, and of injected but unstimulated rats, were fixed and processed for electron microscopy. Fixation was done in glutaraldehyde-osmium tetroxide according to a short schedule (30 min) (31) and postfixation in 1.5% osmium tetroxide for 2 hr.

In the pineal glands of unstimulated animals after five doses of 5-OHDA almost all the vesicles were totally stained (Fig. 4C). After stimulation very few vesicles were totally stained (Fig. 4D); most were unreactive, and only a few showed electron-dense cores. Fixation was done about 1 hr after the last injection.

In pineal glands of unstimulated rats killed after four doses of 5-OHDA (Fig. 4E), almost all the vesicle matrices were electron-lucent whereas in many the cores were stained. After stimulation, almost all the vesicles were electron-lucent (Fig. 4F), and many had flattened or tubular shapes. Fixation was done 5 hr after the last injection.

Large vesicles appeared depleted in stimulated and unstimulated glands. These results seem to indicate that the false transmitter is spontaneously released, mainly from the matrix, and that electrical stimulation influences both compartments.

Controls and stimulated glands from animals without 5-OHDA pretreatment were fixed in the following ways: (a) OsO_4 1% in a balanced solution containing NaCl, KCl, $MgCl_2$, $CaCl_2$, and polyvinylpyrrolidone (M.W. 12,000) for 2 hr and immersion in 2% uranyl acetate in distilled water for 2 hr. (b) Glutaraldehyde 2.5% in collidine buffer (0.1 M, pH 7.2 to 7.4) plus 50 mM $CaCl_2$ overnight, washing in the same buffer with 50 mM $CaCl_2$ for 3 hr, and postfixation in 1.5% OsO_4 in the same buffer plus 50 mM $CaCl_2$ for 2 hr (21). (c) Glutaraldehyde-dichromate, according to Wood (38), with a long (4-hr) (GDL), and a short (30-min) (GDSh) period of fixation. (d) Formaldehyde-glutaraldehyde-dichromate, according to Wood (38), with a long (FGDL) and a short (FGDSh) time of fixation.

It was observed that stimulation produced a decrease in the size and relative proportion of OsO_4-reactive cores (fixation a) (Fig. 5A,B). In glands processed with fixation b, the controls showed small reactive sites, apparently related to the cores (Fig. 6C), which were reduced in size or disappeared after stimulation. Reactive sites in fixation b may correspond to the localization of a calcium-binding protein (21). In both cases—fixations a and b—emptied vesicles after stimulation were distributed in all nerve endings, as in the controls.

After stimulation, GDL-reactive sites, corresponding to the cores, decreased considerably in number and density (Fig. 6A,B), whereas GDSh-reactive sites, corresponding

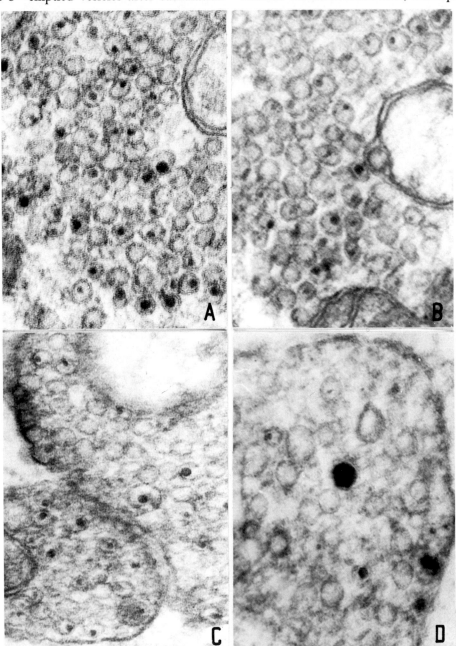

FIG. 5. Nerve endings from rat pineal nerves before **(A,C)** and after **(B, D)** electrical stimulation, processed with different fixatives. See description in the text. ×120,000.

to a matrix reaction, were only slightly reduced (Fig. 6C,D). Similar results were obtained with FGDL (Fig. 6E,F) and FGDSh (Fig. 6G,H). Whereas GD reveals catecholamines and indoleamines, FGD is specific for indoleamines (17).

The results obtained with electrical stimulation show that the monoamines located in

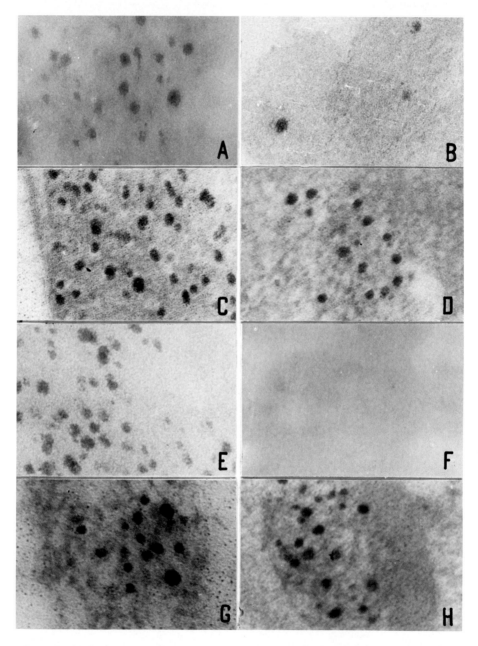

FIG. 6. Nerve endings from rat pineal nerves. Histochemical techniques for catecholamines and indoleamines, and electrical stimulation. A,C,E,G: Before stimulation. B,D,F,H: After stimulation. A, B: GDL. C, D: GDSh. E, F: FGDL. G, H: FGDSh. See description in the text. ×120,000.

both vesicle compartments may be released by electrical stimulation. They also suggest that (a) monoamines are more easily released from the matrix than from the core; (b) electrical stimulation induces the synthesis of monoamines in the matrix; and (c) a Ca-binding protein is related to the core and is depleted or changed by electrical stimulation.

Changes in large vesicles are more difficult to analyze because of their scarcity and the variable morphological changes in the experiments described.

ZIO AND VESICLE COMPARTMENTATION

ZIO reacts with different kinds of synaptic vesicles. It was first used by Akert and Sandri (1) to stain cholinergic vesicles. In monoaminergic vesicles (26,27) ZIO reacts differently with either compartment: the core is slightly stained and the matrix is intensely reactive (Fig. 1C).

Although *in vitro* assays have shown (27) that ZIO reacts with different compounds, giving a black precipitate (which could explain the electron density of different types of synaptic vesicles tested), the ZIO reaction was shown to be due to the presence of reactive -SH groups (22,32), possibly belonging to protein molecules within the vesicles (Fig. 1D). The ZIO reaction is temperature- and time-dependent and varies in different synaptic vesicles (22). Thus in the pineal nerves the reaction takes place at 4°C or above, whereas in other monoaminergic vesicles 20°C or more is needed for a positive reaction. These findings and pharmacological studies (22) suggest that different components of this compartment react with ZIO at different temperatures, or that temperature changes the accessibility of -SH groups. The ZIO reaction in monoaminergic synaptic vesicles is affected by drugs interfering with the uptake, storage, and synthesis of monoamines (24,26–28). The fact that the inhibitors of the synthesis of the monoamines lo-

calized in the vesicle affect the ZIO reactive sites in the vesicle matrix (24) can be taken to suggest that the enzymes involved in the monoamine synthesis are located in that compartment. This interpretation is in line with the speculations by Udenfriend (37) on the integration of norepinephrine-synthesizing enzymes in a subcellular structure (the granular vesicle) that also binds norepinephrine. However, the effect of drugs may be complex; they could act on more than one component, or their effect on the outer compartment of the vesicles could be indirect. It is generally accepted that only dopamine-β-hydroxylase is located in adrenergic vesicles (6), whereas tyrosine hydroxylase and DOPA decarboxylase are considered cytoplasmic enzymes, based mainly on their solubilization after tissue fractionation. However, both enzymes are at least partly particulated (24). Furthermore, the solubilization of a tissue component after fractionation does not exclude the possibility that it has been located in a synaptic vesicle. Arguments for and against the synthesis of serotonin in the pineal nerves have been reported (24). Our experiments with electrical stimulation seem to indicate that catecholamines and indoleamines have the same fate in the granular vesicles of the pineal nerves of the rat and suggest that both amines are synthesized in the vesicles.

MONOAMINERGIC POOLS IN NERVE TERMINALS

On the basis of pharmacological, electrophysiological, and biochemical criteria, at least two pools of norepinephrine have been defined in nerve terminals: an "easily releasable" pool and a "storage" pool (19,39). Most of the authors have placed the easily releasable material in the cytoplasm or in a pool of vesicles located near the plasma membrane and the less releasable material in the granular synaptic vesicles or in vesicles located far from the synaptic cleft. A compartmentation of norepinephrine pools in adrenergic granules or vesicles has also been postulated (13).

Our observations show that the granular vesicles of monoaminergic nerves have two compartments: the matrix and the core. A "loosely bound" and a "tightly bound" pool of monoamines, which could be identified with the "easily releasable" and the "stored material," are located in the matrix and the core, respectively, and may be identified histochemically. The monoamines would be synthesized in the matrix and stored in the core. Thus recently synthesized amines (19) would be located in and released from the matrix. In stimulated nerves, only the matrix persists in most of the vesicles.

The existence of a reserve pool inside the vesicle could provide an efficient mechanism of regulation for the release of the neurotransmitter. Adequate activation of the receptor according to the need of a "tonic" or a "phasic" stimulation of the innervated structures may be thus obtained. Different sizes of the central compartment could be correlated with this bimodal release of the neurotransmitter. A small core is observed in the nerves of organs with a predominant "tonic" functioning (i.e., in small arteries) and large cores in organs with an important "phasic" function (i.e., in the vas deferens or the pineal gland). This bimodal behavior of release would not take place in the large vesicles from which a "matrix" is absent. The core could intervene to maintain an adequate amount of neurotransmitter in the vesicle matrix, thus regulating the synthesis of neurotransmitter by feedback inhibition.

These physiological properties and the life span of synaptic vesicle constituents (5,20,23) seem to indicate that synaptic vesicles are true organelles and not simple secretory granules.

QUANTAL RELEASE AND QUANTAL PACKAGING OF NEUROTRANSMITTERS

The quantal concept of neurotransmitter secretion, first established for the neuromuscular end plate (12), implies that the neurotransmitter is secreted at rest in subthreshold multimolecular packets of preset size which are multiplied by an integer number on nerve stimulation. This concept has been extended to peripheral adrenergic nerves (4) in which a quantal random release has been reported.

The quantal packet has been assumed to be identical with the transmitter content of synaptic vesicles from which it would be released in a quantal all-or-none fashion (18). This assumption has been challenged in peripheral adrenergic nerves in which the quantal packet, at rest, has been estimated as a small fraction, 2 to 39% (35) or 1 to 3% (13) of the vesicle content. According to Folkow and Häggendal (13), the fractionated release could be explained by the compartmentation of norepinephrine in the synaptic granules or vesicles.

Our findings provide anatomical, histochemical, and physiological data which identify the "quantal packet" with a fraction of the transmitter content in the monoaminergic vesicles. They show that the vesicles have two compartments—one constant and the other transient—where the amines are differently bound, possibly in different concentrations, and perhaps released by different mechanisms. In one compartment the amine would be synthesized and easily released; in the other it would be kept in storage. Although their sizes are inversely related, the first compartment is always considerably larger than the second and may be 100% of the vesicle.

If the vesicle represents the quantal packet (18), it could be formed by a homogeneous pool found in the vesicles without a core, or by heterogeneous pools located in both compartments, participating in variable proportions in the quantal packet. If the quantal packet at rest is provided by one compartment, the size of the compartment must be variable if it represents the monoamine content of the vesicle unless only vesicles without a core are involved in the release of the quantal packet at rest.

These considerations and the experimental

results reported previously (see above) led us to believe that the quantal packet at rest corresponds to a fraction of the neurotransmitter contained in the vesicle, which is provided at rest by one compartment, possibly the matrix, and by both compartments during electrical stimulation.

RELEASE OF THE NEUROTRANSMITTER AND VESICLE COMPARTMENTATION

Two possible fates for amine stores in vesicles have been considered: (a) the amines are released into the cytosol from where they diffuse out of the cell, or (b) the amine is released directly into the extracellular space, alone or together with ATP or vesicle proteins. The second alternative implies (34) three possible mechanisms: (a) the vesicle is extruded *in toto;* (b) the vesicle is attached to

the plasma membrane by a "tight" junction; or (c) the vesicle membrane is fused with the plasma membrane, the fused region is lysed, and the vesicle content is extruded to the extracellular space by exocytosis. If a "tight" or a "gap" junction is formed, only small molecules could be extruded. There is no evidence supporting the extrusion of the vesicles *in toto.* Exocytosis has been demonstrated in the adrenal medulla by various methods (34). In the nerve endings exocytosis has been controverted and restricted to large vesicles (6,34) or to fractions of the vesicle contents (13,15,35).

The mechanisms involved in the release of neurotransmitter from monoaminergic synaptic vesicles must be reformulated, taking into account vesicle compartmentation. The possible mechanisms are depicted in Fig. 7.

Although release of an unaltered neurotransmitter into the extracellular space would

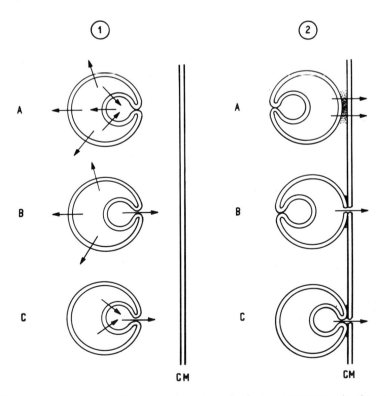

FIG. 7. Possible modes of release of monoamines from both compartments via the cytosol (1) and through the cell membrane (2), as well as the interactions between the compartments (1). (CM) Cell membrane.

16. Hökfelt, T. (1968): In vitro studies on central and peripheral monoamine neurons at the ultrastructural level. *Z. Zellforsch. Mikrosk. Anat.,* 91:1-74.

17. Jaim-Etcheverry, G., and Zieher, L.M. (1968): Cytochemistry of 5-hydroxytryptamine at the electron microscope level. I. Study of the specificity of the reaction in isolated blood platelets. *J. Histochem, Cytochem.,* 16:162-171.

18. Katz, B. (1966): *Nerve, Muscle and Synapse.* McGraw-Hill, New York.

19. Kopin, I.J., Breese, G.R., Krauss, K.R., and Weise, V.K. (1968): Selective release of newly synthesized norepinephrine from the cat spleen during sympathetic nerve stimulation. *J. Pharmacol. Exp. Ther.,* 161:271-278.

20. Lapetina, E.G., Lunt, G.G., and De Robertis, E. (1970): The turnover of phosphatidyl choline in rat cerebral cortex membranes in vivo. *J. Neurobiol.,* 1:295-302.

21. Pappas, G.D., and Rosen, S. (1976) Localization of calcium deposits in the frog neuromuscular junction at rest and following stimulation. *Brain Res.,* 103:362-365.

22. Pellegrino de Iraldi, A. (1977): Significance of the Maillet method (ZIO) for cytochemical studies of subcellular structures. *Experientia,* 33:1-10.

23. Pellegrino de Iraldi, A. (1979): Unpublished.

24. Pellegrino de Iraldi, A., and Cardoni, R. (1979): ZIO staining in synaptic vesicles of the rat pineal nerves after inhibition of serotonin and noradrenaline synthesizing enzymes. *Cell Tissue Res.,* 200:91-100.

25. Pellegrino de Iraldi, A., and Corazza, J.P. (1979): To be published.

26. Pellegrino de Iraldi, A., and Gueudet, R. (1968): Action of reserpine on the osmium tetroxide-zinc iodide reactive sites of synaptic vesicles in the pineal nerves of the rat. *Z. Zellforsch. Mikrosk. Anat.,* 91:178-185.

27. Pellegrino de Iraldi, A., and Suburo, A.M. (1971): Two compartments in the granulated vesicles of the pineal nerves. In: *The Pineal Gland,* edited by G.E.W. Wolstenholme and J. Knight, pp. 177-195. Ciba Foundation Symposium. Churchill Livingstone, Edinburgh.

28. Pellegrino de Iraldi, A., and Suburo, A.M. (1972): Effect of tyramine on the compartments of granulated vesicles of rat pineal nerve endings. *Eur. J. Pharmacol.,* 19:251-259.

29. Pellegrino de Iraldi, A., and Suburo, A.M. (1972): Morphological evidence of a connection between the core of granulated vesicles and their membrane. *Neurobiology,* 2:8-11.

30. Pellegrino de Iraldi, A., Farini Duggan, J., and De Robertis, E. (1963): Adrenergic synaptic vesicles in the anterior hypothalamus of the rat. *Anat. Rec.,* 145:521-531.

31. Pellegrino de Iraldi, A., Gueudet, R., and Suburo, A.M. (1971): Differentiation between 5-hydroxytryptamine and catecholamines in synaptic vesicles. *Prog. Brain Res.,* 34:161-170.

32. Reinecke, M., and Walther, C. (1978): Aspects of turnover and biogenesis of synaptic vesicles at locust neuromuscular junctions as revealed by zinc iodide-osmium tetroxide (ZIO) reacting with intravesicular -SH groups. *J. Cell Biol.,* 78:839-855.

33. Rodriguez de Lores Arnaiz, G., Alberici de Canal, M., and De Robertis, E. (1971): Turnover of proteins in subcellular fractions of rat cerebral cortex. *Brain Res.,* 31:179-184.

34. Smith, A.D., and Winkler, H. (1972): Fundamental mechanisms in the release of catecholamines. In: *Catecholamines,* edited by H. Blaschko and E. Muscholl, pp. 538-617. Springer-Verlag, Berlin.

35. Stjärne, L. (1970): Quantal or graded secretion of adrenal medullary hormone and sympathetic neurotransmitter. In: *New Aspects of Storage and Release Mechanisms of Catecholamines,* edited by H.J. Schümann and G. Kroneberg, pp. 112-127. Bayer Symposium II. Springer-Verlag, Berlin.

36. Tranzer, J.P., and Thoenen, H. (1967): Electron microscopic localization of 5-hydroxydopamine (3,4,5,-trihydroxyphenylethylamine) a new "false" sympathetic transmitter. *Experientia,* 23:743-745.

37. Udenfriend, S. (1968): Physiological regulation of noradrenaline biosynthesis. In: *Adrenergic Neurotransmission,* edited by G.E.W. Wolstenholme and M. O'Connor, pp. 3-11. Ciba Foundation Study Group No. 33. Churchill, London.

38. Wood, J.G. (1967): Cytochemical localization of 5-hydroxytryptamine (5HT) in the central nervous system (CNS). *Anat. Rec.,* 157:343-344.

39. Wurtman, R.J. (1966): *Catecholamines.* Little, Brown, Boston.

Histochemistry and Cell Biology of Autonomic Neurons, SIF Cells, and Paraneurons,
edited by O. Eränkö et al.
Raven Press, New York © 1980.

Cytochemical Investigations on Subcellular Organelles Storing Biogenic Amines in Peripheral Adrenergic Neurons

J.G. Richards and M. Da Prada

Pharmaceutical Research Department, F. Hoffmann-La Roche & Co., Ltd., CH-4002 Basel, Switzerland

A variety of cytochemical, cytopharmacological, and autoradiographic techniques (Fig. 1) can be used to investigate the nature and distribution of intraneuronal storage sites for neurotransmitters such as biogenic monoamines (16). In sympathetic neurons the amine, norepinephrine (NE), is concentrated in an osmotically stable form with 5'-phosphonucleotides [e.g., adenosine-5'-triphosphate (ATP)] in storage organelles whose localization and distribution are best studied by electron microscopy. These investigations provide insight to the biogenesis of synaptic (axonic and dendritic) vesicles from which the neurotransmitter has been shown to be released.

Two cytochemical methods—the chromaffin reaction (21) and the uranaffin reaction (17)—have been used to identify storage sites for NE and ATP, respectively, in adrenergic neurons. These investigations demonstrate that NE is present not only in small (50 nm) and large (100 nm) dense-cored (SDCs and LDCs, respectively) vesicles but also in a tubular endoplasmic reticulum (TER) and in multivesicular bodies (MVBs). Their distribution within the neuron provides some clues to the origin and fate of transmitter-storing organelles. Evidence also points to the presence of ATP in SDC and LDC vesicles, and to a likely role for the TER in the formation of storage organelles. The role of MVBs is

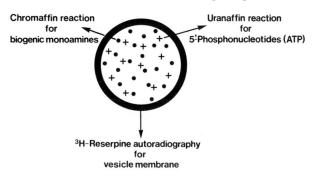

FIG. 1. Labeling amine-storing vesicles. Methods available for the identification of amine-storing vesicles in neurons. Only the cytochemical methods are discussed in detail.

less clear, although they appear to be involved in either the formation or breakdown of SDC vesicles.

CHROMAFFIN REACTION

The original cytochemical reaction of Wood and Barrnett, as modified in our laboratory, produces a highly selective method for detecting biogenic monoamines, including NE. It has since been successfully applied for the identification of amine-storing organelles in peripheral sympathetic neurons and in a few neuronal systems in the brain. The unsubstituted amino group of NE reacts with the aldehyde groups of the glutaraldehyde fixative, leaving the two hydroxyl groups of the aromatic ring of the amine free to react with dichromate ions, thus forming an electron-dense product.

URANAFFIN REACTION

ATP-rich amine storage organelles of aldehyde-fixed tissues have a strong affinity for uranyl ions before their dehydration. An interaction between UO_2^{++} ions and phosphate groups of 5'-phosphonucleotides produces a highly electron-dense reaction product, e.g., in rat adrenal medulla and sympathetic neurons as well as in 5-hydroxytryptamine (5-HT) storing organelles in rabbit platelets and megakaryocytes. The amine-storing organelles of rabbit platelets have been used as a model to study, cytochemically and biochemically, the subcellular distribution of amines and nucleotides (16). Indeed, electron-probe microanalyses of these platelets submitted to the uranaffin reaction demonstrate the extremely high content of phosphorus (2,19) and uranium exclusively in the 5-HT organelles (Figs. 2 and 3).

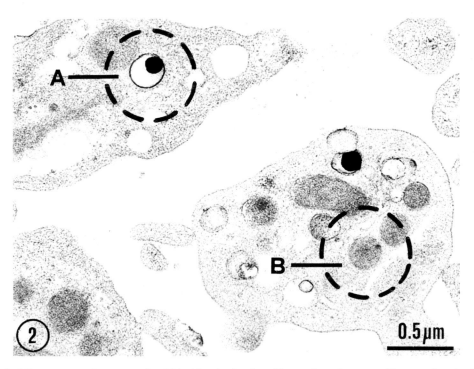

FIG. 2. Ultrastructural aspect of rabbit blood platelets illustrating the uranaffin reaction in highly electron-dense organelles (A) and its absence in adjacent organelles (B). The former are the 5-HT organelles and the latter are the α-granules. The broken circles indicate the area included in the analyses.

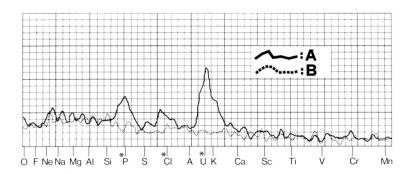

FIG. 3. X-ray microanalyses of two regions of rabbit blood platelets, similar to those illustrated in Fig. 2, reveal that the 5-HT organelle (A) contains much higher quantities of phosphorus (P) and uranium (U) compared to the α-granule (B). Moreover, region A also appears to contain significantly more chlorine (Cl). These quantitative differences lend support to our concept that the uranaffin reaction visualizes 5′-phosphonucleotides in amine-storing organelles.

AMINE-STORING ORGANELLES IN NERVE TERMINALS AND AXONS

Although amine-storing SDC and LDC vesicles have been shown to occur throughout the sympathetic neuron, they are most concentrated in the nerve terminal, where the former represent 90 to 95% of the vesicle population. Cytochemical investigations reveal that both types of vesicle and the TER are highly chromaffin-positive. In rat iris (Fig. 4), but not rat vas deferens, the TER is well developed and easily identified. Both types of vesicle are also uranaffin-positive [e.g., in rat vas deferens (Fig. 5)], although the TER (e.g., in rat iris) is rarely revealed. In myelinated and nonmyelinated nonterminal axons, all three types of organelle are chromaffin-positive, whereas only SDC and LDC vesicles have so far been identified by the uranaffin reaction, e.g., in bovine extrasplenic nerve (Fig. 6). Therefore it is evident that both types of vesicle in nerve terminals and axons contain NE and 5′-phosphonucleotides (probably ATP). Biochemical analyses of vesicles purified by differential and density gradient centrifugation have not only been possible in the case of the LDC vesicles [bovine splenic nerve (11)] but also

more recently of the SDC vesicles [vas deferens of castrated rats (6)].

These analyses revealed, in agreement with the present findings, the presence of NE and ATP in LDC and possibly SDC vesicles, although only the LDC vesicles appear to contain the enzyme dopamine-β-hydroxylase. The concentration of NE in SDC vesicles has been estimated to be at most 40 μg/mg protein (6) and in LDC vesicles 100 μg/mg protein (11). Since the SDC vesicles of rat vas deferens are uranaffin-positive (i.e., they contain ATP), the absence of an NE/ATP storage complex in these vesicles cannot be the explanation for the differences in NE concentration in SDC and LDC vesicles as recently proposed (6). The apparent discrepancy in these values is either real or possibly due to differences in the purification methods. The recent demonstration of the accumulation of [³H]ATP in SDC vesicles (1) nevertheless seems to confirm our findings.

AMINE-STORING ORGANELLES IN PERIKARYA AND DENDRITES

There is now convincing evidence that LDC and SDC vesicles (containing NE and ATP) are present throughout the perikaryon,

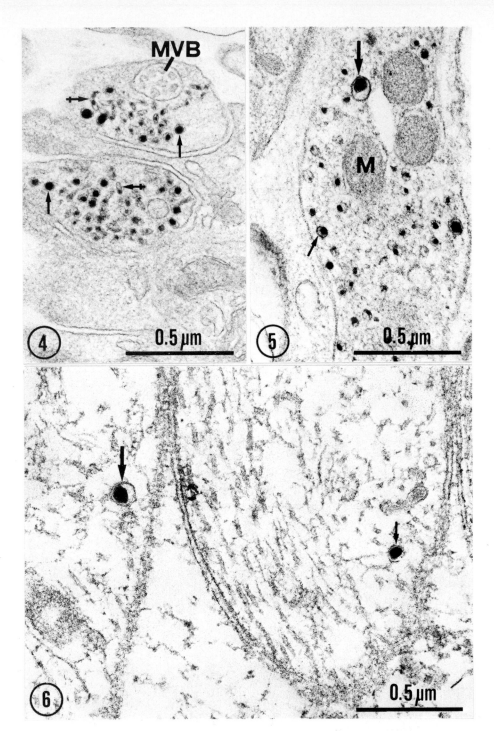

FIG. 4. Rat iris: chromaffin reaction. Note the presence of highly electron-dense SDC vesicles (*arrows*) and TER (*crossed arrows*) in two nerve terminal profiles; LDC vesicles (not shown) are similarly reactive. A rarely observed MVB is chromaffin-negative.

FIG. 5. Rat vas deferens: uranaffin reaction. Note the highly electron-dense reaction product in the matrix and membrane of SDC (*small arrow*) and LDC (*large arrow*) vesicles, whereas adjacent mitochondria (M) are weakly reactive.

FIG. 6. Calf extra splenic nerve: uranaffin reaction. SDC and LDC vesicles are highly electron-dense (*arrows*).

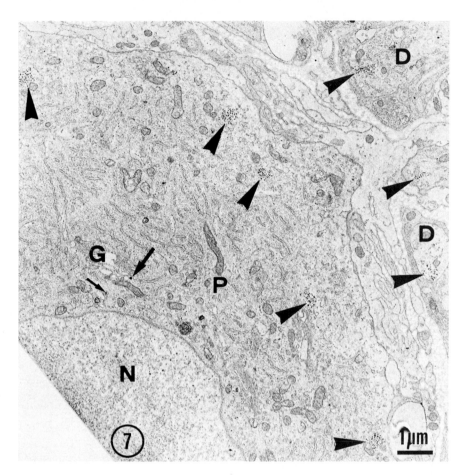

FIG. 7. Rat superior cervical ganglion: chromaffin reaction. The general distribution of electron-dense reaction products in the perikaryon (P) is shown in this survey micrograph. Note the presence of single LDC (*large arrow*) and SDC (*small arrow*) vesicles in a Golgi apparatus (G) and the numerous clusters of SDC vesicles + TER (*arrowheads*) at the cell periphery and in nearby dendritic profiles (D). (N) Nucleus.

e.g., in the Golgi apparatus as individual vesicles, in the periphery of the cell soma, and in dendrites as clusters of SDC vesicles (Figs. 7 and 8). A chromaffin-positive TER is also present in the Golgi and among the clusters of SDC vesicles (Fig. 8), implying the involvement of TER in the formation of storage vesicles. The proliferation of a TER in sympathetic ganglion cells of rats (Fig. 9) 24 hr after two injections of 6-hydroxydopamine (100 mg/kg i.p. within 6 hr) is a further indication of its involvement in vesicle formation under conditions in which demand is very high. Previous studies (8,10,13) have also suggested that the TER [or axonal Golgi elements (15)] is the source of newly formed vesicles whether in the perikaryon or in the nerve terminal. Recent autoradiographic evidence (18) demonstrating a selective accumulation of [3H]reserpine in sympathetic ganglion cells and paracellular regions confirms the presence of amine-storing vesicles in the perikaryon for which this radiolabel is a specific ligand (Fig. 10).

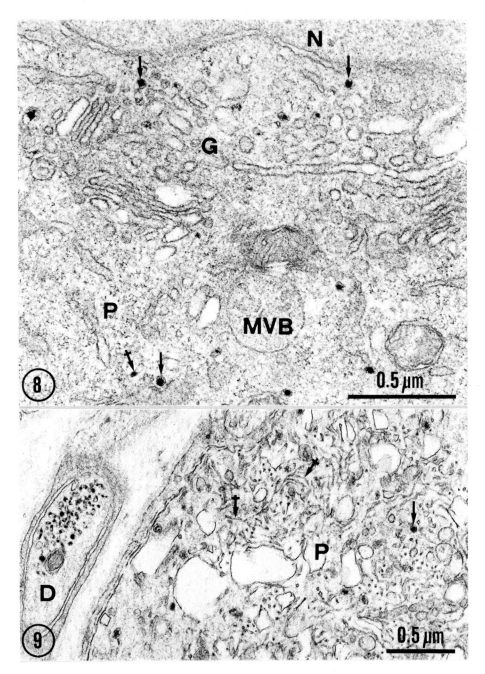

FIG. 8. Rat superior cervical ganglion: chromaffin reaction. Several electron-dense organelles (*arrows*) can be observed in the region of a Golgi apparatus (G). (MVB) Multivesicular body. (N) Nucleus. (P) Perikaryon.

FIG. 9. Rat superior cervical ganglion: chromaffin reaction. In this cell from a rat pretreated with 6-hydroxydopamine, the perikaryon (P) contains numerous profiles of a TER (*crossed arrow*) and the occasional SDC vesicle (*arrow*). The TER illustrated here is possibly proliferating in response to the lack of amine-storing organelles in nerve terminals (due to chemical sympathectomy). (D) Dendrite.

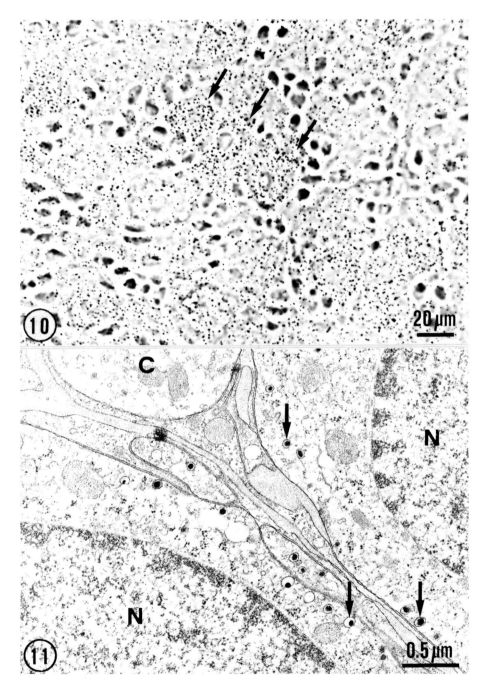

FIG. 10. Rat superior cervical ganglion: [³H]reserpine autoradiography. Note the accumulation of silver grains in cellular (*arrows*) and paracellular regions corresponding to the distribution of amine-storing organelles in this ganglion.
FIG. 11. Superior cervical ganglion: uranaffin reaction. Note the numerous electron-dense granules (*arrows*) in two adjacent small granule-containing cells (= SIF cells) but no reaction product in a nearby cholinergic nerve terminal (C). (N) Nucleus.

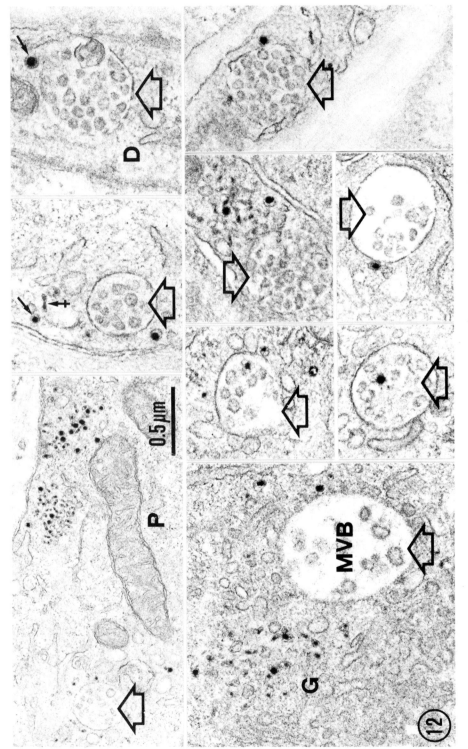

FIG. 12. Superior cervical ganglion: chromaffin reaction. This series of micrographs illustrates the frequent association of MVBs with individual or clustered amine-storing organelles in the perikaryon (P), Golgi apparatus (G), and dendrites (D). In two of the center illustrations, chromaffin-positive SDC vesicles can be observed within the MVB. From these micrographs alone, it cannot be determined whether the amine-storing organelles are being newly formed or sequestered. The magnification of all micrographs (except the one with a calibration bar) is ×56,000.

AMINE-STORING ORGANELLES IN SMALL GRANULE-CONTAINING CELLS

Granule-containing cells, probably identical with SIF cells observed by fluorescence microscopy, are easily identified by the size (100 to 300 nm) and frequency of their granules (20). Their positive reaction to both cytochemical tests suggests that they contain an amine, probably dopamine (4), and ATP. Chromaffin- and uranaffin-positive (Fig. 11) organelles were observed throughout the perikaryon and its processes. Nonreactive, presumably cholinergic, nerve terminals synapsed on these cells.

AMINE-STORING ORGANELLES IN MULTIVESICULAR BODIES

MVBs are frequently found in the vicinity of the Golgi apparatus and near clusters of SDC vesicles and TER in the perikaryon and dendritic processes (Fig. 12). They are occasionally also found in terminal and nonterminal axons (Fig. 4). In the perikaryon and dendrites SDC vesicles are occasionally observed *within* the MVBs (Fig. 12), suggesting that the latter are somehow involved in the formation of SDC vesicles or (perhaps less likely) in their breakdown. MVBs are well characterized morphologically in neuronal (12,14) and nonneuronal (7) tissues. In the latter they have been shown to contain lysosomal enzymes in their matrix but not in their vesicles, which are believed to be derived from the MVB-limiting membrane, which itself may be formed in the Golgi apparatus. The presence of chromaffin-positive SDC vesicles within MVBs confirms the ability of MVBs to accumulate [³H]NE, an observation made by Descarries and Droz (3) using autoradiography. Perhaps the MVBs identified in the present study represent an early stage in the formation of clusters of SDC vesicles and TER. Alternatively, they could indicate the fate of the vesicles after membrane retrieval, in which case it could represent a form of secondary lysosome (see refs. 5,9,22 for a detailed discussion on the origin and fate of amine-storing vesicles).

CONCLUSIONS

Monoamines such as NE are stored (probably as a complex) with ATP in SDC and LDC vesicles. TER also contains NE and might represent a stage in the formation of new SDC vesicles. Although the precise role of MVBs is less clear, it is evident that they are involved in the origin or fate of amine-storing vesicles.

ACKNOWLEDGMENT

The x-ray microanalyses illustrated in Fig. 3 were kindly performed by Mr. D.M. Williams, Department of Metallurgy and Materials Technology, University of Swansea, Wales.

REFERENCES

1. Aberer, W., Stitzel, R., Winkler, H., and Huber, E. (1979): Accumulation of [³H]ATP in small dense core vesicles of superfused vasa deferentia. *J. Neurochem.,* 33:797–801.
2. Costa, J.L., Pettigrew, K.D., and Murphy, D.L. (1979): Electron probe microanalysis of change in dense body phosphorus and calcium content following alterations in platelet serotonin levels. *Biochem. Pharmacol.,* 28:23–27.
3. Descarries, L., and Droz, B. (1968): Incorporation de noradrénaline-³H (NA-³H) dans le système nerveux central du rat adulte: Etude radio-autographique en microscopie électronique. *C R Acad. Sci. Paris,* 266:2480–2482.
4. Elfvin, L., Hökfelt, T., and Goldstein, M. (1975): Fluorescence, microscopical, immunohistochemical and ultrastructural studies on sympathetic ganglia of guinea pig, with special reference to SIF cells and their catecholamine content. *Ultrastruct. Res.,* 51:377–396.
5. Fillenz, M. (1977): The factors which provide short-term and long-term control of transmitter release. *Prog. Neurobiol.,* 8:251–278.
6. Fried, G., Lagercrantz, H., and Hökfelt, T. (1978): Improved isolation of small noradrenergic vesicles from rat seminal ducts following castration: A density gradient centrifugation and morphological study. *Neuroscience,* 3:1271–1291.

7. Friend, D.S. (1969): Cytochemical staining of multivesicular body and Golgi vesicles. *J. Cell Biol.*, 41:270–279.

8. Hökfelt, T. (1973): On the origin of small adrenergic vesicles: evidence for local formation in nerve endings after chronic reserpine treatment. *Experientia*, 29:580–582.

9. Holtzman, E. (1977): The origin and fate of secretory packages, especially synaptic vesicles. *Neuroscience*, 2:237–355.

10. Kanerva, L., Hervonen, A., Rechardt, L., Hervonen, H., and Partanen, M. (1978) Ultrastructural localization of granules and smooth endoplasmic reticulum with permanganate fixation in monoamine cells. In: *Peripheral Neuroendocrine Interaction*, pp. 15–28. Springer, Heidelberg.

11. Lagercrantz, H. (1977): On the composition and function of large dense cored vesicles in sympathetic nerves. *Neuroscience*, 1:81–92.

12. Matthews, M.R., and Raisman, G. (1972): A light and electron microscopic study of the cellular response to axonal injury in the superior cervical ganglion of the rat. *Proc. R. Soc. Lond. [Biol]*, 181:43–79.

13. Mercer, L., Del Fiacco, M., and Cuello, A.C. (1979): The smooth endoplasmic reticulum as a possible storage site for dendritic dopamine in substantia nigra neurones. *Experientia*, 35:101.

14. Peters, A., Palay, S.L., and Webster, H. (1970): *The Fine Structure of the Nervous System*, pp. 18–20. Harper & Row, New York.

15. Quatacker, J., and De Potter, W. (1978): Relationship between Golgi apparatus and axonal reticulum in sympathetic ganglion cells. In: *9th International Congress on Electron Microscopy, Toronto*, Vol. II, pp. 598–599.

16. Richards, J.G. (1978): Cytochemistry and autoradiography in the search for transmitter-specific neuronal pathways. In: *Peripheral Neuroendocrine Interaction*, pp. 1–14. Springer, Heidelberg.

17. Richards, J.G., and Da Prada, M. (1977): Uranaffin reaction: A new cytochemical technique for the localization of adenine nucleotides in organelles storing biogenic amines. *J. Histochem. Cytochem.*, 25:1322–1336.

18. Richards, J.G., Da Prada, M., Würsch, J., Lorez, H.P., and Pieri, L. (1979): Mapping monoaminergic neurons with [3H]reserpine by autoradiography. *Neuroscience*, 4:937–950.

19. Skaer, R.J. (1975): Elemental composition of platelet dense bodies. In: *Biochemistry and Pharmacology of Platelets*, pp. 239–259. Ciba Foundation Symposium 35. Elsevier, Amsterdam.

20. Taxi, J., and Mikulajova, M. (1976): Some cytochemical and cytological features of the so-called SIF cells of the superior cervical ganglion of the rat. *J. Neurocytol.*, 5:283–295.

21. Tranzer, J.P., and Richards, J.G. (1976): Ultrastructural cytochemistry of biogenic amines in nervous tissue: Methodologic improvements. *J. Histochem. Cytochem.*, 24:1178–1193.

22. Winkler, H. (1977): The biogenesis of adrenal chromaffin granules. *Neuroscience*, 2:657–683.

*Histochemistry and Cell Biology of
Autonomic Neurons, SIF Cells, and
Paraneurons,*
edited by O. Eränkö et al.
Raven Press, New York © 1980.

Observations on the Ultrastructural Localization of Monoamines with Permanganate Fixation

Lasse Kanerva, *Antti Hervonen, and Mats Grönblad

*Departments of Anatomy and of Dermatology, University of Helsinki, Helsinki, Finland; and
Department of Biomedical Sciences, University of Tampere, Tampere, Finland

In neurocytological fine structural research, one of the main tasks has been to localize transmitter substances at certain cytoplasmic locations. Richardson (25) was the first to discover that permanganate fixation can be used to differentiate between adrenergic and cholinergic terminals in the peripheral nervous system. Hökfelt (14) also localized small granular vesicles (SGVs) in adrenergic terminals within the central nervous system. During the following decade, permanganate was considered a useful tool in localizing monoamine-containing structures in electron microscopy (21). Furthermore, permanganate demonstrates well the endoplasmic reticulum, as originally reported by Luft (23). In the present chapter we summarize our recent observations with this fixative.

ADRENOMEDULLARY CELLS AND SIF CELLS

Small intensely fluorescent (SIF) cells contain large amounts of monoamines (4,5). Thus it was expected that permanganate fixation (25) would reveal granular vesicles (GVs) characteristic of monoamines in these cells. They were observed in two types of SIF cell in the paracervical ganglion of the rat. In one type the GVs were about 100 nm in diameter (Fig. 1B); the other type contained, in addition, larger GVs, 200 to 300 nm in diameter (12,18) (Fig. 1C). In the corresponding ganglion of the male rat, the hypogastric ganglion, two types of SIF cell were also observed (21) after permanganate fixation (Fig. 1A). SIF cells with granular vesicles were abundantly found also in the newborn rat after permanganate fixation (15,16).

It was therefore surprising to find that using the same fixation method as for the SIF cells above (i.e., 3% KMnO$_4$ buffered with Krebs-Ringer glucose to pH 7.0 or with 0.1 M phosphate buffer to pH 7.0 at 4°C for 30 to 60 min) no GVs were observed in the adrenal medulla of the rat (19) (Fig. 2A). We also tried 9% LiMnO$_4$ and used buffers at 0.2 M, but GVs were again absent. This applied to perinatal rats as well (19). By lowering the pH of the permanganate fixative to pH 5.0, distinct amine granules were observed in about one-third of the adrenomedullary cells, probably the norepinephrine cells (24) (Fig. 2C). In this modification of the permanganate fixative, we used ice-cold 3% KMnO$_4$ solution buffered with 0.1 or 0.2 M acetate buffer at pH 5.0. The specimens were fixed for 30 min and, after a brief rinse in the acetate buffer, block-stained with uranyl acetate. The epinephrine (E) cells showed only occasional precipitates in their vesicles. The rea-

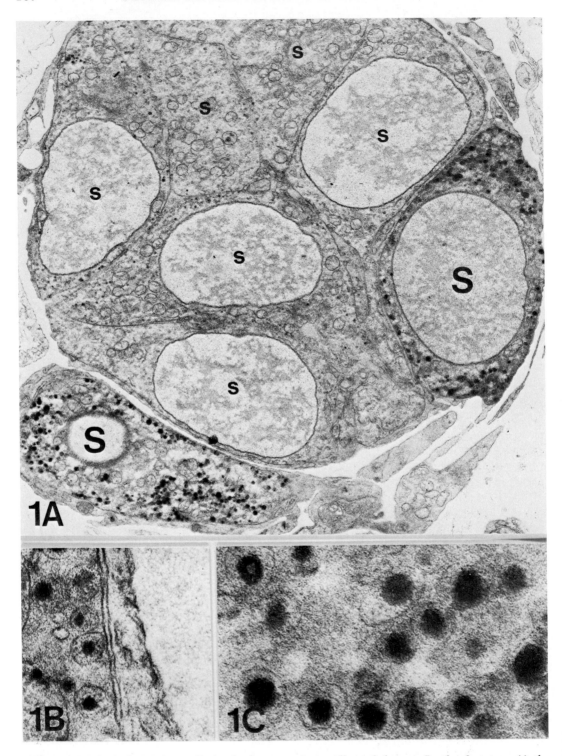

FIG. 1. A. Two types of SIF cells in the hypogastric ganglion of the rat. In the first type (s) the granulated vesicles are about 100 nm in diameter, and in the other type (S) granulated vesicles 200 to 300 nm in diameter are also found. **1B** and **C**: Two types of SIF cell from the paracervical ganglion of the rat at larger magnification than in **A.** KMnO₄ 3% immersion fixation, 0.1 M phosphate buffer, pH 7.0, no staining. **(A)** ×7,100. **(B)** ×75,000. **(C)** ×105,000. (From ref.21.)

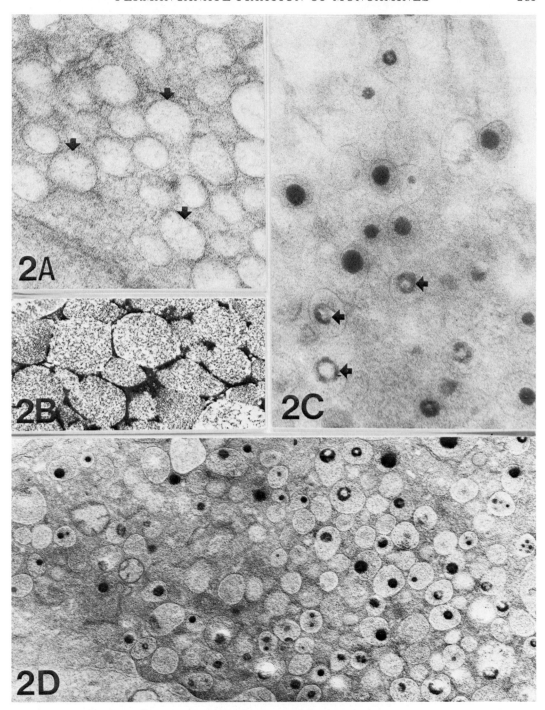

FIG. 2. A. Adrenomedullary cell from adult rat. All the chromaffin vesicles are clear (*arrows*). Fixation as in Fig. 1A, staining *en bloc* (19). ×66,400. **B:** Mast cell in the hypogastric ganglion of adult rat. The vesicles contain a reticular structure but no granules. Fixation as in Fig. 1A, no staining. ×35,000. **C:** Cytoplasm from adrenomedullary cell of adult rat containing electron-dense cores in the chromaffin vesicles. Some of the vesicles contain ring-like precipitates (*arrows*) and some are clear. KMn4 3% immersion fixation at pH 5.0, staining *en bloc* (24). ×52,000. **D:** Extra-adrenal chromaffin tissue cell (organ of Zuckerkandl) from a human fetus. A dark granular core is present in many vesicles. Ring-like and scattered precipitates are seen in some vesicles. Fixation as in Fig. 1A. ×30,100.

son why acid permanganate fixation favors the immediate precipitation reaction compared to fixation at a neutral pH was previously discussed (21,24). The different reaction after permanganate fixation in morphologically and developmentally closely related cells, the adrenomedullary cells and the SIF cells, might be that (a) the amine in the SIF cells is differently bound to the protein carrier complex, and/or (b) the protein carrier complex of the SIF cells is less soluble than in the adrenomedullary cells, thus allowing the formed precipitate to stay in the granule after the oxidation-reduction reaction.

MAST CELLS

Rat mast cells are known to contain considerable amounts of 5-hydroxytryptamine (5-HT) in their granules, but after permanganate fixation mast cells in the paracervical and the hypogastric ganglion of the rat were devoid of a dense core in their vesicles (19,21) (Fig. 2B). This applied also to rat mast cells isolated from the peritoneal fluid (19). It was suggested that the reactive ending of the 5-HT-protein complex does not react with permanganate, or that the reaction product is loosely bound and lost during tissue dehydration and embedding.

CAROTID BODY

Previously, Duncan and Yates (3) reported a lack of GVs in the glomus cells of the carotid body of the cat after neutral permanganate fixation. In the rat about 10% of the glomus cells contained GVs, as noted by Grönblad and Eränkö (7). The GVs were smaller (90 to 130 nm) than after glutaraldehyde-osmium tetroxide fixation.

By lowering the pH to 6.4 (0.1 M acetate buffer, fixation for 30 min at 4°C), an increased number of GVs were found in the glomus cells (Fig. 3). At this pH the chromaffin vesicles of the rat adrenomedullary cells were clear. By further lowering the pH to 5.0,

more GVs were found in approximately one-third of the glomus cells. Small granular vesicles about 50 nm in diameter were also observed in the glomus cells. Adrenergic nerve terminals in the carotid body with small dense-cored vesicles were easily demonstrated with neutral permanganate fixation even in the deeper parts of the organ.

EXTRA-ADRENAL CHROMAFFIN TISSUE

Abdominal paraganglia (organ of Zuckerkandl) from human fetuses contain large amounts of monoamines (MAs) (8,11). After permanganate fixation at neutral pH, a dense precipitate was formed inside the MA vesicles (Fig. 2D). The granular core was smaller than after glutaraldehyde-osmium tetroxide fixation, and some of the cores were circular with a clear center. Exocytosis-like openings were sometimes observed at the plasma membrane.

SYMPATHETIC GANGLIA

The cell bodies in adult rats contain a moderate number of large granular vesicles (LGVs) and varying numbers of small granular vesicles (SGVs) after neutral permanganate fixation (14,15). The dendrites also contain SGVs and LGVs (12). During development the situation is different. MAs are present in the neuroblasts of human fetal sympathetic ganglia (9), but after permanganate fixation no granular vesicles were observed (10). This suggests a loose nonparticulate binding of MAs during development, perhaps in the endoplasmic reticulum. When studying human fetal sympathetic ganglia in culture and maintaining these for up to 5 weeks *in vitro,* no granular vesicles developed after permanganate fixation, although formaldehyde-induced monoamine fluorescence was observed. This indicates also that we are not dealing with a permanganate penetration problem, since monolayers of tissue culture specimens were used (13).

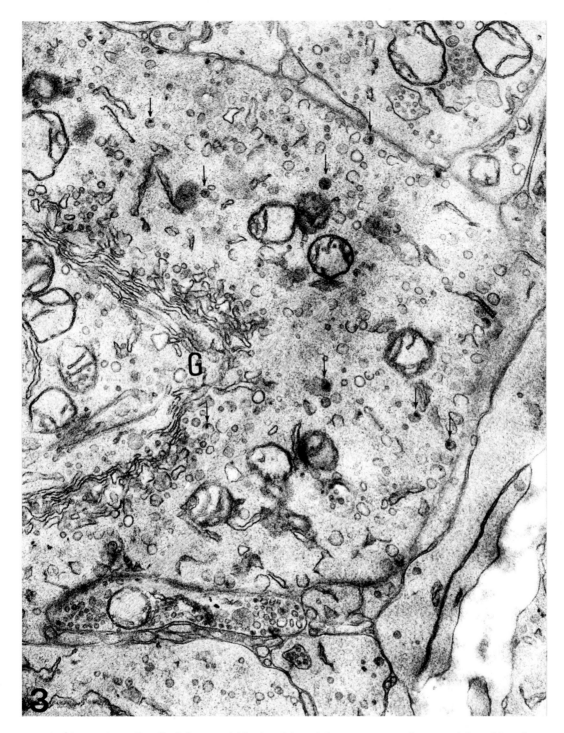

FIG. 3. Glomus (type I) cell of the carotid body of the adult rat. *Arrows* point to vesicles with a dense core. (G) Golgi apparatus. Many of the vesicles in the cytoplasm are devoid of a dense core. KMnO$_4$ 3% immersion fixation, 0.1 M acetate buffer, pH 6.4, no staining. ×31,360.

PERIPHERAL ADRENERGIC NERVES

The axons contain LGVs and SGVs, both of which can be demonstrated with permanganate fixation (14,15,21). The ratio of LGVs to SGVs is much higher in the axon than in the terminals (15). Permanganate fixation has also been used with success to study the endoplasmic reticulum (23). Axonic endoplasmic reticulum (ASER) is involved in axonic transport (2) and is especially prominent during development. Dixon and Gosling (1) suggested that monoamines in peripheral adrenergic nerves are stored in ASER during development since they found no SGV but an extensive amount of ASER. ASER was also reported by us (20,22) in developing MA synaptosomes. MA synaptosomes from neonatal rats differed morphologically from adult synaptosomes after permanganate fixation: (a) The developing MA synaptosomes contained neurotubule-like structures and relatively more smooth endoplasmic reticulum. (b) The number of synaptic vesicles was smaller than in the adult. (c) The dense core of LGVs was smaller than in the adult. (d) Large agranular vesicles were found more often than in the adult. (e) Dense cores were occasionally seen inside the tubules of the endoplasmic reticulum.

CONCLUDING REMARKS

Richardson (25) first demonstrated that concentrated permanganate salts fix especially well the monoamine storage vesicles. Since then much information on the fine structural distribution of monoamines in the peripheral and central nervous systems has been obtained, and permanganate fixation has been considered a specific means of demonstrating monoamines. Our recent observations demonstrate that, surprisingly, the monoamines in the adrenomedullary cells of the rat are not visualized with the conventional neutral permanganate fixation, and an acid permanganate demonstrated the norepinephrine cells.

During development, granular vesicles are not found in all nervous structures where glutaraldehyde-osmium tetroxide reveals such vesicles. Thus care must be taken when interpreting a negative finding after permanganate fixation.

In our hands, the ease with which granular vesicles in monoamine-containing paraneurons (6) after neutral pH have been demonstrated is of the following order (from the easiest to the most difficult): SIF cells > extra-adrenal chromaffin tissue > carotid body > adrenal medulla. This finding suggests that the granular vesicles in different paraneurons have different amine-binding properties.

REFERENCES

1. Dixon, J.S., and Gosling, J.A. (1976): Extravesicular noradrenaline in developing peripheral adrenergic nerves. *Experientia,* 32:1052–1053.
2. Droz, B., Rambourg, A., and Koening, H.L. (1975): The smooth endoplasmic reticulum: Structure and role on the renewal of axonal membrane and synaptic vesicles by fast axonal transport. *Brain Res.,* 93:1–13.
3. Duncan, D., and Yates, R. (1967): Ultrastructure of the carotid body of the cat as revealed by various fixatives and the use of reserpine. *Anat. Rec.,* 157:667–682.
4. Eränkö, O., editor (1976): *SIF Cells, Structure and Function of the Small Intensely Fluorescent Cells.* Fogarty International Center Proceedings, No. 30., DHEW Publication No. (NIH) 76-942. Government Printing Office, Washington, D.C.
5. Eränkö, O., and Eränkö, L. (1971): Small intensely fluorescent granule-containing cells in the sympathetic ganglion of the rat. *Prog. Brain Res.,* 34:39–51.
6. Fujita, T. (1976): The gastro-enteric endocrine cell and its paraneuronic nature. In: *Chromaffin, Enterochromaffin and Related Cells,* edited by R.E. Coupland and T. Fujita, pp. 191–208. Elsevier, Amsterdam.
7. Grönblad, M., and Eränkö, O. (1978): Fine structure of dense-cored vesicles in glomus cells of rat carotid body after fixation with permanganate or glutaraldehyde. *Histochemistry,* 57:305–312.
8. Hervonen, A. (1971): Development of catecholamine-storing cells in human paraganglia and adrenal medulla: A histochemical and electron microscopical study. *Acta Physiol. Scand. [Suppl.],* 368:1–94.
9. Hervonen, A., and Kanerva, L. (1972): Cell types of human fetal superior cervical ganglion. *Z. Anat. Entwicklungsgesch.,* 137:257–269.

10. Hervonen, A., Kanerva, L., and Hervonen, H. (1977): Ultrastructure of human fetal sympathetic ganglia after permanganate fixation—absence of granular vesicles. *Neurosci. Lett.,* 5:303-308.

11. Hervonen, A., Kanerva, L., Partanen, S., and Vaalasti, A. (1978): The histochemistry and fine structure of the paraganglia of man. In: *Peripheral Neuroendocrine Interaction,* edited by R.E. Coupland and W.G. Forssmann, pp. 48-59. Springer, Berlin.

12. Hervonen, A., Kanerva, L., and Teräväinen, H. (1972): The fine structure of the paracervical ganglion of the rat after permanganate fixation. *Acta Physiol. Scand.,* 85:506-510.

13. Hervonen, H., Hervonen, A., and Kanerva, L. (1978): Light and electron microscopic monoamine histochemistry of the human fetal sympathetic ganglion in culture. *Histochem. J.,* 10:271-286.

14. Hökfelt, T. (1968): In vitro studies on central and peripheral monoamine neurons at the ultrastructural level. *Z. Zellforsch.,* 91:1-74.

15. Hökfelt, T. (1971): Distribution of noradrenaline storing particles in peripheral adrenergic neurons as revealed by electron microscopy. *Acta Physiol. Scand.,* 76:427-440.

16. Kanerva, L. (1972): Ultrastructure of sympathetic ganglion cells and granule-containing cells in the paracervical (Frankenhäuser) ganglion of the newborn rat. *Z. Zellforsch.,* 126:25-40.

17. Kanerva, L. (1972): Development, histochemistry and connections of the paracervical (Frankenhäuser) ganglion of the rat uterus: A light and electron microscopical study. *Acta Inst. Anat. (Helsinki) (Suppl.),* 2.

18. Kanerva, L., and Hervonen, A. (1976): SIF cells, short adrenergic neurons and vacuolated nerve cells of the paracervical (Frankenhäuser) ganglion. In: *SIF Cells. Structure and Function of the Small Intensely Fluorescent Sympathetic Cells,* edited by O. Eränkö, pp. 19-34. Fogarty International Center Proceedings No. 30, DHEW Publication No. NIH 76-942, Government Printing Office, Washington, D.C.

19. Kanerva, L., Hervonen, A., and Rechardt, L. (1977): Permanganate fixation demonstrates the monoamine-containing granular vesicles in the SIF cells but not in the adrenal medulla or mast cells. *Histochemistry,* 52:61-72.

20. Kanerva, L., Hervonen, A., and Tissari, A.H. (1976): Ultrastructural identification of monoaminergic synaptosomes from one-day-old rat brain. *Histochemistry,* 48:233-240.

21. Kanerva, L., Hervonen, A., Rechardt, L., Hervonen, H., and Partanen, M. (1978): Ultrastructural localization of granules and smooth endoplasmic reticulum with permanganate fixation in monoamine cells. In: *Peripheral Neuroendocrine Interaction,* edited by R.E. Coupland and W.G. Forssmann, pp. 15-28. Springer, Berlin.

22. Kanerva, L., Tissari, A.H., Suurhasko, B.V.A., and Hervonen, A. (1977): Ultrastructural characterization of synaptosomes from neonatal and adult rats with special reference to monoamines. *J. Comp. Neurol.,* 174:631-658.

23. Luft, J.H. (1956): Permanganate—a new fixative for electron microscopy. *J. Biophys. Biochem. Cytol.,* 2:799-802.

24. Rechardt, L., Kanerva, L., and Hervonen, H. (1977): Ultrastructural demonstration of amine granules in the adrenal medullary cells of the rat using acidic permanganate fixation. *Histochemistry,* 54:339-343.

25. Richardson, K.C. (1966): Electron microscopic identification of autonomic nerve endings. *Nature,* 210:756.

Histochemistry and Cell Biology of Autonomic Neurons, SIF Cells, and Paraneurons,
edited by O. Eränkö et al.
Raven Press, New York © 1980.

The Fate of Adrenergic Fibers Which Enter the Superior Cervical Ganglion

William G. Dail, Susanne Khoudary, Celia Barraza, Heather M. Murray, and Carol Bradley

Department of Anatomy, University of New Mexico, School of Medicine, Albuquerque, New Mexico 87131

A role for catecholamines in the functions of autonomic ganglia has been proposed by several investigators (2,9,24,27). The effect of catecholamines, originally described as a depression of transmission (1,27), may take place at two sites in the sympathetic ganglion. First, catecholamines may act postsynaptically to result in a hyperpolarizing potential in the postganglionic sympathetic neuron. This pathway seems to involve an interneuron which, when activated by preganglionic nerve impulses, releases a catecholamine onto adrenergic receptor sites on the postganglionic cell (14,24–26). In a second proposal, catecholamines may act on adrenergic receptors on preganglionic terminals to inhibit acetylcholine release (3,4,12, 13).

In support of an important role of catecholamines in the mechanisms of autonomic ganglia, several studies have revealed an extensive plexus of adrenergic varicose fibers in close proximity to ganglion cells (18,21,32, 36). It is apparent that, depending on ganglia and species, small intensely fluorescent cells (SIF cells, autonomic interneurons) and the principal neurons may contribute to this plexus. In the superior cervical ganglion of the rabbit, an important animal model, it has

been suggested that the autonomic ground plexus is due largely to the processes of SIF cells. Electron microscopic studies have shown that in some species SIF cells form efferent contacts with the principal neurons (6,28,37); and some evidence indicates that synapses occur between catecholamine-containing dendrites of ganglion cells (15).

In comparison, hypotheses of the possible presynaptic effects of catecholamines in autonomic ganglia have received relatively little study. It is probable that direct contacts between adrenergic terminals and cholinergic preganglionic terminals is not required in these models (nor have such contacts been identified in ganglia). Whether free endings of ganglion cells or SIF cells (e.g., those in close relationship to blood vessels) could serve as a source of catecholamines, which in turn would inhibit the release of acetylcholine, is not known.

Common to all current models of the roles of catecholamines in the functions of sympathetic ganglia is the general belief that the source of the catecholamines is endogenous, i.e., from the population of principal neurons and SIF cells within the ganglion. This view has been challenged in a series of investigations which purport to show that the norepinephrine (NE) released as a result of preganglionic nerve stimulation is due to adrenergic fibers in the cervical sympathetic nerve (30,31). The authors cite histofluorescence data which show that adrenergic fibers enter the caudal margin of the superior cervical ganglion (SCG) in some species, including the SCG of the rabbit (20,21). Presumably, the proposed system of Noon et al. (31) accounts for a proportion of the adrenergic ground plexus seen in the SCG, although the authors are careful to point out that it is not known where the fibers terminate in the ganglion. Studies which have used deafferentation (decentralization, denervation, section of the preganglionic nerve) have consistently pointed to an origin of the autonomic ground plexus from nerve elements within the ganglion. Histofluorescence (17,21) and electron microscopic studies (16,34,35,38) have failed to detect a loss of adrenergic terminals after section of the preganglionic input. However, many of these studies were not quantitative or were performed on ganglia which have relatively few adrenergic endings. Moreover, many of the studies were done on ganglia which had been chronically decentralized; and it is not known to what degree possible sprouting may have affected the results (7,34).

Electron microscopic studies have shown that the SCG of the rabbit has a remarkably high percentage of adrenergic terminals (8,35). It is not known if the adrenergic terminals described in these studies might, in part, arise from peripherally located ganglion cells. Another correlate of the hypothesis of Noon et al. has also not been adequately investigated: that the release of NE might be due to vascular fibers derived from the cervical sympathetic trunk.

The purpose of this investigation is first to determine in a quantitative manner if decentralization results in a loss of adrenergic terminals in the SCG of the rabbit, and, second, to define the source of the adrenergic fibers present in the cervical sympathetic trunk.

MATERIALS AND METHODS

Young adult male New Zealand rabbits (3 to 5 kg) were used in this study. In six rabbits ligatures were placed on the cervical sympathetic trunk (CST) to determine the presence and course of adrenergic fibers in the trunk. In each of these animals a thread was tied on the CST 2 to 3 mm from the caudal margin of the SCG. Four days later the ligated portion of the trunk was removed and treated for the demonstration of catecholamines according to the method of de la Torre and Surgeon (10). In all catecholamine histofluorescence studies, rabbits received a single injection of nialamide (100 mg/kg) 4 hr before sacrifice. In a retrospective study, ligatures were tied around the external carotid nerve (ECN) branch of the SCG in four rabbits to study the possible buildup of catecholamines in the cervical sympathetic chain.

The cervical sympathetic trunk of several rabbits (six in the final quantitative study) was cut 2 to 3 mm from the SCG to determine if adrenergic fibers in the trunk terminated in the SCG. The left cervical sympathetic chain served as the control. On the operated side care was taken to avoid injury to the small blood vessels adherent to the CST. Initial studies were aimed at selecting an appropriate time at which to examine the SCG for possible loss of adrenergic terminals. With a survival time of 24 hr, many preganglionic terminals still appeared normal; whereas at a longer survival period (48 hr), degeneration was so advanced that it became impossible to determine the type of degenerating ending on the basis of its content of transmitter vesicles. A survival time of 36 hr was finally chosen since some degenerative changes could be seen in the majority of preganglionic terminals. However, even at this interval, there was considerable variation in the stage of degeneration among terminals, with some preganglionic terminals still normal in appearance. Thirty minutes prior to removing the ganglia, or prior to the denervation, 5-hydroxydopamine (5-OHDA)

100 mg/kg was given in an ear vein. Ganglia were fixed by perfusion with glutaraldehyde and paraformaldehyde as previously described (8). To ensure adequate sampling, the SCG was divided into caudal and rostral halves at its normal constriction above the origin of the external carotid nerve. Thin sections were taken from two blocks in each half of the ganglion and placed on grids. The entire thin section, which usually had portions of about 100 ganglion cells, was surveyed for terminals. In control ganglia, terminals were categorized into two groups: adrenergic endings, distinguished by their content of small vesicles labeled with 5-OHDA, and endings containing small clear vesicles (presumed cholinergic preganglionic terminals). Further distinction was made on the basis of whether the profile showed features of a synapse (synaptic densities, polarity of vesicles toward the membrane, etc.) or demonstrated no efferent contacts (adrenergic varicosity). This information was gathered in an effort to determine if denervation caused a shift in the ratio of adrenergic synapses to adrenergic varicosities since one of the effects of deafferentation may be the lifting off of synapses from the postsynaptic structure (29). In denervated and intact ganglia, records were also kept of the viewing time in the electron microscope in an effort to compare the incidence of adrenergic terminals encountered in experimental and control ganglia in a unit of time. In control ganglia, a total of 1,800 terminals were counted and classified. By comparison, a count of 1,800 endings 36 hr after denervation revealed that about 1,381 of the 1,800 terminals were in various stages of degeneration.

Since it is also possible that the release of NE elicited by stimulation of the preganglionic nerve might be coming from vascular nerves (31), the perimeter of all vessels found in the sections in the quantitative study was surveyed for adrenergic fibers in experimental and control ganglia.

The obvious processes of granule-containing cells (SIF cells) were not included in this quantitative study, nor were those neurites which contained scattered collections of small dense-cored vesicles (presumed dendrites) (8).

Horseradish Peroxidase Studies

Based on our results from ligation of the external carotid nerve, it was suspected that ganglion cells caudal to the SCG might send their axons through this efferent nerve. To locate the cell bodies of these ganglion cells along the cervical sympathetic chain, horseradish peroxidase (HRP) was applied to the cut end of the external carotid nerve. In six rabbits, a 4- to 6-mm segment of the external carotid nerve was dissected free from the external carotid artery. This distal end of the cut nerve was placed in a small boat containing a solution of HRP (25%, Sigma type VI). The nerve was exposed to the solution for about 90 min. Twenty-four hours later the rabbits were again anesthetized and perfused with saline followed by cold glutaraldehyde (2.5% in Sorenson's phosphate buffer). The SCG, segments of the cervical sympathetic trunk, the middle cervical ganglion, and the stellate ganglion were removed and treated for the demonstration of peroxidase activity with the *o*-dianisidine method (5).

RESULTS

Effect of Ligation of the Cervical Sympathetic Trunk

A ligature placed on the CST resulted in a large buildup of catecholamines proximal to the tie (Figs. 1 and 2). No increase was seen on the distal side of the ligature. Swollen axons, some of which were varicose, and reactive ganglion cells could be traced for approximately 1 cm down the sympathetic trunk 4 days following the ligation. Other sites such as the area of the accessory cervical ganglion (ACG) and more caudal collections of ganglion cells were not examined for possible reactions to this procedure. The pattern

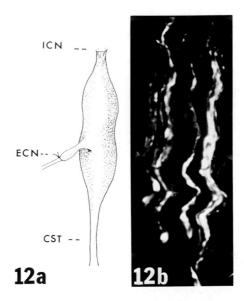

ICN

ECN

CST

12a   **12b**

FIG. 12. Ligation of the external carotid nerve **(a)** and the effect on adrenergic nerve fibers in the rostral part of the cervical sympathetic trunk **(b)**. See Fig. 1 for abbreviations.

tributed with the external carotid nerve. Similar experiments with other efferent nerves of the SCG have not been attempted.

HRP Studies

In an attempt to locate the ganglion cells which send adrenergic fibers up the cervical sympathetic chain, HRP was applied to the cut end of the ECN (Fig. 13a). HRP-labeled neurons were distributed in the SCG in a manner similar to that found by ligating the ECN; i.e., they were mostly located in the caudal portion of the ganglion. Figure 13b illustrates the very distal end of the CST as it enters the SCG. Some HRP-filled neurons are present on either side of a caudally directed bundle of HRP-labeled fibers. At about 3 cm from the caudal margin of the SCG can be found a collection of ganglion cells, many of which are labeled with HRP (the accessory cervical sympathetic ganglion). More dramatic labeling of ganglion cells, however, occurs in the middle cervical sympathetic ganglion, where as many as 40% of the cells may be filled with HRP (Fig. 13c). A few labeled cells could always be

found in the stellate ganglion, a full 8 cm from the SCG (Fig. 13d).

CONCLUDING REMARKS

The failure of deafferentation to result in degeneration of adrenergic profiles indicates that the SCG may not be a target tissue for adrenergic fibers in the CST. On the contrary, we found that many of the adrenergic fibers in the CST are distributed with the ECN. This observation agrees with the work of Douglas and Ritchie (11), who are able to obtain a C potential in the CST by stimulating efferent nerves of the rabbit SCG. They ascribed the potential to a collection of neurons about 3 cm caudal to the SCG and, taking the lead of Langley (23), termed the cells the accessory cervical sympathetic ganglion. However, Noon et al. (31) showed that decentralization below the ACG abolished the release of NE for the SCG. Therefore it was reasoned that ganglion cells somewhere caudal to the ACG were responsible for the release of NE elicited by stimulation of the CST. Our studies with HRP indeed showed that some ganglion cells in the middle and

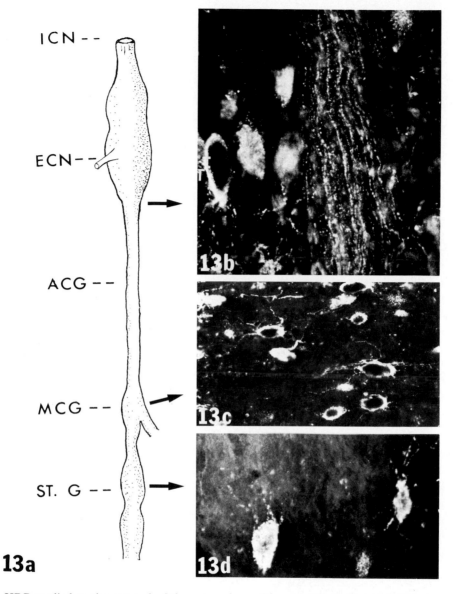

FIG. 13. HRP applied to the cut end of the external carotid nerve **(a)** results in labeled neurons and nerve fibers in the rostral part of the cervical sympathetic trunk **(b)** and the middle cervical ganglion **(c)**, and a few nerve cells in the stellate ganglion **(d)**. (ICN) Internal carotid nerve. (ECN) External carotid nerve. (ACG) Accessory cervical ganglion. (MCG) Middle cervical ganglion. (ST.G) Stellate ganglion. **b:** ×200. **c:** ×120. **d:** ×120.

stellate ganglia send their fibers up the CST to enter the SCG; however, these fibers seem to continue on into the ECN. Based on our studies with HRP, a model of the organization of the cervical sympathetic chain in the rabbit is presented in Fig. 14.

As of now, we can offer no sound explanation to fit our data with those of Noon et al. (31) unless it could be speculated that the purported adrenergic fibers terminate in a singular area of the SCG not sampled in our study. It is well to point out that a recent

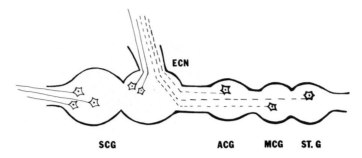

FIG. 14. Organization of the cervical portion of the sympathetic trunk in the rabbit. Ganglion cells as far caudal as the middle and stellate ganglia send their axons up the cervical sympathetic trunk to join the external carotid nerve. (SCG) Superior cervical ganglion. (ECN) External carotid nerve. (ACG) Accessory cervical ganglion. (MCG) Middle cervical ganglion. (ST.G) Stellate ganglion.

study offered evidence which suggests that cells in the celiac ganglion of cats may innervate cells in the opposite celiac ganglion (22). However, this has yet to be verified by electron microscopy.

An explanation for the presence of adrenergic fibers in the CST might be found in the embryology of the cervical portion of the sympathetic nervous system. Pick and Sheehan (33) suggest that primordia of sympathetic ganglia may variously split and fuse, resulting in connections of several ganglia to one cervical nerve. It seems plausible to suggest that some ganglion cells which form a portion of the primordium of the SCG may be drawn downward in the CST. The axons of these displaced ganglion cells therefore would course in the CST to reach their normal pathway for distribution, efferent nerves of the SCG.

REFERENCES

1. Bulbring, E. (1944): The action of adrenalin on transmission in the superior cervical ganglion. *J. Physiol. (Lond)*, 103:55–67.
2. Bulbring, E., and Burn, J.H. (1942): An action of adrenaline on transmission in sympathetic ganglia which may play a part in shock. *J. Physiol. (Lond)*, 101:289–303.
3. Christ, D.D., and Nishi, S. (1971): Site of adrenalin blockage in the superior cervical ganglion of the rabbit. *J. Physiol. (Lond)*, 101:289–303.
4. Christ, D.D., and Nishi, S. (1971): Effects of adrenaline on nerve terminals in the superior cervical ganglion of the rabbit. *Br. J. Pharmacol.*, 41:331–338.

5. Colman, D.R., Scalia, F., and Cabrales, E. (1976): Light and electron microscopic observations on the anterograde transport of HRP in the optic pathway in the mouse and rat. *Brain Res.*, 102:156–163.
6. Dail, W.G. (1977): Histochemical and fine structural studies on SIF cells in the major pelvic ganglion of the rat. In: *SIF Cells: Structure and Function of the Small Intensely Fluorescent Cells*, edited by O. Eränkö, pp. 8–18. Fogarty International Proceedings No. 30. DHEW Publication No. (NIH) 76-942. Government Printing Office, Washington, D.C.
7. Dail, W.G., and Evan, A.P. (1978): Effects of chronic deafferentation on adrenergic ganglion cells and small intensely fluorescent cells. *J. Neurocytol.*, 7:25–37.
8. Dail, W.G., and Evan, A.P. (1978): Ultrastructure of adrenergic terminals and SIF cells in the superior cervical ganglion of the rabbit. *Brain Res.*, 148:469–477.
9. DeGroat, W.C., and Volle, R.L. (1966): The actions of catecholamines on transmission in the superior cervical ganglion of the cat. *J. Pharmacol. Exp. Ther.*, 154:211–218.
10. De la Torre, J.C., and Surgeon, J.W. (1977): A methodological approach to rapid and sensitive monoamine histofluorescence using a modified glyoxylic acid technique: The SPG method. *Histochemistry*, 49:81–93.
11. Douglas, W.W., and Richie, J.M. (1956): The conduction of impulses through the superior cervical ganglia of the rabbit. *J. Physiol. (Lond)*, 133:220–231.
12. Dun, N., and Karczmar, A.G. (1977): The presynaptic site of action of norepinephrine in the superior cervical ganglion of the guinea pig. *J. Pharmacol. Exp. Ther.*, 200:328–335.
13. Dun, N., and Nishi, S. (1974): Effects of dopamine on the superior cervical ganglion of the rabbit. *J. Physiol. (Lond)*, 239:155–164.
14. Eccles, R.M., and Libet, B. (1961): Origin and blockage of the synaptic responses of curarized sympathetic ganglia. *J. Physiol. (Lond)*, 157:484–503.

15. Elfvin, L-G. (1971): Ultrastructural studies on the synaptology of the inferior mesenteric ganglion of the cat. *J. Ultrastruct. Res.,* 37:432–448.

16. Grillo, M.A. (1966): Electron microscopy of sympathetic tissues. *Pharmacol. Rev.,* 18:387–399.

17. Hamberger, B., and Norberg, K-A. (1965): Studies on some systems of adrenergic synaptic terminals in the abdominal ganglia of the cat. *Acta Physiol. Scand.,* 65:235–242.

18. Hamberger, B., Norberg, K-A., and Sjoqvist, T. (1964): Evidence for adrenergic nerve terminals and synapses in sympathetic ganglia. *Int. J. Neuropharmacol.,* 2:279–282.

19. Hamori, J., Land, E., and Simon, L. (1968): Experimental degeneration of the preganglionic fibers in the superior cervical ganglion of the cat. *Z. Zellforsch.,* 90:37–52.

20. Jacobowitz, D. (1970): Catecholamine fluorescence studies of adrenergic neurons and chromaffin cells in sympathetic ganglia. *Fed. Proc.,* 29:1929–1944.

21. Jacobowitz, D., and Woodward, J.K. (1968): Adrenergic neurons in the cat superior cervical ganglion and cervical sympathetic nerve trunk: A histochemical study. *J. Pharmacol. Exp. Ther.,* 162:213–226.

22. Kelts, K.A., Whitlock, D.G., Ledbury, P.A., and Reese, B.H. (1979): Postganglionic connections between sympathetic ganglia in the solar plexus of the cat demonstrated autoradiographically. *Exp. Neurol.,* 63:120–134.

23. Langley, J.N. (1893): On an 'accessory' cervical ganglion in the cat and notes on the rami of the superior cervical ganglion. *J. Physiol. (Lond),* 14:i–ii.

24. Libet, B. (1970): Generation of slow inhibitory and excitatory postsynaptic potentials. *Fed. Proc.,* 29:1945–1956.

25. Libet, B. (1976): The role that SIF cells play in ganglionic transmission. *Adv. Biochem. Psychopharmacol.,* 16:541–546.

26. Libet, B., and Owman, C.H. (1974): Concomitant changes in formaldehyde-induced fluorescence of dopamine interneurons and in slow inhibitory postsynaptic potentials of the rabbit superior cervical ganglion, induced by stimulation of the preganglionic nerve or by a muscarinic agent. *J. Physiol. (Lond),* 237:635–662.

27. Marrazzi, A.S. (1939): Electrical studies on the pharmacology of autonomic synapses. II. The action of a sympathomimetic drug (epinephrine) on sympathetic ganglia. *J. Pharmacol. Exp. Ther.,* 65:395–404.

28. Matthews, M.R., and Raisman, G. (1969): The ultrastructure and somatic efferent synapses of small granule-containing cells in the superior cervical ganglion. *J. Anat.,* 105:255–282.

29. Matthews, M.R., and Raisman, G. (1972): A light and electron microscopic study of the cellular response to axonal injury in the superior cervical ganglion of the rat. *Proc. R. Soc. Lond. [Biol],* 181:43–79.

30. Noon, J.D., and Roth, R.H. (1975): Some physiological and pharmacological characteristics of the stimulus induced release of norepinephrine from the rabbit superior cervical ganglion. *Naunyn Schmiedebergs Arch. Pharmacol.,* 291:163–174.

31. Noon, J.P., McAfee, D.A., and Roth, R.H. (1975): Norepinephrine release from nerve terminals within the rabbit superior cervical ganglion. *Naunyn Schmiedebergs Arch. Pharmacol.,* 291:139–162.

32. Norberg, K-A., and Hamberger, B. (1964): The sympathetic adrenergic neuron: Some characteristics revealed by histochemical studies on the intraneuronal distribution of the transmitter. *Acta Physiol. Scand. [Suppl. 238],* 62:1–42.

33. Pick, J.D., and Sheehan, D. (1946): Sympathetic rami in man. *J. Anat.,* 80:12–20.

34. Quilliam, J.D., and Tamarind, J.P. (1972): Electron microscopy of degenerative changes in decentralized rat superior cervical ganglia. *Micron,* 3:454–472.

35. Tamarind, D.L., and Quilliam, J.P. (1971): Synaptic organization and other ultrastructural features of the superior cervical ganglion of the rat, kitten and rabbit. *Micron,* 2:204–234.

36. Watanabe, H. (1971): Adrenergic nerve elements in the hypogastric ganglion of the guinea pig. *Am. J. Anat.,* 130:305–330.

37. Williams, T.H., and Palay, S.L. (1969): Ultrastructure of the small neurons in the superior cervical ganglion. *Brain Res.,* 15:17–34.

38. Yokota, R., and Yamauchi, A. (1974): Ultrastructure of the mouse superior cervical ganglion, with particular reference to the pre- and postganglionic elements covering the soma of its principal neurons. *Am. J. Anat.,* 140:281–298.

Histochemistry and Cell Biology of Autonomic Neurons, SIF Cells, and Paraneurons,
edited by O. Eränkö et al.
Raven Press, New York © 1980.

Ultrastructural Studies on Rat Sympathetic Ganglia After 5-Hydroxydopamine Administration

R. M. Santer, J. D. Lever, *K. S. Lu, and S. A. Palmer

*Department of Anatomy, University College, Cardiff, CF1 1XL, Wales; and *Department of Anatomy, College of Medicine, National Taiwan University, Taipei, Taiwan*

Fluorescence histochemical studies for the localization of biogenic amines have demonstrated an intraganglionic plexus of norepinephrine (NE) containing varicosities within autonomic ganglia of several species (9,11). The nature of these varicosities, which pass among the principal ganglion cells and in some cases form calyces around them, has attracted attention from researchers with a view to understanding more fully the architecture of autonomic ganglia in general, and of sympathetic ganglia in particular.

It has been suggested that a proportion, at least, of these varicose fibers may originate from small intensely fluorescent (SIF) cells (4,7), some of which are undoubtedly interneuronal in position (13,15) and which produce varicose processes. However, the difficulty of characterizing the vesicle populations of such processes at an ultrastructural level (3,6) has hindered attempts at establishing their relationship with principal ganglion cells.

Another possible origin of intraganglionic varicose fibers may be the dendritic arborizations and/or recurrent collaterals (21) of the principal neurons themselves. Certainly, the demonstration of biogenic amine storage in dendrites of neurons of the rat superior cervical ganglion (17) and of substantia nigra (1), as well as the demonstration of amine release from dendrites of these last neurons (8), supports the concept of a functional, integrative role for dendrites. The possibility of a noradrenergic pathway originating outside autonomic ganglia and terminating within them was suggested by Noon et al. (16) and cannot be excluded.

The present study reports ultrastructural observations on the celiac-mesenteric ganglion (CMG) complex of the rat following the administration of the false transmitter 5-hydroxydopamine (5-OHDA) to mark sites of vesicular storage of NE in the intact ganglion and after postganglionic section.

MATERIALS AND METHODS

White Wistar rats 4 months of age were used in this study. 5-OHDA 100 mg/kg (Sigma) dissolved in 0.5 ml saline containing ascorbic acid 0.2 μg/ml was administered under ether anesthesia via the cannulated femoral vein. After 30 min the thorax was opened and the animal perfused for 10 min via the descending thoracic aorta with ice-cold 3% glutaraldehyde in 0.1 M sodium cacodylate buffer at pH 7.2. The CMG complex was then dissected out, diced into pieces less than 1 mm³, and fixed for 2 hr in fresh fixative at 4°C. Following a buffer wash overnight and osmication in 1% OsO₄ buffered with 0.1 M sodium cacodylate, the tissue pieces were rinsed in buffer and dehydrated through a graded ethanol series prior to embedding in Spurr resin. Fine sections showing pale gold–silver interference colors were mounted on copper grids, stained with lead citrate and uranyl acetate, and viewed in Siemens Elmiskop and AE1 801 electron microscopes at 40 to 60 kV.

Thirty minutes before sacrifice the false transmitter (5-OHDA) was administered (as described above) to six intact rats and four neurectomized rats. Laparotomy was performed on these rats under ether anesthesia with aseptic precautions in order to expose the CMG complex. Nerve bundles from the CMG along the celiac and anterior mesenteric arteries were cut with fine scissors approximately 0.5 cm from the ganglion. Animals were sacrificed either at 1 day (two rats) or 3 days (two rats) after ganglionectomy. Controls, intact and operated, were similarly injected with saline containing ascorbic acid 0.2 μg/ml.

RESULTS

Axodendritic and axosomatic preganglionic cholinergic-type synaptic contacts were frequently observed, and in structural appearance were similar to those previously described (5). These nerve terminals were consistently unlabeled by 5-OHDA, although they were often in the vicinity of sites labeled by this marker.

Dendrites were distinguished from axonal processes by a content of ribosomes and in some sections were observed in continuity with nerve cell bodies. Vesicle collections showing 5-OHDA loading were superficially situated along dendrites and in obvious dendritic spines.

5-OHDA Labeling at Sites of Neuronal Contact

Synaptic contacts between neuronal processes containing 5-OHDA-labeled vesicles and typical dendrites were commonly observed (Fig. 1). Some of the vesicles within the presynaptic processes were circular, and others were oval or irregular in outline with a mean diameter of 45 nm. Positive labeling with 5-OHDA was found in some of these vesicles. The population density of vesicles within such terminals was invariably high.

Terminal sites were recognized by specializations of pre- and postsynaptic membranes. Although both membranes were thickened and of increased electron density, these features were usually more prominent in the postsynaptic component (Fig. 1). The synaptic cleft was consistently 21 nm wide and contained a material of moderate electron density.

At these terminal sites the postsynaptic components presented all the features characteristic of dendrites as well as occasional 5-OHDA-labeled vesicles (Fig. 1) with a mean diameter of 47 nm.

Schwann-Invested 5-OHDA-Labeled Neuronal Profiles

Labeled neuronal processes not in synaptic contact with other elements were observed throughout the ganglion fields between well-defined groups of neurons. They were invariably enclosed by Schwann cells, being covered by mesaxons of one or more concentric turns. Other neuronal elements were also enclosed by the same Schwann cells but not in direct contact with the labeled processes. These other elements were interpreted as (a) near-terminal cholinergic axons because of their characteristic vesicle content; (b) axons without near-terminal characteristics; and (c) putative but unlabeled dendrites.

Labeled processes were characterized by clusters of pleomorphic vesicles, many of which were circular in outline and averaged

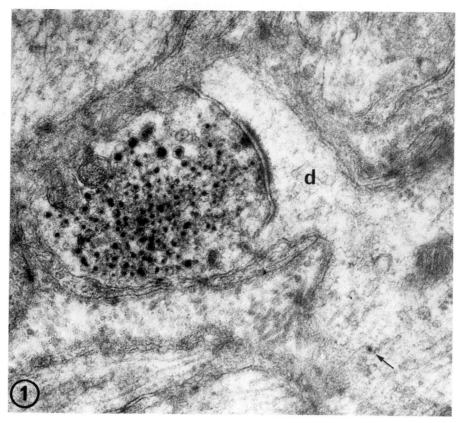

FIG. 1. A 5-OHDA-labeled neuronal process in synaptic relationship with the dendrite (d) of a principal neuron. Note the loaded vesicle (*arrow*) in this dendrite. ×48,000.

50 nm in diameter. Although some of these vesicles were loaded with 5-OHDA, others were not. Narrow intervesicular tubular connections were sometimes observed, as were connections with larger membrane profiles, which might be regarded as smooth reticulum.

Neuronal Relationships Within Neurectomized CMG

Swollen unmyelinated neuronal processes were frequently encountered throughout the ganglion enclosed by Schwann cell coverings together with unaltered neuronal profiles. The cytoplasm of these swollen processes was occupied by large numbers of 5-OHDA-labeled and clear vesicles—rounded, oval, or tubular in shape (mean cross section 85 nm)—and by prominent secondary lysosomes (many with myelinic figures), multivesicular bodies, mitochondria, neurotubules, and smooth endoplasmic reticulum. Evidence of the endocytosis of surface membrane was provided by the finding of numerous rounded, coated vesicular inclusions (110 nm in diameter) some of which were seen in surface contact or communication (Fig. 2).

Swollen myelinated neuronal processes were infrequently encountered. These elements were characterized by an intact myelin sheath and neurocytoplasmic concentrations of mitochondria, smooth endoplasmic reticulum, secondary lysosomes, and rounded vesicles (105 nm diameter), some with moderately dense cores but none labeled by 5-

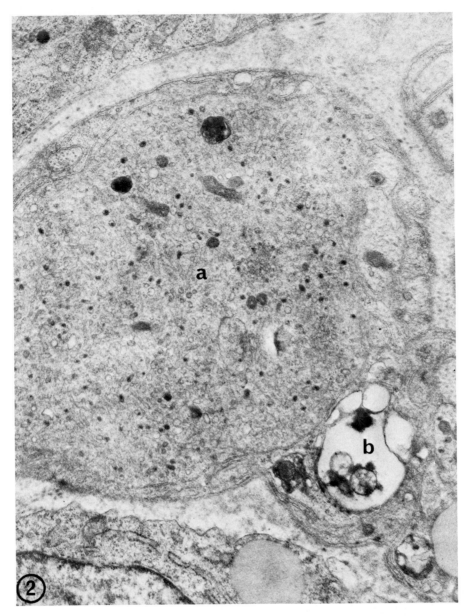

FIG. 2. Twenty-four hours after post-ganglionectomy, nerve fascicles are observed in which unaffected, swollen (a) and degenerated (b) nerves can be seen within the same Schwann cell covering. ×21,000.

OHDA. These processes were interpreted as either sensory nerves or preganglionic parasympathetic nerves *en passage.*

Unmyelinated degenerated neuronal processes were observed in all low-magnification (× 4,000) electron micrographs of postgan-glionectomized CMG (Figs. 2 and 3). These processes were contained within the same Schwann cell coverings as unaffected cholin-ergic-type nerves, putative dendrites, and swollen labeled neuronal processes. Degener-ated profiles were not swollen and had well-defined mesaxonal coverings. In some of these profiles eccentrically placed axoplasmic remnants, often of high electron density and of indiscernible composition, were surrounded by electron-lucent vacuous areas. In a few other less-degenerate processes, (Fig. 3), a complement of small (50 nm) clear vesicles and a few larger (100 nm) dense-cored vesicles indicated a possible cholinergic identity. No 5-OHDA labeling was observed in any of these degenerated nerve fibers.

Although these degenerated nerves were not infrequently observed in close proximity to dendrites and neuron somata, evidence of any synaptic relationship with these elements has not yet been observed, presumably be-

cause of degeneration changes that had taken place prior to fixation.

DISCUSSION

NE storage, as visualized by 5-OHDA labeling in the present study, within pleomorphic structures at sites in sympathetic neurons of the rat CMG was previously demonstrated in other sympathetic neurons *in vitro* (20) and is in agreement with the ultrastructural studies of Richards and Tranzer following aldehyde/dichromate fixation (17). All the above workers detected biogenic amine storage in vesicular as well as smooth reticular components of axonal and dendritic

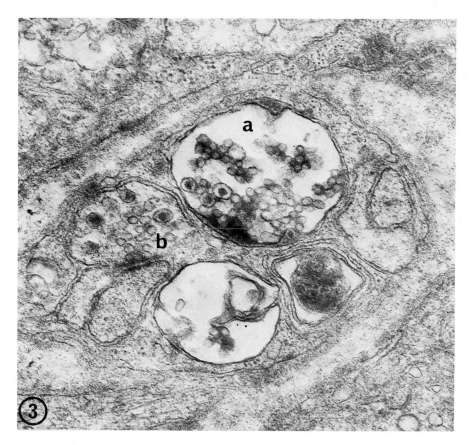

FIG. 3. A fascicle of unmyelinated nerves 24 hr following post-ganglionectomy in which three degenerated nerves can be seen, one of which (a) contains a cholinergic-type vesicle complement. Also shown is a presumed preganglionic (b) synapse on a putative dendritic spine. ×48,000.

cytoplasm. The relationship between the vesicular and reticular storage compartments is unclear, but Richards and Tranzer (17) suggested that small (50 nm diameter) dense-cored vesicles are budded from the smooth reticulum at dendritic locations.

In the context of the present investigation, it is appropriate to question the possible existence of recurrent collaterals from sympathetic postganglionic axons. Early silver impregnation studies (2) did not reveal any identifiable recurrent collaterals. Ultrastructural identification of dendrites as opposed to axons (in cross section) is aided by a ribosomal content in dendrites which is not present in axons (20) and thus unlikely to be present in axonal derivatives such as recurrent collaterals. Although one cannot be certain that some of the neuronal profiles containing 5-OHDA-loaded dense-cored vesicles, but lacking demonstrable ribosomes, were *not* recurrent collaterals, sufficient instances of these two organelles being observed within the same neuronal profile suggests that these are dendrites and, further, that dendrites of CMG neurons not only store NE but also make synaptic contact with one another. As a result of this study, we think it likely that at least some of the intraganglionic fluorescent varicosities observable by light microscope histochemistry can be attributed to dendritic stores of NE. Because of the favorable architecture of the substantia nigra, it has been possible to demonstrate the release of dopamine from dendrites (8) but not as yet from the dendrites of sympathetic neurons. In sympathetic ganglia the principal neurons and their processes are too closely intermingled to identify specifically any NE which might be released from dendrites. Thus the functional significance of the observed contacts between postganglionic neurons remains speculative.

Ultrastructural changes we observed in sympathetic postganglionic axons following axotomy were similar to those observed by others following constriction of the postganglionic branches of the SCG (14). The presence of 5-OHDA-loaded vesicles in swollen axons and in dendrites following postganglionectomy indicated that damaged sympathetic neurons retain their ability to take up exogenous amines for at least 3 days after operation. This is in agreement with the earlier work of Lever et al. (12).

The extensive terminal ramifications and many axodendritic contacts of preganglionic sympathetic axons would account in part for the presence of unloaded (by 5-OHDA) processes located within the same Schwann cell covering as 5-OHDA-loaded processes. Further, the many peptide-containing nerve fibers revealed immunohistochemically by Hökfelt et al. (10) within the rat CMG may well be included within these fascicles of unmyelinated nerves. In this connection it is interesting to speculate that degenerated, unmyelinated nerves frequently encountered by us in the CMG after ganglionectomy could represent centripetal fibers of neurons located at visceral sites peripheral to the CMG. This finding is probably complementary to recent reports by Schultzberg et al. (18) of peptide-containing nerve cell bodies within the gut and to evidence for a cholinergic input from the guinea pig colon to the inferior mesenteric ganglion (19).

ACKNOWLEDGMENT

The valuable photographic assistance of Mr. P.F. Hire is gratefully acknowledged.

REFERENCES

1. Bjorklund, A., and Lindvall, O. (1975): Dopamine in dendrites of substantia nigra neurons: Suggestions for a role in dendritic terminals. *Brain Res.,* 83:531–537.
2. Cajal, S.R. (1911): *Histologie du Systeme Nerveux de L'Homme et des Vertebres,* Vol. II. Maloine, Paris.
3. Dail, W.G., and Evan, A.P. (1978): Ultrastructure of adrenergic terminals and SIF cells in the superior cervical ganglion of the rabbit. *Brain Res.,* 148:469–477.
4. Dail, W.G., and Evan, A.P. (1978): Effects of chronic deafferentation on adrenergic ganglion cells and small intensely fluorescent cells. *J. Neurocytol.,* 7:25–37.

5. Elfvin, L-G. (1971): Ultrastructural studies on the synaptology of the inferior mesenteric ganglion of the cat. II. Specialised serial neuronal contacts between pre-ganglionic end fibers. *J. Ultrastruct. Res.*, 37:426–431.

6. Elfvin, L-G., Hökfelt, T., and Goldstein, M. (1975): Fluorescence microscopical, immunohistochemical and ultrastructural studies on sympathetic ganglia of the guinea pig with special reference to the SIF cells and their catecholamine content. *J. Ultrastruct. Res.*, 51:337–396.

7. Eränkö, O., and Eränkö, L. (1971): Small, intensely fluorescent granule-containing cells in the sympathetic ganglia of the rat. *Prog. Brain Res.*, 34:39–51.

8. Geffen, L.B., Jessell, T.M., Cuello, A.G., and Iversen, L.L. (1976): Release of dopamine from dendrites in rat substantia nigra. *Nature*, 260:258–260.

9. Hamberger, B., Norberg, K-A., and Sjoqvist, F. (1963): Evidence for adrenergic nerve terminals and synapses in sympathetic ganglia. *Int. J. Neuropharmacol.*, 2:279–282.

10. Hökfelt, T., Schultzberg, M., Elde, R., Nilsson, G., Terenius, L., Said, S., and Goldstein, M. (1978): Peptide neurons in peripheral tissues including the urinary tract: Immunohistochemical studies. *Acta Pharmacol. Toxicol. (Kbh)*, 43:79–89.

11. Jacobowitz, D. (1970): Catecholamine fluorescence studies of adrenergic neurons and chromaffin cells in sympathetic ganglia. *Fed. Proc.*, 29:1929–1944.

12. Lever, J.D., Spriggs, T.L.B., Graham, J.D.P., and Ivens, C. (1970): The distribution of ³H-noradrenaline and acetylcholinesterase (AChE) proximal to constrictions of hypogastric and splenic nerves in the cat. *J. Anat.*, 107:407–419.

13. Lu, K-S., Lever, J.D., Santer, R.M., and Presley, R. (1976): Small granulated cell types in rat superior cervical and coeliac-mesenteric ganglia. *Cell Tissue Res.*, 172:331–343.

14. Matthews, M.R. (1973): An ultrastructural study of axonal changes following constriction of post-ganglionic branches of the superior cervical ganglion in the rat. *Philos. Trans. R. Soc. Lond. [Biol]*, 246:479–508.

15. Matthews, M.R., and Raisman, G. (1969): The ultrastructure and somatic efferent synapses of small granule-containing cells in the superior cervical ganglion. *J. Anat.*, 105:255–282.

16. Noon, J.P., McAfee, D.A., and Roth, R.A. (1975): Norepinephrine release from nerve terminals within the rabbit superior cervical ganglion. *Arch. Pharmacol.*, 291:139–162.

17. Richards, J.G., and Tranzer, J.P. (1975): Localization of amine storage sites in the adrenergic cell body. *J. Ultrastruct. Res.*, 53A:204–216.

18. Schultzberg, M., Dreyfus, C.F., Gershon, M.D., Hökfelt, T., Elde, R., Nilsson, G., Said, S., and Goldstein, M. (1978): VIP-, enkephalin-, substance P- and somatostatin-like immunoreactivity in neurones intrinsic to the intestine: Immunohistochemical evidence from organotypic tissue cultures. *Brain Res.*, 155:239–248.

19. Szurszewski, J.A. (1977): Towards a new view of prevertebral ganglion. In: *Nerves and the Gut*, edited by F.P. Brookes and P.W. Evers, pp. 244–260. Charles B. Slack, New Jersey.

20. Teichberg, S., and Holtzman, E. (1973): Axonal agranular reticulum and sympathetic neurons. *J. Cell Biol.*, 57:88–108.

21. Yokota, R., and Yamauchi, A. (1974): Ultrastructure of the mouse superior cervical ganglion, with particular reference to the pre- and post-ganglionic elements covering the soma of its principal neurons. *Am. J. Anat.*, 140:281–298.

Histochemistry and Cell Biology of Autonomic Neurons, SIF Cells, and Paraneurons,
edited by O. Eränkö et al.
Raven Press, New York © 1980.

Extraneuronal Catecholamine as an Index for Sympathetic Activity in the Iris of the Rat: A Scanning Microfluorimetric Study

J. Schipper, F.J.H. Tilders, and *J.S. Ploem

*Department of Pharmacology, Medical Faculty, Free University, 1081 BT Amsterdam, The Netherlands; and *Department of Histochemistry and Cytochemistry, Medical Faculty, Leiden University, 2333 AL Leiden, The Netherlands*

The formaldehyde-induced fluorescence (FIF) method has been frequently used for the visualization of catecholamines and indolamines in histochemical preparations (4). In addition, measurements on FIF intensity have shown a good correlation with total biogenic amine content in various tissue preparations (1,7,14). Recently we developed a two-dimensional microfluorimetric scanning method which provides detailed information on the fluorescence intensity in neuronal and extraneuronal elements selectively (11,13). In studies on the noradrenergically innervated dilator muscle of the iris of the rat, we found a decrease in extraneuronal fluorescence after surgical or pharmacological depletion of neuronal catecholamine (CA) stores (10). We concluded that CA is present extraneuronally in the iris of untreated rats, which might reflect norepinephrine released *in vivo* from the sympathetic nerve fibers in the preparation. In the present investigation, we attempted to obtain further information on the potential value of extraneuronal FIF as a quantitative histochemical parameter for sympathetic neurotransmission.

EXPERIMENTAL PROCEDURES

Male Wistar rats (140 to 160 g) were used. Drugs were injected intraperitoneally. The diameter of the pupil was measured in conscious rats using an operation microscope with ocular micrometer at a light intensity of 600 lux (12). The increase in pupillary diameter was calculated from the difference between drug- and saline-treated groups. Rats were sacrificed by decapitation, and the eyes were enucleated immediately. The iris was dissected and whole mount preparations were treated with formaldehyde vapor as described earlier (11). Scanning microfluorimetry was performed with an automated microfluorimeter (MPVII, Leitz) equipped with a 0.5 μm scanning stage and controlled by a PDP 12 computer (11). Areas of 50×50 μm were scanned under standardized conditions with a scan speed of 100 spots/sec. With the aid of a PDP 11/60 computer, the data of each scan (10,000) were converted into a fluorescence histogram comprising all measured values.

A mathematical analysis was used to discriminate between populations of measurements on nerve fibers and background (smooth muscle) according to objective criteria (13). By this approach, two frequency distributions were generated: a histogram of measurements on nerve fiber (neuronal) fluorescence and a histogram of smooth muscle (extraneuronal) measurements. Total integrated fluorescence of the nerve fiber histogram was used as a parameter for neuronal fluorescence. This parameter showed a good correlation with changes in norepinephrine content as determined by a radioenzymatic method (Schipper et al., *in preparation*). The mean fluorescence intensity of the smooth muscle was used as a parameter for "extraneuronal" fluorescence. This parameter is independent of drug-induced changes in the number of fluorescent nerve fibers in the scan area (e.g., reserpine). Data presented in the figures show the mean ± SEM of group values expressed as the percentage of sham-treated controls and are compared by Student's *t*-test; N refers to the number of irides per group (three to four scans per iris).

EFFECT OF RESERPINE AND DENERVATION ON EXTRANEURONAL FLUORESCENCE

A decrease in extraneuronal fluorescence to about 60% of control is obtained after administration of reserpine (Fig. 1). A similar reduction was observed 7 days after extirpation of the superior cervical ganglion result-

ing in a sympathetic denervation. These observations suggest that CA is present extraneuronally. Indeed, the emission spectrum of this "reserpine-sensitive" component in the smooth muscle showed the characteristics of norepinephrine (10).

Since the extraneuronal fluorescence after surgical sympathectomy was reduced to a similar extent as after general depletion of CA from the sympathetic nervous system by reserpine, it is likely that extraneuronal CA in the iris is derived from the intraocular sympathetic nerve fibers rather than from the circulation. We conclude that about 40% of the extraneuronal fluorescence in the iris of untreated rats is due to extraneuronal CA. The remaining extraneuronal fluorescence represents "noncatecholamine" fluorescence (e.g., autofluorescence of smooth muscle cells, etc.). The difference in extraneuronal fluorescence in irides of untreated and reserpinized (3 mg/kg, 6 hr) rats is denoted as "extraneuronal CA fluorescence" (100%).

RELATIONSHIP BETWEEN EXTRANEURONAL AND NEURONAL CA FLUORESCENCE

It could be argued that extraneuronal CA fluorescence might be an artifact rather than a potentially interesting parameter. If so, one would expect a parallel between changes in neuronal and extraneuronal CA fluorescence. As illustrated in Fig. 2A, the time course of

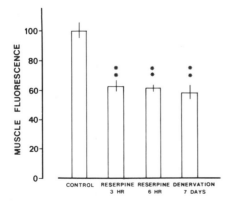

FIG. 1. Effect of reserpine (3 mg/kg i.p.) and sympathetic denervation on the fluorescence intensity of the smooth muscle of the rat iris. Muscle fluorescence is expressed as the percentage of that of saline-treated controls (N=4). The difference in smooth muscle fluorescence between saline-treated and reserpinized rats is denoted as "extraneuronal CA fluorescence" (c.f. Fig. 2, etc.). (**)p<0.01. (Data from ref. 12.)

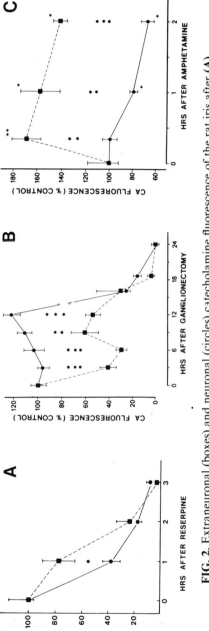

FIG. 2. Extraneuronal (boxes) and neuronal (circles) catecholamine fluorescence of the rat iris after (**A**) reserpine (3 mg/kg) treatment, (**B**) extirpation of the superior cervical ganglion, and (**C**) *d*-amphetamine (10 mg/kg) treatment. Data are expressed as the percentage of that of sham-treated controls ($N = 5$). (*)$p < 0.05$. (**)$p < 0.01$. (***)$p < 0.001$: level of significance between neuronal and extraneuronal CA values, whereas * indicates similar level of significance as compared with controls. (Data from ref. 12.)

the decline in neuronal and extraneuronal CA fluorescence after reserpine treatment did *not* indicate such a parallel. Also after extirpation of the superior cervical ganglion, a pronounced dissociation was observed between neuronal and extraneuronal CA fluorescence. The acute blockade of the impulse flow resulted in a decrease of extraneuronal CA fluorescence which was prominent as early as after 3 hr (Fig. 2B). In contrast, the nerve fiber fluorescence showed a gradual increase to $122.6 \pm 6.5\%$ after 12 hr, probably induced by a disturbance of the balance between synthesis and release. The rapid decline in neuronal CA fluorescence between 12 and 16 hr is in agreement with degeneration of nerve terminals reported by others (9).

Amphetamine (10 mg/kg) also induced a nonparallel change in neuronal and extraneuronal CA fluorescence (Fig. 2C). In addition, administration of the CA reuptake inhibitors [imipramine, desmethylimipramine (DMI), and cocaine] resulted in a strong increase in extraneuronal CA fluorescence which was not reflected in neuronal CA fluorescence (Fig. 3). The reverse was found af-

ter inhibition of monoamine oxidase (MAO) activity. Administration of nialamide (300 mg/kg), which is known to increase CA content, resulted in a strong increase in neuronal CA fluorescence. Although this drug evokes specifically an elevation of nongranular localized CA within the nerve terminals (5), which might diffuse most easily from the nerve fibers, no changes were observed in the extraneuronal CA fluorescence (Fig. 3).

The absence of a correlation between neuronal and extraneuronal CA fluorescence under these various experimental conditions strongly suggests that extraneuronal CA in the iris is not a simple artifact caused by release of CA from nerve fibers during preparation, diffusion during the histochemical reaction, etc.

RELATIONSHIP BETWEEN EXTRANEURONAL CA FLUORESCENCE AND PUPILLARY DIAMETER

In rats subjected to superior cervical ganglionectomy, miosis and ptosis were apparent directly after recovery from the anesthesia. A

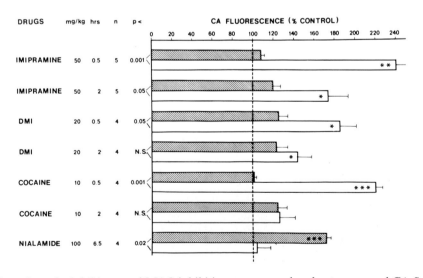

FIG. 3. Effect of uptake inhibitors and MAO inhibition on neuronal and extraneuronal CA fluorescence in the rat iris. Open bars represent extraneuronal CA fluorescence; cross-hatched bars represent neuronal CA fluorescence. Data are expressed as the percentage of the results in saline-injected controls ($N = 5$). (*)$p < 0.01$. (***) $p < 0.001$: level of significance as compared with controls. (Data from ref. 12.)

similar parallel between the decrease in extraneuronal CA fluorescence and miosis was found after reserpine treatment (3 mg/kg). On the other hand, according to the CA-releasing effect of amphetamine, this drug induced a marked increase in extraneuronal CA fluorescence, accompanied by an increase in pupillary diameter. Furthermore, the CA reuptake inhibitors imipramine, DMI, and cocaine strongly elevated extraneuronal CA fluorescence and induced mydriatic response. A quantitative relationship between the increase in pupillary diameter and the increase in extraneuronal CA fluorescence is expected only for those drugs that produce a mydriatic effect fully mediated by the intraocular sympathetic nerve fibers. A good correlation ($r = 0.98$) was found after administration of various indirect sympathomimetics (Fig. 4). However, imipramine and DMI induced a more pronounced mydriatic effect than expected according to the change in extraneuronal CA fluorescence.

Mydriatic effects of drugs, not mediated by the sympathetic nerve fibers, were studied in the superior cervical ganglionectomized rat. It appeared that imipramine and DMI evoked a pronounced mydriasis in intact and sympathectomized rats, whereas amphetamine, cocaine, and nialamide failed to show mydriatic effects in sympathetically denervated rats (12). Therefore the mydriatic effect of imipramine and DMI is to a large extent *not* mediated by inhibition of CA reuptake mechanisms but, more likely, by relaxation of the iris sphincter due to the pronounced antimuscarinic activity of these drugs (8).

EXTRANEURONAL CA FLUORESCENCE: A CONSEQUENCE OF "NONSYNAPTIC" NEUROTRANSMISSION?

Based on these results, we speculate that the extraneuronal CA fluorescence reflects CA which is directly involved in the interaction with α-adrenoreceptors. Interestingly, the estimated concentration of this extraneuronal CA is within the effective dose range for the α-adrenoreceptors (Schipper et al., *in preparation*). This is in agreement with the observation that normal rats show an acute pupillary constriction (miosis) after sympathectomy, indicating that under basal conditions a sympathetic tone is already present. Electron microscopical studies revealed that most of the varicosities in sympathetically innervated organs, including the iris, lack the "synaptic membrane complex," which frequently serves as a morphological criterion for the presence of synapses. Actually, the

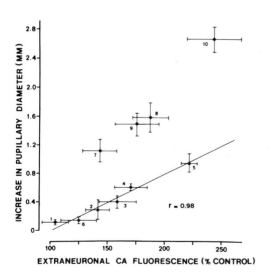

FIG. 4. Relationship between extraneuronal CA fluorescence and pupillary diameter in the rat iris after (1) nialamide (100 mg/kg, 6.5 hr); (2) amphetamine (10 mg/kg, 2 hr); (3) amphetamine (10 mg/kg, 1 hr); (4) amphetamine (10 mg/kg, 20 min); (5) cocaine (10 mg/kg, 30 min); (6) cocaine (10 mg/kg, 2 hr); (7) desmethylimipramine (20 mg/kg, 2 hr); (8) desmethylimipramine (20 mg/kg, 30 min); (9) imipramine (50 mg/kg, 2 hr); (10) imipramine (50 mg/kg, 30 min). (Data from ref. 12.)

majority of the nerve terminals is localized at a considerable distance from the nearest postsynaptic membrane, which may be up to some microns (3). These observations suggest a "nonsynaptic" rather than a classical "synaptic" neurotransmission (2). As a consequence of "nonsynaptic" neurotransmission, CA might be released in an extended extracellular space to reach the effector cells. In this respect it should be noted that the extracellular space in the iris represents as much as 40% of the total volume (6). However, the diffuse extraneuronal CA fluorescence in the iris might also suggest the presence of CA within smooth muscle cells, since CA, even at low concentrations, are able to penetrate smooth muscle cells. Therefore these microfluorimetric results cannot discriminate between an extracellular or an intracellular localization of extraneuronal CA.

We conclude that: (a) in the iris of untreated rats, CA is present extraneuronally; (b) this extraneuronal CA originates from intraocular sympathetic nerve fibers; and (c) extraneuronal CA fluorescence reflects CA released *in vivo* and can be used as a histochemical parameter for sympathetic activity.

ACKNOWLEDGMENTS

The authors thank Miss A. van Reeven for typing the manuscript and Mr. H.W. Nordsiek for preparing the figures. This investigation was supported by the Netherlands Foundation for Medical Research (FUNGO).

REFERENCES

1. Bacopoulos, N.G., Bhatnagar, R.K., Schnute, W.J., and Van Orden, L.S., III (1975): On the use of the fluorescence histochemical method to estimate catecholamine content in brain. *Neuropharmacology,* 14:291-299.
2. Beaudet, A., and Descarries, L. (1978): The monoamine innervation of rat cerebral cortex: Synaptic and nonsynaptic axon terminals. *Neuroscience,* 3:851-860.
3. Ehinger, B., Falck, B., and Sporrong, B. (1970): Possible axo-axonal synapses between peripheral adrenergic and cholinergic nerve terminals. *Z. Zellforsch.,* 107:308-321.
4. Falck, B., Hillarp, N.A., Thieme, G., and Torp, A. (1962): Fluorescence of catecholamines and related compounds with formaldehyde. *J. Histochem. Cytochem.,* 10:348-354.
5. Fillenz, M., Howe, P.R.C., and West, D.P. (1976): Vesicular noradrenaline in nerve terminals of rat following inhibition of monoamine oxidase and administration of noradrenaline. *Neuroscience,* 1:113-116.
6. Jonsson, G., Hamberger, B., Malmfors, T., and Sachs, C. (1969): Uptake and accumulation of ^{3}H-noradrenaline in adrenergic nerves of rat iris: Effect of reserpine, monoamine oxidase and tyrosine hydroxylase inhibition. *Eur. J. Pharmacol.,* 8:58-72.
7. Löfström, A., Jonsson, G., Wiesel, F.A., and Fuxe, K. (1976): Microfluorimetric quantitation of catecholamine fluorescence in rat median eminence. II. Turnover changes in hormonal states. *J. Histochem. Cytochem.,* 24:430-442.
8. Rehavi, M., Maayani, S., and Sokolovsky, M. (1977): Tricyclic antidepressants as antimuscarinic drugs: In vivo and in vitro studies. *Biochem. Pharmacol.,* 26:1559-1567.
9. Sachs, C., and Jonsson, G. (1973): Quantitative microfluorimetric and neurochemical studies on degenerating adrenergic nerves. *J. Histochem. Cytochem.,* 21:902-911.
10. Schipper, J., and Tilders, F.J.H. (1979): On the presence of extraneuronal catecholamine in the iris of the rat: a scanning microfluorimetric study. *Neurosci. Lett.,* 12:229-234.
11. Schipper, J., Tilders, F.J.H., and Ploem, J.S. (1978): Microfluorimetric scanning of sympathetic nerve fibres: An improved method to quantitate formaldehyde induced fluorescence of biogenic amines. *J. Histochem. Cytochem.,* 26:1057-1066.
12. Schipper, J., Tilders, F.J.H., and Ploem, J.S. (1979): Extraneuronal catecholamine as an index for sympathetic activity: A scanning microfluorimetric study on the iris of the rat. *J. Pharmacol. Exp. Ther.,* 211:265-270.
13. Schipper, J., Tilders, F.J.H., Groot Wassink, R., Boleij, H., and Ploem, J.S. (1980): Microfluorimetric scanning of sympathetic nerve fibres: Quantification of neuronal and extraneuronal fluorescence with aid of histogram analysis. *J. Histochem. Cytochem.,* 28:124-132.
14. Tilders, F.J.H., Ploem, J.S., and Smelik, P.G. (1974): Quantitative microfluorimetric studies on formaldehyde induced fluorescence of 5-hydroxytryptamine in the pineal gland of the rat. *J. Histochem. Cytochem.,* 22:967-975.

Histochemistry and Cell Biology of
Autonomic Neurons, SIF Cells, and
Paraneurons,
edited by O. Eränkö et al.
Raven Press, New York © 1980.

Extensive Sympathetic Denervation of the Uterus During Pregnancy as Evidenced by Tyrosine Hydroxylase Determinations in the Guinea Pig

Ch. Owman, P. Alm, A. Björklund, and G. Thorbert

Department of Histology, University of Lund, S-22362 Lund, Sweden

The guinea pig uterus receives a rich supply of adrenergic nerves that innervate blood vessels as well as the myometrium. The innervation has a complex origin: It is derived from the short adrenergic neurons in the paracervical ganglia and from the pre- and paravertebral ganglia via the hypogastric nerve as well as via nerves running in the suspensory ligament to the tubal ends of the uterine horns, i.e., probably classical long adrenergic neurons (22). In contrast to other parts of the peripheral nervous system, the uterine adrenergic innervation is highly influenced by endocrine factors (e.g., sex steroids) and pregnancy (16). For example, it was recently shown for the guinea pig that the level of norepinephrine (NE) transmitter varies during the estrous cycle (25). Even more pronounced are the changes seen during the course of pregnancy, when the neuronal NE almost totally disappears (15). The latter phenomenon may be due to changes in the synthesis, release, or degradation of NE transmitter induced by humoral and/or local factors.

In view of this, the capacity for uterine adrenergic transmitter synthesis was investigated during advancing pregnancy and following delivery by measuring the activity of tyrosine hydroxylase (TH), the rate-limiting enzyme in NE synthesis (13). This was carried out in various uterine regions and, for comparison, in the submandibular gland, representing a peripheral nongenital organ richly supplied with sympathetic nerves (2). The results show a disappearance of TH in the uterine horns but not in the submandibular gland during pregnancy, supporting our idea that a functional sympathetic denervation occurs in the myometrium during pregnancy.

MATERIALS AND METHODS

Animals

Female guinea pigs of a mixed strain were used. Animals from three periods of pregnancy (20 to 25, 30 to 40, and 60 to 65 days; in the text and illustrations these periods are referred to as 25, 40, and 65 days of pregnancy) were selected by palpation or by known mating dates; after killing under light ether anesthesia, the day of pregnancy was determined by measuring the mean weight and crown–rump length of the fetuses (8,11). Puerperal animals were studied at 3 weeks, 6 weeks, 3 months, and 6 months after delivery. To check whether pregnancy had been unilateral or bilateral, the abdomen was opened under halothane anesthesia (Halo-than, Hoechst AG) and aseptic conditions 2 to 5 days postpartum. After closure of the incision, 10 mg oxytetracycline (Terramycin, Pfizer) was given by a single intramuscular injection.

The nonpregnant material consisted of sexually mature virgin animals (400 to 500 g body weight) in the diestrous stage, used on days 7 to 12 of the estrous cycle, which was determined as previously described (22).

Tissues

Immediately after killing, various tissues were dissected out, weighed, frozen, and stored in a deep-freezer ($-90°C$) until further processing. One submandibular gland was taken from each animal. The type of uterine preparations removed varied according to the reproductive state. In virgin animals one whole uterine horn was used. In the case of unilateral pregnancy, the whole empty horn was always analyzed separately. At 25 or 40 days of pregnancy the entire fetus-containing horn was used, or after separation into those parts not distended (parafetal uterus) and those distended (perifetal uterus) by the conceptus. At term pregnancy (65 days) the entire fetus-containing horn was taken as it consisted almost completely of perifetal uterine tissue. After delivery the whole uterine horn that had contained the conceptus was taken.

Determination of Tyrosine Hydroxylase Activity

The various tissue specimens were homogenized in distilled water containing 0.1% Triton X-100. TH (L-tyrosine-3-monooxygenase, E.C.1.14.16.2) activity was determined according to the method of Hendry and Iversen (10). Briefly, an aliquot of the homogenate was incubated with ^{3}H-L-tyrosine (L-[side chain-2,3]-^{3}H-tyrosine; 22 Ci/mM of tyrosine, Radiochemical Centre, Amersham,

England) and cofactors, and the ^{3}H-DOPA (3,4-dihydroxyphenylalanine) formed was isolated and its radioactivity counted in a liquid scintillator. TH activity was expressed as picomoles of ^{3}H-DOPA formed per hour per milligram of wet tissue (specific TH activity) or nanomoles per whole organ (total TH activity). To remove possible contaminations of ^{3}H-DOPA or ^{3}H-dopamine, the isotope was purified with alumina prior to each experiment.

Blank values were obtained by replacing the homogenate with an equal volume of water. Determinations having counts below the blank values were considered to have nondetectable activity.

Statistics

Observed differences in mean values were evaluated using Student's t-test.

RESULTS

Fetus-Containing Uterine Horn

The changes in weight during pregnancy and after delivery are shown in Fig. 1. At 25 days of pregnancy the total organ activity of TH was reduced by approximately 50% compared to that in virgin animals ($p<0.01$). At 40 days of pregnancy it had further decreased, being only 30% of that in virgin animals (Fig. 1). By full term, about 65 days of pregnancy, no TH activity was detected. After parturition the total TH activity increased slowly. Thus at 3 and 6 weeks and 3 months it was about 10%, and after 6 months postpartum it had increased to about 40% of that in virgin animals ($p<0.005$).

In a similar way, the specific values for TH activity were gradually reduced compared to those in virgin animals. The mean values at 25 and 40 days of pregnancy were 18% and 4%, respectively ($p<0.01$ and $p<0.001$) compared to virgin animals, and at full term no detectable activity could be demonstrated.

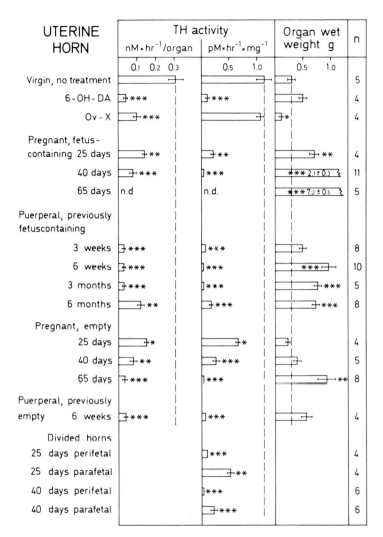

FIG. 1. Changes in TH activity and weight of the uterine horns before, during, and after pregnancy. For comparison, the effect of treatment with 6-OHDA (two injections of 100 mg/kg i.v. with a 1-day interval) and bilateral ovariectomy (Ov-X; 4 weeks) are included (1). The level of the mean values of virgin, untreated control animals is extended by interrupted lines. The statistical evaluation of observed differences in means between, on the one hand, various treatments, pregnancy, and puerperal groups and, on the other, control animals is indicated: (*)$p < 0.05$, (**)$p < 0.01$, (***)$p < 0.001$. (n) Number of observations. (n.d.) Nondetectable enzyme activity.

After delivery it slowly increased, but at 6 months postpartum it was still only 14% of that in virgin animals. Further, at increasing times of pregnancy the TH activity per milligram of tissue was lower in the perifetal than in the parafetal uterus, this difference being significant at 25 and 40 days of pregnancy ($p < 0.001$ and $p < 0.01$, respectively).

Sterile Uterine Horn

The empty horn in animals with unilateral pregnancy also showed distinct changes in TH activity during pregnancy compared to that in the uterine horn from virgin animals (Fig. 1). At 25 days of pregnancy the organ weight was unchanged ($p > 0.05$), but the total and specific TH activities were reduced, being about 50% and 40%, respectively, of the corresponding values in virgin animals ($p<0.02$). The total TH activity level did not significantly differ from that in the fetus-containing horn ($p>0.05$). At 40 days of pregnancy the weight of the empty horn was still unchanged ($p>0.05$), but the total and specific TH activities were further reduced, the levels being 26% and 22%, respectively, of those in virgin animals ($p<0.01$ and $p<0.001$). The reduction in the total TH activity was thus approximately the same as in the fetus-containing horn.

At full term the weight of the empty horn had increased about three times ($p<0.01$). The total and the specific TH activities were reduced by about 90% ($p<0.001$ and $p<0.001$), which should be compared with the fetus-containing horn, where the TH activity had fallen to nondetectable amounts. At 6 weeks after delivery the organ weight did not differ from that in virgin animals ($p>0.05$). The total and specific TH activities remained at the same level as at full term: about 10% of that in virgin animals ($p<0.001$). Further, at this time there was no difference in total and specific TH activities between uterine horns that had contained fetuses or not ($p>0.05$).

Submandibular Gland

During pregnancy the organ weight and total TH activity increased gradually, reaching levels approximately 50% above virgin controls at full term. The total TH activity in the submandibular gland generally tended to follow the changes in the weight of the organ, and consequently the specific activity did not significantly deviate from that in virgin animals ($p>0.05$) (Fig. 2). After parturition these increases progressed further, and at 6 months postpartum the organ weight and total TH activity were about three times higher than in the controls.

DISCUSSION

The enzyme TH is known to be the rate-limiting in the biosynthesis of the adrenergic nerve transmitter NE (13). The TH activity is supposed to be regulated in two principal ways. One is short-term regulation (e.g., after short periods of increased nerve activity) when there is an unchanged tissue level of the enzyme but an increased TH activity due to modifications in tissue constituents or in the enzyme molecule (14). On the other hand, in long-term regulation, variations in the synthesis of catecholamines depend on

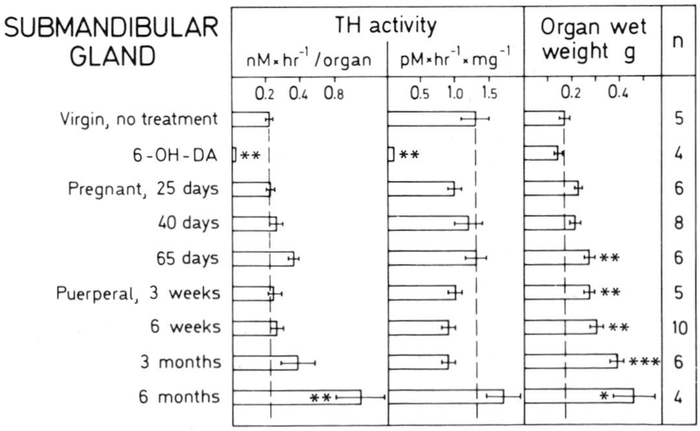

FIG. 2. Changes in TH activity and weight of the submandibular gland in some of the situations described in Fig. 1, where the symbols are explained.

changes in the amount of enzyme protein. Changes in tissue levels of the enzyme can be elicited by hormones such as adrenocorticotropic hormone (ACTH), glucocorticoids, and insulin; by long-term increase in sympathetic tone; through the destruction of adrenergic nerve terminals by 6-hydroxydopamine (6-OHDA); or by an increased or decreased growth of adrenergic nerves evoked by nerve growth factor or its antibodies (17). Since TH is localized intraneuronally in the adrenergic nerves (10), it is often used as a specific marker for the demonstration of such nerves. This is supported by the findings of a reduction by about 90% of myometrial TH activity after chemical sympathectomy induced by 6-hydroxydopamine (1). The remaining enzyme activity is probably derived from persisting adrenergic preterminals as judged from fluorescence microscopic findings (18).

The present findings in guinea pigs demonstrate a dramatic reduction of the TH activity in the fetus-containing uterine horn during advancing pregnancy, and at term no activity could be found. There is reason to believe that these changes represent a real reduction in TH activity and not simply a "spacing effect" of the adrenergic nerve plexus during the extreme increase in organ size. The reduction was pronounced when expressed in terms of whole organ content and on a weight basis; in the latter case the reduction at 25 and 40 days was even more pronounced than could be accounted for by the weight increase. Thus the TH activity was also lowered at a stage when the increase in uterine size was still only modest. The reduced TH activity of the fetus-containing horn is associated with a gradual, and finally a complete, loss of neuronal NE (15), fluorescence microscopic signs of terminal degeneration (23), and an inability of the uterus at full term to form ^{3}H-NE from ^{3}H-tyrosine *in vivo* (24). The fluorescence microscopic absence of uterine adrenergic nerves is evident even after loading the tissue with α-methyl-

NE (23), in accordance with a complete loss of neuronal ^{3}H-NE uptake as measured in uterine slices *in vitro* (3). Moreover, electron microscopic studies have shown characteristic signs of axonal degeneration in perifetal regions of the uterus during pregnancy (21). Taken together, these observations provide strong evidence for the view that the adrenergic nerves in the fetus-containing uterine horn undergo structural degeneration which is almost complete at term pregnancy, thus explaining the observed loss of TH.

Pregnant guinea pigs often carry fetuses in only one of the uterine horns. In our study the total TH activity in the empty horn decreased during progressing pregnancy, reaching only 10% of virgin levels at term. In this horn the changes evidently reflect true reductions in neuronal TH activity, although the underlying mechanisms appear to be different from the degenerative changes in the fetus-containing horn (see further below). Thus at 25 and 40 days of pregnancy the weight of the empty uterine horn was not significantly increased, whereas the total and specific TH activities were reduced by about 50 to 60% and 70 to 80%, respectively. Further, at term the weight of the horn had increased only threefold, but the TH activity expressed on a weight basis was reduced by as much as 96% and the total TH activity of the organ, independent of weight changes, was lowered by 90%. In the empty horn the endogenous neuronal NE shows no significant decrease until 50 days of pregnancy (15) despite a highly reduced total TH activity within 25 days postcoitum. Moreover, fluorescence microscopic findings demonstrate that in the empty horn the adrenergic innervation is essentially intact structurally (as compared with the contralateral fetus-containing horn), which is also reflected in the persistence of a substantial ^{3}H-NE uptake capacity (3). Thus in the empty horn adrenergic nerve degeneration would explain only a part of the observed reduction in the TH activity with progressive pregnancy. Rather, in the empty horn the major mechanism seems to be a reduced enzyme activity in structurally intact adrenergic nerves. The observations on the empty horn may be interpreted in terms of a functional adrenergic inactivation due to a much reduced transmitter release. This may be reflected in the earlier stages of pregnancy by a maintained level of transmitter (15) despite reduced TH activity and reduced neuronal uptake capacity for ^{3}H-NE (3). Near term the rate-limiting enzyme and the nerve transmitter levels reach very low levels, suggesting a more-advanced impairment of the adrenergic nerves.

It is interesting to compare the changes in TH activity in the uterine horns with those observed in the submandibular gland, which, like the uterus, increases in weight and size during pregnancy. In the submandibular gland the specific TH activity remained relatively constant despite the marked increase in organ weight (about 50% increase by the end of pregnancy and 150% at 12 weeks postpartum), indicating that the innervation kept pace with the changes in the size of the target tissue. Such an adaptive change is in line with the alterations that occur during normal growth of the organism and reflects a preserved neuron–target organ relation (9).

It is suggested that the progressive decrease in uterine TH activity with advancing pregnancy, in combination with the almost total loss of uterine NE transmitter, is caused by degeneration of the adrenergic nerve plexus in the fetus-containing horn, whereas in the empty horn the nerves are to a larger extent structurally intact (23). Since the pronounced reductions in TH activity and NE level occur in uterine horns regardless of whether fetuses are present, it seems unlikely that they are induced solely by a local action of the conceptus on the uterine muscle. There is some evidence, though, that the magnitude of the changes in the fetus-containing uterine horn may be related to the location and size of the conceptus (23). On the other hand, from numerous previous studies it seems likely that hormonal mechanisms are also in-

volved (16). In this respect, the system of short adrenergic neurons supplying the female reproductive tract appears to have a peculiar sensitivity to the action of certain sex steroids. However, since the adrenergic nerves do not react uniformly throughout the uterine and cervical region (1), it seems unlikely that the nerve degeneration is caused only by systemic humoral factors acting directly on the adrenergic neurons. Rather, the neuronal changes may be caused by effects, probably of hormonal nature, induced locally in the target organ. Neuron–target cell interactions are known to be of fundamental importance in the establishment and maintenance of nerve connections (for review, see refs. 9,20). The target organ has thus been shown to greatly influence the development of those sympathetic (and parasympathetic) neurons by which it is innervated (12). In the sympathetic nervous system it has been shown that removal of the effector organ (7) or a hormonally induced increase in the target cell mass (6) causes a reduction or increase, respectively, in the TH activity of the superior cervical ganglion. Likewise, culturing the superior cervical ganglion in the presence or absence of a target (iris) has shown that the target tissue induces an increased axonal proliferation (19). These effects were proposed to be mediated via the production of trophic substances in the target tissue; in the case of the superior cervical ganglion, there is some evidence that nerve growth factor could play such a role (9,19). This is in agreement with the observation that ovariectomy, which reduces the weight of the cervix and the uterine horns, also reduces total organ TH activity (1). On these grounds it may be hypothesized that the uterus is able to alter its state of innervation through a hormonally controlled variation in the production of such neurotropic material essential for the maintenance of a normal innervation. This implies that the functional state of the uterus as a target organ is highly important for the regulation of the adrenergic neurotransmission in this organ during pregnancy.

After delivery there was a very slow return of the total TH activity in the fetus-containing uterine horn. At 6 months postpartum the total TH activity was only about 40% of that in virgin animals, which is in good accordance with the level of endogenous NE, being about 50% of virgin levels at this point of time (4). In parallel, there was a very slow restoration of the adrenergic nerve plexus as judged by fluorescence microscopy (23). At 3 months postpartum only few terminals were seen; the number was increased, however, after incubation in the presence of α-methyl-NE, although the density of nerve terminals in virgin animals was not reached. In comparison, after chemical sympathectomy with 6-OHDA, which produces an almost total disappearance of uterine NE, the transmitter level had recovered to about 50% of control values by 6 weeks after injection (4). The present findings might therefore suggest that the return of the TH activity and the restoration of the uterine adrenergic nerve plexus is markedly slower than after chemically induced axotomy. A corresponding difference is seen with regard to the speed of regeneration of adrenergic nerves after the degeneration of axon terminals induced by 6-OHDA compared with that caused by antiserum to nerve growth factor (5).

SUMMARY

The capacity for neuronal NE synthesis in the guinea pig uterus was studied during pregnancy and postpartum utilizing animals with bilateral or unilateral pregnancies. The activity of the NE-synthesizing enzyme TH was measured in various parts of the uterus. The submandibular gland was used for comparison. During advancing pregnancy the TH activity in the fetus-containing uterine horn was progressively reduced, reaching undetectable levels at term. This change was not seen in the submandibular gland. In parallel with previous fluorescence microscopic observations on uterine adrenergic innervation, the pronounced reduction in TH activity in the fetus-containing uterine horn during progressive pregnancy is suggested to be due to degeneration of the terminal network of adrenergic nerves. In the contralateral empty uterine horn of unilateral pregnancy, the TH activity was decreased by about 90% at term, despite evidence for a relatively intact adrenergic nerve plexus. During the puerperal period the TH activity increased very slowly in the previously empty as well as the fetus-containing horns. In the latter tissue, even at 6 months after delivery, the activity was only 14% of that in uterine horns of virgin animals. In view of the marked regional heterogeneity in the changes, also related to the position and size of the conceptus, it is assumed that they are caused mainly by local humoral and probably also mechanical factors within the uterus.

ACKNOWLEDGMENTS

This work was supported by grants from the Ford Foundation (680-0383 A), the Swedish Medical Research Council (04X-56; 04X-3874), and "Förenade Liv" Insurance Company, Sweden. 6-Hydroxydopamine was supplied through courtesy of Astra Pharmaceuticals Ltd., Sweden.

REFERENCES

1. Alm, P., Björklund, A., Owman, Ch., and Thorbert, G. (1979): Tyrosine hydroxylase and DOPA decarboxylase activities in the guinea-pig uterus: Further evidence for functional adrenergic denervation in association with pregnancy. *Neuroscience (in press).*

2. Alm, P., Bloom, G.D., and Carlsöö, B. (1973): Adrenergic and cholinergic nerves of bovine, guinea-pig and hamster salivary glands. *Z. Zellforsch.,* 138:407–420.

3. Alm, P., Owman, Ch., Sjöberg, N-O., and Thorbert, G. (1979): Uptake and metabolism of [³H]norepinephrine in uterine nerves of pregnant guinea pig. *Am. J. Physiol.,* 236:C277–C285.

4. Alm, P., Owman, Ch., Sjöberg, N-O., and Thorbert, G. (1980): Restoration of uterine sympathetic transmitter after pregnancy. *Submitted for publication.*

5. Bjerre, B., Wiklund, L., and Edwards, D.C. (1975): A study of the de- and regenerative changes in the

sympathetic nervous system of the adult mouse after treatment with the antiserum to nerve growth factor. *Brain Res.,* 92:257-278.

6. Dibner, M.D., and Black, I.B. (1976): Elevation of sympathetic ganglion tyrosine hydroxylase activity in neonatal and adult rats by testosterone treatment. *J. Neurochem.,* 27:323-324.

7. Dibner, M.D., Mytlineou, C., and Black, I.B. (1977): Target organ regulation of sympathetic neuron development. *Brain Res.,* 123:301-310.

8. Draper, R.L. (1920): The prenatal growth of the guinea-pig. *Anat. Rec.,* 18:369-392.

9. Hendry, J.A. (1976): Control in the development of the vertebral sympathetic nervous system. In: *Reviews of Neurosciences,* Vol. 2, edited by S. Ehrenpreis and I.J. Kopin, pp. 149-194. Raven Press, New York.

10. Hendry, J.A., and Iversen, L.L. (1971): Effect of nerve growth factor and its antiserum on tyrosine hydroxylase activity in mouse superior cervical sympathetic ganglion. *Brain Res.,* 29:159-162.

11. Kaufmann, P. (1969): Die Meerschweinchen-placenta und ihre Entwicklung. *Z. Anat. Entwicklungsgesch.,* 129:83-101.

12. Landmesser, L., and Pilar, G. (1974): Synapse formation during embryogenesis in ganglion cells lacking a periphery. *J. Physiol. (Lond),* 241:715-736.

13. Levitt, M., Spector, S., Sjoerdsma, A., and Udenfriend, S. (1965): Elucidation of the rate-limiting step in norepinephrine biosynthesis in the guinea-pig heart. *J. Pharmacol. Exp. Ther.,* 148:1-8.

14. Lovenberg, W., Bruckwick, E., and Hanbauer, I. (1975): Protein phosphorylation and regulation of catecholamine synthesis. In: *Chemical Tools in Catecholamine Research,* Vol. II, edited by O. Almgren, A. Carlsson, and J. Engel, pp. 37-44. North-Holland Publishing Co., Amsterdam.

15. Owman, Ch., Alm, P., Rosengren, E., Sjöberg, N-O., and Thorbert, G. (1975): Variations in the level of uterine norepinephrine during pregnancy in the guinea-pig. *Am. J. Obstet. Gynecol.,* 122:961-964.

16. Owman, Ch., and Sjöberg, N-O. (1977): Influence of pregnancy and sex hormones on the system of short adrenergic neurons in the female reproductive tract. *Exerpta Medica Int. Congr. Ser.,* 402:205-209.

17. Pletscher, A. (1972): Regulations of catecholamine turnover by variations of enzyme levels. *Pharmacol. Rev.,* 24:225-232.

18. Sachs, Ch., and Jönsson, G. (1975): Mechanisms of action of 6-hydroxydopamine. *Biochem. Pharmacol.,* 24:1-8.

19. Silberstein, S.D., Johnsson, D.G., Jacobowitz, D.M., and Kopin, I.J. (1971): Sympathetic reinnervation of the rat iris in organ culture. *Proc. Natl. Acad. Sci. USA,* 68:1121-1124.

20. Smith, B.H., and Kreutzberg, G.W. (1976): Neuron-target cell interaction. *Neurosci. Res. Prog. Bull.,* 14:215-453.

21. Sporrong, B., Alm, P., Owman, Ch., Sjöberg, N-O., and Thorbert, G. (1978): Ultrastructural evidence for adrenergic nerve degeneration in the guinea pig uterus during pregnancy. *Cell Tissue Res.,* 195:189-193.

22. Thorbert, G., Alm, P., Owman, Ch., and Sjöberg, N-O. (1977): Regional distribution of autonomic nerves in guinea-pig uterus. *Am. J. Physiol.,* 233:C25-C34.

23. Thorbert, G., Alm, P., Owman, Ch., Sjöberg, N-O., and Sporrong, B. (1978): Regional changes in structural and functional integrity of myometrial adrenergic nerves in pregnant guinea-pig, and their relationship to the localization of the conceptus. *Acta Physiol. Scand.,* 103:120-131.

24. Thorbert, G., Alm, P., Owman, Ch., and Sjöberg, N-O. (1979): Pregnancy-induced alterations in the turnover rate of ³H-noradrenaline formed from ³H-tyrosine in guinea-pig uterus. *Acta Physiol. Scand.,* 105:428-436.

25. Thorbert, G., Alm, P., and Rosengren, E. (1978): Cyclic and steroid-induced changes in adrenergic neurotransmitter level of guinea-pig uterus. *Acta Obstet. Gynecol. Scand.,* 57:45-48.

Histochemistry and Cell Biology of Autonomic Neurons, SIF Cells, and Paraneurons,
edited by O. Eränkö et al.
Raven Press, New York © 1980.

Paraneuron Concept and Its Current Implications

Tsuneo Fujita, Shigeru Kobayashi, and Ryogo Yui

Department of Anatomy, Niigata University School of Medicine, Niigata, 951 Japan

During our studies on various endocrine cells, especially gastro-entero-pancreatic (GEP) endocrine cells, we noted that these cells are closely akin to neurons, and so we coined the name "paraneurons" (3). The criteria of a paraneuron are as follows (8): (a) a cell which possesses neurosecretion-like and/or synaptic vesicle-like granules; (b) a cell which may produce substances identical with, or related to, neurosecretions or neurotransmitters; (c) a cell that is receptosecretory in function, receiving adequate stimuli and releasing its secretory substances in response to them.

The category "paraneurons" includes, in addition to amine/peptide-producing endocrine cells, such internuncial cells as small intensely fluorescent (SIF) cells and various types of sensory or neuroepithelial cells (4). Recent studies have provided convincing evidence that several types of neuron and paraneuron comprise a continuous spectrum of grading cell biological features, as illustrated in Fig. 1.

The implications of the paraneuron concept suggest various topics for discussion. However, this chapter concentrates on the GEP endocrine cells—their neuron-like features and new aspects about neuroendocrine coordination that have been developed through the study of these cells.

GEP ENDOCRINE CELLS AS PARANEURONS

Gastroenteric and pancreatic endocrine cells are homologous and comprise a typical category of paraneurons (3,6). Immunohistochemical and electron microscopic studies indicate that the open-type basal granulated cells in the gut epithelium, which are chemoreceptor cells for food information (5,7), represent the ancestral form of the pancreatic islet cells, which are more differentiated in order to extract information from the blood (2,10).

As is known for other neurons and paraneurons, the secretory granules of the GEP endocrine cells contain: (a) peptide(s), which usually have hormonal, paracrine, or transmitter activities; (b) ATP and other adenine nucleotides; (c) a carrier or large complex protein; and (d) sometimes monoamines or acetylcholine. Substances 'b' and 'd' may also exert biological activities (3,7). ATP was biochemically found by Leitner et al. (14) to be released in a constant ratio to insulin from glucose-stimulated islets and can now be localized in the islet cell granules by the quinacrine method (1) under the light microscope (Fig. 2, left) and by the uranaffin reaction (16) under the electron microscope (Fig. 2, right). The so-called carrier protein includes an acidic substance responsible for the meta-

FIG. 1. Representative neurons and paraneurons comprising a graded spectrum of cells.

chromatic and other stain-technology reactions, including aldehyde fuchsin (6). It may also include converting or splitting enzymes to produce final secretory substances from their precursors and prohormones.

On the basis of the morphological and physiological findings now available, it is likely that the four substances mentioned above are extruded at the same time from the neuronal or paraneuronal cell by exocytosis of either synaptic vesicles or secretory granules (3).

The secretion component(s) thus released may be called transmitters when they act on a synaptically juxtaposed target. Figure 3 shows somatostatin-containing D cells in the canine gastric fundus which are in synapse-like contact with parietal cells by a cytoplasmic process (13). This seems to be the morphological basis for a direct inhibitory control of gastric acid secretion by somatostatin, which may now be called a paraneurotransmitter. When the substance can move a short distance through the extracellular space to its

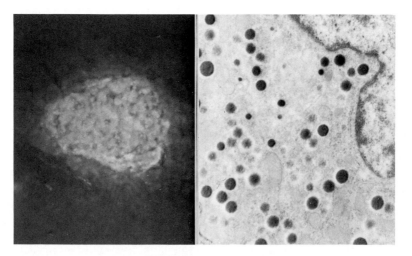

FIG. 2. Light and electron micrographs suggesting the occurrence of ATP in pancreatic islet cell granules. **Left**: Quinacrine gathered in rat islet. Quinacrine 5 mg/kg was given intraperitoneally before the rat was sacrificed, and the pancreas was frozen-sectioned. **Right**: Uranaffin reaction in dog islet A cells. The granule cores are stained dark, suggesting the occurrence of ATP. ×13,000.

target, it may be called paracrine substance. Next comes the transport of local hormones by diffusion, by blood or lymph capillaries, and by special small portal vessels. Only a portion of the secretions released from neurons and paraneurons is brought to remote targets by the general circulation.

In the case of pancreatic islet cells, mutual paracrine actions between A (glucagon), B (insulin), and D (somatostatin) cells are believed to occur by their juxtaposition to each other (15) and by the intrainsular microcirculation (11). Furthermore, the local action of islet hormones on the exocrine pancreas is effected by a microvascular route called the insuloacinar portal system (11).

It may be unnecessary to add that the secretory substances discussed here are largely common to neurons and paraneurons. It does not seem an exaggeration to say that every peptide and amine produced in the GEP paraneurons may be found in the brain and vice versa.

NEURONS AND PARANEURONS IN THE PANCREATIC ISLET

It has long been known that the autonomic nerve fibers in the pancreas are concentrated in the endocrine portion, and that in some species, including the dog, the islets of Langerhans contain a large number of nerve fibers derived partly from extrapancreatic and partly from intrapancreatic ganglia.

In a recent electron microscopic study of the dog pancreas (9), we found that the insular nerve fibers are only partly involved in the control of the endocrine cells, and that most of them end in the pericapillary space (Fig. 4). According to the morphology of synaptic vesicles in their end portions, the fibers are divided into at least four types,

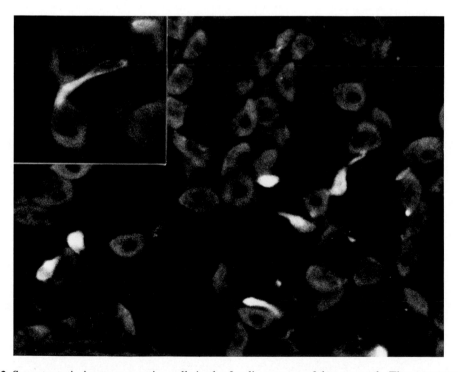

FIG. 3. Somatostatin immunoreactive cells in the fundic mucosa of dog stomach. The somatostatin cells (*bright*) are in contact with parietal cells (*slightly bright*), either with their cell body or with a slender cytoplasmic process.

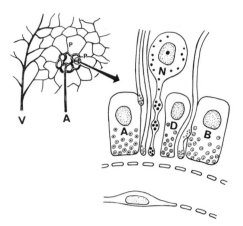

FIG. 4. The new aspect on the pancreatic islet. At the upper left is the vascular design of the pancreas. (P) Insuloacinar portal vessels. At right is the portion of the islet that is encircled at left. Together with the endocrine cells (A, B, D), intrinsic (N) and extrinsic nerve fibers end in the pericapillary space to release their secretions.

presumably corresponding to neurons containing acetylcholine, norepinephrine, dopamine, and some peptides including vasoactive intestinal polypeptide (VIP). That VIP or a similar substance occurs in the canine intrapancreatic neuronal somata and fibers, as well as, in a few islet endocrine cells can be confirmed by the immunohistochemical technique. Occurrence of a large number of monoamine fibers in the dog is confirmed by the Falck-Hillarp method.

The new concept thus reached is that most of the nerve fibers gathered in the islet are not there to control the islet cells or "supply" the blood vessels, as hitherto suggested, but to release their secretions into the capillary blood. Conveyed by the insuloacinar portal vessels, the insular neurosecretions, together with the insular hormones, exert their actions on the exocrine pancreas in high concentrations (Fig. 4). It has been proved that dopamine and VIP are secretin-like secretagogues to the exocrine pancreas of the dog. The marked effects of acetylcholine and norepinephrine, as well as of vagal and splanchnic stimulation, on the canine exocrine pancreas have long been known to physiologists, and they have wondered why the exocrine pancreas receives so little nerve supply. This discrepancy is accounted for by the present view

that the "neurotransmitters" involved in the pancreatic secretion are distributed by the blood as islet-born neurohormones.

Neurons and paraneurons in the canine pancreatic islet are literally juxtaposed in parallel to release their secretions into the blood. Noteworthy also is that the Schwann cells invest not only the neurons but also parts of the endocrine cells in the same attitude. This means that the endocrine cells or paraneurons of the islet are so neuron-like in their nature that the nurse cells specific to neurons cannot discriminate them from the neurons.

INSULIN-CONTAINING NEURONS

The parallel and unsegregative nature of neurons and paraneurons may be symbolized by the fact that even the most classical hormone, insulin, has been extracted from the mammalian brain (12) and from insect neurosecretory tissues, corpus allatum, and corpus cardiacum (17). Using antiporcine insulin guinea pig antiserum, we demonstrated that certain neurons in the brain and their terminals in the corpus allatum of the silkworm larva *(Bombicus mori)* are heavily laden with insulin-like immunoreactivity (Fig. 5); a detailed report appears elsewhere (18).

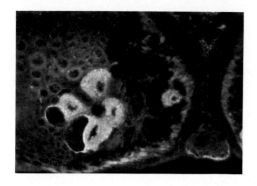

FIG. 5. Insulin immunoreactive neurons in the brain of the silkworm larva.

Insulin has been a marvelously stable molecule throughout the long history of animal evolution and seems to represent a useful tool to elucidate the phylogeny of neurons and paraneurons.

ACKNOWLEDGMENTS

The anti-VIP antiserum R502 for the immunohistochemical technique was provided by Dr. N. Yanaihara. The anti porcine insulin guinea pig antiserum was produced by Dr. S. Ito.

REFERENCES

1. Crowe, R., and Whitear, M. (1978): Quinacrine fluorescence of Merkel cells in *Xenopus laevis. Cell Tissue Res.,* 190:273–283.
2. Falkmer, S., and Östberg, Y. (1977): Comparative morphology of pancreatic islets in animals. In: *The Diabetic Pancreas,* edited by B.W. Volk and K.F. Wellmann, pp. 15–59. Plenum, New York.
3. Fujita, T. (1976): The gastro-enteric endocrine cell and its paraneuronic nature. In: *Chromaffin, Enterochromaffin and Related Cells,* edited by R.E. Coupland and T. Fujita, pp. 191–208. Elsevier, Amsterdam.
4. Fujita, T. (1977): Concept of paraneurons. *Arch. Histol. Jpn. (Suppl.),* 40:1–12.
5. Fujita, T., and Kobayashi, S. (1971): Experimentally induced granule release in the endocrine cells of dog pyloric antrum. *Z. Zellforsch.,* 116:52–60.
6. Fujita, T., and Kobayashi, S. (1973): The cells and hormones of the GEP endocrine system—the current of studies. In: *Gastro-Entero-Pancreatic Endocrine System: A Cell-Biological Approach,* edited by T. Fujita, pp. 1–16. Igaku Shoin, Tokyo.
7. Fujita, T., and Kobayashi, S. (1977): Structure and function of gut endocrine cells. *Int. Rev. Cytol. (Suppl.),* 6:187–233.
8. Fujita, T., and Kobayashi, S. (1979): Current views

on the paraneurone concept. *Trends Neurosci.,* 2:27–40.
9. Fujita, T., and Kobayashi, S. (1979): Proposal of a neurosecretory system in the pancreas: An electron microscope study in the dog. *Arch. Histol. Jpn.,* 42:277–295.
10. Fujita, T., Kobayashi, S., Yui, R., and Iwanaga, T. (1980): Evolution of neurons and paraneurons. In: *Proceedings of the International Symposium on Hormones and Evolution,* edited by H. Kobayashi. Japan Scientific Society Press, Tokyo *(in press).*
11. Fujita, T., Yanatori, Y., and Murakami, T. (1976): Insulo-acinar axis, its vascular basis and its functional and morphological changes caused by CCK-PZ and caerulein. In: *Endocrine Gut and Pancreas,* edited by T. Fujita, pp. 347–357. Elsevier, Amsterdam.
12. Havrankova, J., Schmechel, D., Roth, J., and Brownstein, M. (1978): Identification of insulin in rat brain. *Proc. Natl. Acad. Sci. USA,* 75:5737–5741.
13. Kusumoto, Y., Iwanaga, T., Ito, S., and Fujita, T. (1979): Juxtaposition of somatostatin cell and parietal cell in the dog stomach. *Arch. Histol. Jpn.,* 42:459–465.
14. Leitner, J.W., Sussman, K.E., Vatter, A.E., and Schneider, F.H. (1975): Adenine nucleotides in the secretory granule fraction of rat islets. *Endocrinology,* 96:662–677.
15. Orci, L., Malaisse-Lagae, F., Ravazzola, M., Rouiller, D., Renold, A.E., Perrelet, A., and Unger, R. (1975): A morphological basis for intercellular communication between α- and β-cells in the endocrine pancreas. *J. Clin. Invest.,* 56:1066–1070.
16. Richards, J.G., and Da Prada, M. (1977): Uranaffin reaction: A new cytochemical technique for the localization of adenine nucleotides in organelles storing biogenic amines. *J. Histochem. Cytochem.,* 25:1322–1336.
17. Tager, H.S., Markese, J., Kramer, K.J., Speirs, R.D., and Childs, C.N. (1976): Glucagon-like and insulin-like hormones of the insect neurosecretory system. *Biochem. J.,* 156:515–520.
18. Yui, R. (1980): Evolution of GEP endocrine cells. In: *Proceedings of the Symposium on Paraneurons, Their Features and Functions,* edited by T. Kanno. *Biomed. Res. Suppl. 1 (in press).*

Histochemistry and Cell Biology of Autonomic Neurons, SIF Cells, and Paraneurons,
edited by O. Eränkö et al.
Raven Press, New York © 1980.

Unidirectional Cellular Processes of Stimulus–Secretion Coupling in a Cell Secreting Cholecystokinin–Pancreozymin

Tomio Kanno, Atsushi Saito, and Hidetoshi Yonezawa

Department of Physiology, Faculty of Veterinary Medicine, Hokkaido University, Sapporo 060, Japan

Paraneurons are cells which should be regarded as closely related to neurons. The criteria of the paraneuron are: (a) a cell which possesses neurosecretion-like and/or synaptic vesicle-like granules; (b) a cell that may produce substances identical with or related to neurosecretions or neurotransmitters; and (c) a cell that receives adequate stimuli and releases its secretory substances in response to them (3). These criteria correspond exactly to the steps involved in a sequence of events in outward membrane-mobile (M-M) transport. The term M-M transport has been proposed in place of "cytosis" (1) to describe the dynamic transformation of cell membrane during a sequence of events (7–9). The outward M-M transport consists of eight steps: in the steady state: ingestion of raw materials, synthesis of secretory substances, intracellular transport of the secretory substances, and storage of the secretory granules or vesicles; and in the activated state: reception of adequate stimuli, excitation, conduction, and extrusion (10). The steady state corresponds to the first and second criteria of the paraneuron, and the activated state coincides the third criterion. The activated state is also equivalent to the stimulus–secretion coupling proposed by Douglas and Rubin (2). The paraneurons may behave characteristically in the activated state, which is analyzed in an enteric paraneuron, a cell secreting cholecystokinin–pancreozymin (CCK-PZ cell) in the present study.

METHODS

The isolated and perfused rat pancreas with attached duodenum was prepared as reported previously (13). The rate of vascular flow was kept constant at 2.0 ml/min. The composition of the standard Krebs–Henseleit solution used for perfusing and bathing the preparation was the same as in a previous paper (8). Dextran T-70 (Pharmacia, Uppsala) was added to the solution at a final concentration of 5% (w/v). A synthetic trypsin inhibitor [N,N-dimethylcarbamoylmethyl-4-(4-guanidinobenzoyloxy)-phenylacetate] methansulfonate (FOY 305), was dissolved in distilled water and administered into the isolated duodenum to stimulate the CCK-PZ cells. Pure natural CCK-PZ (3,500 units/mg; GIH Research Unit, Karolinska Institutet, Stockholm) was injected through a cannula into the inlets of the vascular perfusion. Its concentration is expressed in Ivy dog units (6). The flow rate of pancreatic juice and the total protein and amylase contents were estimated as reported previously (8).

RESULTS

CCK-PZ Release

The isolated rat duodenum was perfused with the standard Krebs-Henseleit solution containing 8 vol% dog erythrocytes. After infusion of FOY 305 (100 mg/kg body weight) into the duodenal lumen, the vascular perfusate was collected from the portal vein for 30 min. The perfusate was oxygenated and its biological activity examined on the isolated rat pancreas as follows. The isolated rat pancreas was perfused with the standard Krebs-Henseleit solution containing 8 vol% dog erythrocytes for 20 min, and then with the solution containing erythocytes and CCK-PZ 10 mU/ml for 5 min. The entire perfusion sequence was repeated two times with the standard and the CCK-PZ-containing solution in the presence of erythrocytes.

Figure 1A shows a series of responses (mean ± SE) obtained in this way. Initial perfusion with CCK-PZ resulted in a large increase in pancreatic juice flow and protein and amylase outputs. The initial responses were followed by a slight diminution in the subsequent responses. The protein output in response to the second and third stimuli was 83% and 63%, respectively, of the initial response. The biological activity of the perfusate collected from the portal vein was examined by replacing the second CCK-PZ injection with the perfusate injection (Fig. 1B). The perfusate collected from the portal vein after the FOY 305 infusion induced secretory responses lower than those induced by the CCK-PZ (10 mU/ml) stimulus. By reference to the dose–response relation in the isolated rat pancreas (5), we calculated that the perfusate contained about 7 mU CCK-PZ/ml. Only small increases in the secretory

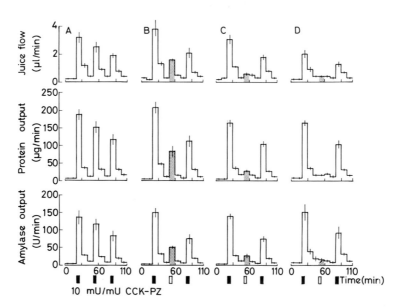

FIG. 1. Bioassay of CCK-PZ activity of venous perfusate collected from the portal vein of pancreatico-duodenal preparation of the rat. **A:** Secretory responses (pancreatic juice flow, protein output, and amylase output) of the isolated rat pancreas induced by successive 5-min perfusions with CCK-PZ 10 mU/ml at regular intervals of 25 min. The biological activity of the perfusate collected from the portal vein was examined by replacing the second CCK-PZ stimulus with the perfusate. **B:** Biological activity of the perfusate collected after infusion of FOY 305 (100 mg/kg body weight) into the duodenum. **C:** Activity of the perfusate collected after infusion of water. **D:** Activity of the perfusate collected after infusion of FOY 305 (100 mg/kg body weight) during vascular perfusion with the Ca-deficient solution. Each value represents the mean (± SE) of five experiments.

responses were observed during perfusion with the venous perfusate collected after duodenal infusion of the distilled water that was used as a solvent of FOY 305 (Fig. 1C).

Inhibition of CCK-PZ Release in Ca-Deficient Medium

The isolated duodenum was perfused with the Ca-deficient Krebs–Henseleit solution containing 8 vol% dog erythrocytes. After a 10-min perfusion, FOY 305 (100 mg/kg body weight) was infused into the duodenum, and the vascular perfusion was continued with the Ca-deficient solution. The Ca concentration of the Ca-deficient perfusate collected from the portal vein was adjusted to 2.5 mM by addition of CaCl$_2$ before the bioassay of CCK-PZ. As shown in Fig. 1D, no increase in secretory responses was detected during perfusion of the pancreas with the Ca-ad-

justed perfusate, which contained less than 0.2 mU CCK-PZ/ml by reference to the sensitivity of the preparation.

Lack of Effect of Tetrodotoxin on CCK-PZ Release

One of the possible mechanisms that mediate the reception of the stimulus at the apical portion of the cell membrane and the CCK-PZ release from the basal portion seems to be generation and conduction of action potential. To examine this possibility, we adopted tetrodotoxin (TTX), a specific inhibitor of the Na-dependent action potential (14). Figure 2A is the control and shows no interference of 1 μg TTX/ml with the CCK-PZ-induced secretion in the exocrine pancreas. The perfusate was collected from the portal vein during perfusion with the standard solution containing erythrocytes and 1 μg TTX/

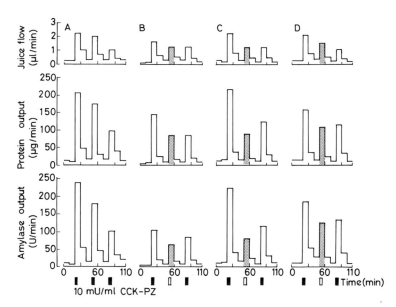

FIG. 2. Influence of TTX on the FOY 305-induced CCK-PZ release. The procedure for examining the biological activity is similar to that used in Fig. 1. **A:** Secretory responses of the isolated rat pancreas induced by CCK-PZ 10 mU/ml in the presence of TTX 1 μg/ml. **B:** Biological activity of the perfusate collected after the intraduodenal infusion of FOY 305 (100 mg/kg body weight) during vascular perfusion with the standard solution containing TTX 1 μg/ml. **C:** Biological activity of the perfusate collected after infusion of FOY 305 and TTX 1 μg/ml into the duodenal lumen during vascular perfusion with the standard solution. **D:** Biological activity of the perfusate collected after infusion of FOY 305 and TTX 10 μg/ml into the duodenal lumen during vascular perfusion with the standard solution. Each value represents the result of a single experiment.

ml after the infusion of FOY 305 (100 mg/kg body weight) into the duodenum. The perfusate induced secretory responses (Fig. 2B) comparable to the corresponding responses depicted in Fig. 1B. In the next experiments, the perfusates were collected during perfusion with the standard solution containing erythrocytes after intraduodenal infusion of FOY 305 (100 mg/kg body weight) and 1 or 10 μg TTX/ml. The perfusates also induced the secretory responses comparable to the corresponding responses depicted in Fig. 1B (Fig. 2C,D). These results show that luminal and vascular applications of TTX have nothing to do with the mechanism of FOY-induced CCK-PZ release.

Effect of High $[K^+]_o$

The preceding experiments showed that the TTX-sensitive Na spike potential may not be involved in the stimulus–secretion coupling of the cells secreting CCK-PZ. There is a possibility that a TTX-insensitive slow depolarization may mediate the reception of stimulus and the CCK-PZ release.

The possibility was examined by infusing a high-K solution into the duodenal lumen. An increase in $[K^+]_o$ in the duodenal lumen induced a dose-dependent release of CCK-PZ, and the release was completely nullified when the vascular perfusion was switched from the standard to the Ca-deficient solution (Fig. 3).

DISCUSSION

The present experiments confirmed the previous findings (11,12) that the trypsin inhibitors cause CCK-PZ release in the isolated perfused rat duodenum, and that the release is nullified when the blood vessels are perfused with the Ca-deficient solution. The present experiments showed further that the CCK-PZ release was not inhibited by intraduodenal and intravenous application of TTX, and that intraduodenal infusion of high-K medium caused the dose-dependent and Ca-dependent CCK-PZ release without any other stimulus in the duodenal lumen. Scratcherd et al. (16) also showed that tryptophan and phenylalanine induced the Ca-

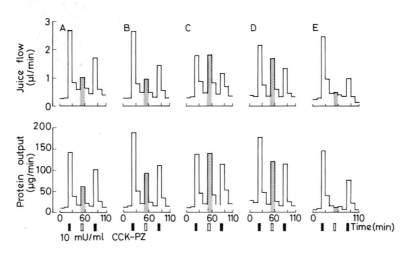

FIG. 3. CCK-PZ released by infusion of a high-K solution into the duodenal lumen. The procedure of examining the biological activity is similar to that used in Fig. 1. The biological activities of the perfusates collected after infusion of 5.6, 16, 30, and 56 mM KCl solution into the duodenal lumen are shown in **A, B, C,** and **D,** respectively. **E:** Activity of the perfusate collected after infusion of 56 mM KCl solution during vascular perfusion with the Ca-deficient solution. Each value represents the mean of two experiments.

dependent CCK-PZ release in the isolated everted jejunum of the ferret, but they could not show the secretory effect of high-K medium on the CCK-PZ release.

The previous and present experiments give a fresh insight into the cellular mechanism of CCK-PZ release. In the steady state of outward M-M transport, raw material (e.g., amino acids, sugar, fatty acids, H_2O, ions) may be ingested from the blood capillary by the cell secreting CCK-PZ. The synthesis step may be similar to that presented comprehensively by Palade (15) for the pancreatic acinar cell. The secretory granules containing CCK-PZ may be translocated toward the basal portion and stored in the basal portion of the intracellular space. The active state may be initiated by interaction between adequate stimuli (e.g., trypsin inhibitors, amino acids) and the specific receptor on the apical part of the membrane of the cell. Penetration of adequate stimuli into the cell secreting CCK-PZ may not be necessary for initiation of the active state, since there is no indication that the natural trypsin inhibitor of higher molecular size can penetrate the luminal (apical) surface of the cell. The intestinal proteolytic activity that is inhibited by the trypsin inhibitors has been regarded as an important factor controlling CCK-PZ release in the rat (4). The duodenal proteolytic activity in the isolated rat duodenum, however, may be very low even in the resting state, despite the fact that very little CCK-PZ is released in the steady state. Thus the view of Green and Lyman (4) may be insufficient to explain the present results. The most possible explanation may be that the trypsin inhibitors per se are sufficient to activate the receptor on the apical part of the membrane of the CCK-PZ cell. The interaction of the secretagogues and the receptor may in turn induce a TTX-insensitive slow depolarization, which may propagate to the basal part of the membrane and induce an increase in $[Ca^{2+}]_i$. The increase in $[Ca^{2+}]_i$ is generally regarded to be a direct cause of exocytosis to the secretory granules in a wide variety of secretory cells.

ACKNOWLEDGMENTS

The investigation was supported by grants to one of the authors (T.K.) from the Ministry of Education, Science, and Culture, Japan. FOY 305 was a gift of Ono Pharmaceutical Co., Osaka, Japan.

REFERENCES

1. De Duve (1963): Footnote. In: *Lysosomes,* edited by A.V.S. de Reuck and M.P. Cameron, p. 126. Churchill, London.
2. Douglas, W.W., and Rubin, R.P. (1961): The role of calcium in the secretory response of the adrenal medulla to acetylcholine. *J. Physiol. (Lond),* 159:40-57.
3. Fujita, T., and Kobayashi, S. (1979): Current views on the paraneurone concept. *Trends Neurosci.,* 2:27-30.
4. Green, G.M., and Lyman, R.L. (1972): Feedback regulation of pancreatic enzyme secretion as a mechanism for trypsin inhibitor-induced hypersecretion in rats. *Proc. Soc. Exp. Biol. Med.,* 140:6-13.
5. Habara, Y., Kanno, T., and Saito, A. (1979): Cold acclimation in the secretory responses of the isolated exocrine pancreas of the rat. *J. Physiol. (Lond) (in press).*
6. Ivy, A.C., and Jenecek, H.M. (1959): Assay of Jorpes-Mutt secretin and cholecystokinin. *Acta Physiol. Scand.,* 45:220-230.
7. Kanno, T. (1974): Absorptive and secretory function of cellular membrane. *Heredity,* 28:30-36 (Japanese).
8. Kanno, T. (1976): The relationship between changes in electrophysiological properties of adrenal chromaffin cells and catecholamine release. In: *Chromaffin, Enterochromaffin and Related Cells,* edited by R.E. Coupland and T. Fujita, pp. 47-57. Elsevier, Amsterdam.
9. Kanno, T. (1977): Membrane-mobile transport. *Membrane,* 2:98-108 (Japanese).
10. Kanno, T. (1977): Physiology of paraneurons. *Arch. Hist. Jpn. (Suppl.)* 40:13-29.
11. Kanno, T., and Imai, S. (1976): Stimulus-secretion coupling in the cell secreting cholecystokinin-pancreazymin. In: *Endocrine Gut and Pancreas,* edited by T. Fujita, pp. 245-254. Elsevier, Amsterdam.
12. Kanno, T., Saito, A., Yonezawa, H., Sato, H., Yanaihara, C., and Yanaihara, N. (1979): Calcium-dependent release of the gut hormones, kitten VIP and rat CCK-PZ. In: *Gut Peptides,* edited by A. Miyoshi, K. Abe, and Z. Itoh. Kodansha Scientific, Tokyo.

13. Kanno, T., Suga, T., and Yamamoto, M. (1976): Effects of oxygen supply on electrical and secretory responses of humorally stimulated acinar cells in isolated rat pancreas. *Jpn. J. Physiol.,* 26:101–115.

14. Narahashi, T., Deguchi, T., Urakawa, N., and Ohkubo, Y. (1960): Stabilization and rectification of muscle fiber membrane by tetrodotoxin. *Am. J. Physiol.,* 198:934–938.

15. Palade, G.E. (1959): Functional changes in the structure of cell components. In: *Subcellular Particles,* edited by T. Hayashi, pp. 64–80. Ronald Press, New York.

16. Scratcherd, T., Syme, G.B., and Wynne, R.D'A. (1976): The release of cholecystokinin-pancreozymin from the isolated jejunum of the ferret. In: *Stimulus-Secretion Coupling in the Gastrointestinal Tract,* edited by R.M. Case and H. Goebell, pp. 341–353. MTP Press, Lancaster.

Immunohistochemistry of Opioid and Other Peptide Neurons in the Autonomic and Central Nervous Systems

Histochemistry and Cell Biology of
Autonomic Neurons, SIF Cells, and
Paraneurons,
edited by O. Eränkö et al.
Raven Press, New York © 1980.

Morphological Studies on Central and Peripheral Connections of Sympathetic Ganglia

Lars-G. Elfvin

Department of Anatomy, Karolinska Institutet, Stockholm, Sweden

During the last two decades electron microscopy and the use of histochemical fluorescence techniques for visualizing biogenic amines have significantly increased our knowledge of the cell bodies and nerve fiber systems in sympathetic ganglia. The application of newly developed immunohistochemical methods for dealing with the cellular localization of various peptide hormones to the study of sympathetic ganglia has revealed additional important facts concerning the organization of the ganglia. It has been possible to demonstrate several types of small peptide hormones in nerve fiber networks and terminals as well as in neuronal cell bodies and SIF cells.

Although the functional role(s) of the peptide-containing neurons is still uncertain, it may be assumed that the peptides act as modulators or transmitters at synaptic sites. The putative peptidergic neurons may be involved in certain reflex arcs, relating the ganglia to either the peripheral target organ or the spinal cord. To clarify this point, it is important to obtain as much information as possible about the origin and projections of the peptide-containing neurons as well as of other nervous connections of sympathetic ganglia.

The present chapter summarizes results dealing with the localization of the various peptides in neuron systems and small intensely fluorescent (SIF) cells of the sympathetic ganglia. This is followed by a description of recent studies on the central and peripheral connections of sympathetic ganglia, particularly as revealed by the horseradish peroxidase axonal tracing method (17).

ULTRASTRUCTURE STUDIES

A fairly comprehensive picture of the principal ganglion cells and their processes has been obtained by analyzing serial sections through sympathetic ganglia in the electron microscope (6). The relationship between the pre- and postganglionic neurons has also been analyzed, and SIF cells and their projections have been studied with this technique (8,9).

A common finding during an electron microscope analysis of sections through the neuropil of the ganglia is that several nerve endings contain different types of granules or vesicles and form specialized contacts with one another. This is particularly evident in the prevertebral ganglia. Figure 1 shows a cross section of what appears to be four types of such closely apposed terminals in the celiac ganglion of the guinea pig. Even if we analyze long series of consecutive sections, it is impossible to find the origin of some of the fibers that form these terminals. The significance of these nerve structures may be related to their storage and release of different types of synaptic transmitters or modulators (e.g., biologically active peptide hormones), and they may be of extrinsic as well as intrinsic origin. In the latter case they may come from other postganglionic cells or possibly SIF cells.

PEPTIDE-CONTAINING NEURONS AND SIF CELLS

Various peptide-containing neuron systems were recently revealed in sympathetic ganglia by means of immunohistochemistry. We studied the occurrence of substance P, vasoactive intestinal polypeptide (VIP), enkephalin, and somatostatin in ganglia from the cat, rat, and guinea pig (12–14,22). These peptides have been shown to be particularly prevalent in the prevertebral ganglia of the guinea pig, which, as mentioned above, were found by electron microscopy to have a complex fiber pattern as well.

To sum up, substance P occurs in fibers, VIP in fibers and some ganglion cell bodies, enkephalin in fibers and a few ganglion cell bodies as well as in SIF cells, and somatostatin in the perikarya of the principal gan-

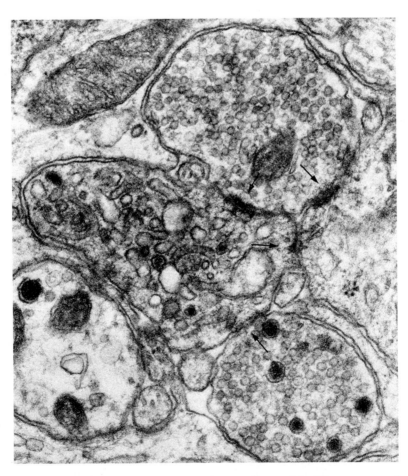

FIG. 1. Section through the celiac ganglion. Several nerve fibers with different types of granules and vesicles form specialized contacts with one another (*arrows*). ×62,000 (Electron micrograph by C. Forsman.)

glion cells. The fibers, which contain peptides, closely surround the postganglionic nerve cell bodies and their processes. Some VIP fibers and, in particular, enkephalin-containing fibers have been observed in the SIF cell groups. No substance P fibers were observed among the SIF cells.

ORIGIN AND PROJECTIONS OF THE PEPTIDE NEURONS

The functional significance of the peptide neurons in sympathetic ganglia is not known. It may be assumed that the substance P fibers represent, at least in part, branches of primary sensory neurons, since cells containing substance P were recently demonstrated in spinal ganglia (15). This assumption is further supported by our horseradish peroxidase studies (see below). It may be suggested that the substance P fibers belong to reflex arcs, which at least in part are connected to the transmission of visceral pain sensation (10).

With regard to the origin of the VIP-positive nerve terminals in the prevertebral ganglia, it is of great interest to note that numerous VIP-containing cell bodies have been demonstrated in the submucous and in the myenteric plexus of the gut (Schultzberg et al., *this volume*). These cells may constitute a source of the intraganglionic VIP-containing nerve network (see below).

The nerve fibers and terminals containing enkephalin-like immunoreactivity and forming a dense plexus in prevertebral ganglia probably originate mainly in the central nervous system. Hökfelt et al. (11) described neuronal cell bodies with enkephalin-like immunoreactivity in the spinal cord of the rat, which may be assumed to be the source of many of these fibers. The presence of enkephalin in nerve terminals of possibly a central origin is interesting as it indicates the occurrence of the opioid peptide in nerves which traditionally have been classified as cholinergic. It is not known, however, whether the enkephalin-containing fibers represent a special population of nerves or if

the peptide occurs together with acetylcholine in these fibers.

The noradrenergic ganglion cells which contain somatostatin-like immunoreactivity seem to send their axons mainly to the gut. This assumption is justified since ligation of the perivascular mesenteric nerves coming from the IMG causes accumulation of somatostatin on the proximal side of the ligature (Lundberg et al., *unpublished*).

STUDIES OF SYMPATHETIC PATHWAYS BY THE USE OF HORSERADISH PEROXIDASE

The development of the horseradish peroxidase (HRP) tracing technique for studies of neuronal connectivity was recently used by several investigators in analyses of peripheral sympathetic pathways. By injecting the enzyme into sympathetic ganglia, where it is taken up by nerve endings and transported to the cell bodies, neurons sending their axons to the sympathetic ganglia can be identified. The cell bodies can subsequently be analyzed by light and electron microscopy.

Prevertebral Ganglia

Earlier morphological (18) and physiological (4,16,19) work indicated that certain neuronal cell bodies located in the periphery may project onto prevertebral ganglia. This assumption has received support from the above-cited studies on the peptide neurons. This problem is further clarified by using the HRP technique.

Our investigations utilized mainly the inferior mesenteric ganglion (IMG) of the guinea pig. After application of HRP into the IMG, the enzyme was found to be transported to nerve cell bodies in at least four regions. In preliminary experiments labeled cell bodies were found in the celiac-superior mesenteric ganglion complex and in ganglia of the pelvic plexus (Elfvin and Dalsgaard, *unpublished*). Furthermore, cells were labeled in spinal ganglia (7) and the spinal cord (5).

In the celiac–superior mesenteric ganglion complex, the labeled cells are sparse and occur singly, apparently distributed at random (Fig. 2). Labeled cells are sometimes present also in the small ganglia located along the aorta. Some cells are clearly binucleated, as are many sympathetic postganglionic neurons in the guinea pig. The functional significance of the connections between the superior abdominal ganglia and the IMG is obscure. It may be that the labeled cells are the origin of some of the slow C-fibers described by McLennan and Pascoe (19) going caudally to the IMG in the intermesenteric nerves connecting the solar plexus with the IMG. The labeled neurons may in part correspond to the VIP-positive cells described in the solar plexus ganglia (13) and contribute to the VIP-containing fibers observed in the IMG. Alternatively, some of the labeled neurons may participate in forming the noradrenergic network surrounding the ganglion cells in the IMG (8).

The HRP-positive cells in the pelvic plexus also occur singly and are sparse, although occasional small groups of labeled cells are seen (Fig. 3). In earlier physiological studies (16) it was demonstrated that neurons located along the hypogastric nerve may project to the IMG. The present findings are in agreement with these earlier results. Possibly the labeled neurons correspond to the VIP-positive cells recently observed in rather large numbers in the pelvic plexus (Schultzberg et al., *unpublished*) and contribute to the VIP-positive fibers observed in the IMG.

Approximately 5% of the cells in the lumbar spinal ganglia are labeled, particularly at the L2–L3 levels. The cells are small and may in part correspond to cells containing substance P (see above).

The preganglionic neurons related to the IMG have also been studied with the HRP method. Labeled cells are located bilaterally in the intermediolateral (IML) and intercalated (IC) nuclei in the T13–L4 segments. The majority of cells are seen at the L2–L3 level. The IC cells represent about one-third of the total number of labeled cells. The cells of the IC nucleus are located medially of the zona intermedia, with most cells dorsal and dorsolateral of the central canal. It appears justified to assume that some of these cells correspond to the enkephalin-containing cells observed dorsal to the central canal (11), and that they constitute the source of the enkephalin-containing fibers observed in the IMG (see above).

The present findings thus give direct morphological evidence for the presence of complex peripheral connections of the prevertebral ganglia, in agreement with the view expressed earlier by several authors (1,18,23, 24).

Paravertebral Ganglia

The preganglionic neurons related to the superior cervical ganglion (SCG) and the stellate ganglion (SG) in the guinea pig have been studied (5). The HRP-positive cells are located ipsilaterally in the IML and IC nuclei (Fig. 4). After injection of HRP into the SCG, labeled cells are found in the C8–T7 segments, with most IML cells at the T1–T3 level and most IC cells at the T3–T5 level. Following injection into the SG, HRP-positive cells are located in the C8–T11 segments, with most IML cells at the T1–T5 level and most IC cells at the T3–T7 level. The IC neurons represent about one-fifth and one-fourth of the total number of labeled cells, respectively.

The difference in segmental accumulation peaks of the IML cells compared to the IC cells has not previously been described. In a recent physiological study on the guinea pig, Njå and Purves (20) demonstrated a relationship between preganglionic fibers from various cord levels and specific effects relayed in the SCG. There is a striking correspondence between the segmental distribution of the IML and IC cell groups and these peripheral sympathetic effects. It is tempting to suggest that there is a relationship between the two cell groups and the tested functions.

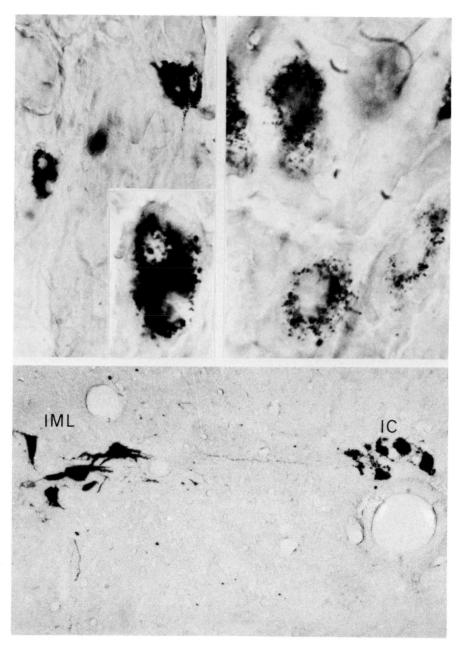

FIG. 2. Labeled cells in the celiac ganglion after HRP application into the IMG. ×510. **Inset:** A binucleated cell. ×1,300.
FIG. 3. Labeled ganglion cells in the pelvic plexus after HRP application into the IMG. ×1,300.
FIG. 4. Labeled preganglionic neurons in the spinal cord (T5) after HRP application into the stellate ganglion. Several IC cells are seen dorsal to the central canal. ×210. (Light micrograph by C. J. Dalsgaard.)

The present study shows that in the guinea pig not only prevertebral but also paravertebral ganglia are innervated by comparatively large numbers of IC neurons, in contrast to what has been described for other animals studied with the HRP technique (2,3,21). This anatomical feature may facilitate further studies on this animal on the specific functions mediated by these cells.

Peroxidase injected into the SCG and SG of the guinea pig is also transported to spinal ganglion cells at certain cervical and upper thoracic levels. A detailed account of these findings must await a more extensive presentation.

ACKNOWLEDGMENTS

This work was supported by grants from the Swedish Medical Research Council (Project 12x5189) and the Karolinska Institutet.

REFERENCES

1. Bulygin, I.A. (1976): *Reflex Function in Vegetative Ganglia.* Nauka i Tekhnika, Minsk. (Russian)
2. Chung, J.M., Chung, K., and Wurster, R.D. (1975): Sympathetic preganglionic neurons of the cat spinal cord: Horseradish peroxidase study. *Brain Res.,* 91:126–131.
3. Chung, K., Chung, J.M., Lavelle, F.W., and Wurster, R.D. (1979): Sympathetic neurons in the spinal cord projecting to the stellate ganglion. *J. Comp. Neurol.,* 185:23–30.
4. Crowcroft, P.J., Holman, M.E., and Szurszewski, J.H. (1971): Excitatory input from the distal colon to the inferior mesenteric ganglion of the guinea pig. *J. Physiol. (Lond),* 219:443–461.
5. Dalsgaard, C-J., and Elfvin, L-G. (1979): Spinal origin of preganglionic fibers projecting onto the superior cervical ganglion and inferior mesenteric ganglion of the guinea pig, as demonstrated by the horseradish peroxidase technique. *Brain Res.,* 172:139–143.
6. Elfvin, L-G. (1971): Ultrastructural studies on the synaptology of the inferior mesenteric ganglion of the cat. *J. Ultrastruct. Res.,* 37:411–448.
7. Elfvin, L-G., and Dalsgaard, C-J. (1977): Retrograde axonal transport of horseradish perioxidase in afferent fibers of the inferior mesenteric ganglion of the guinea pig: Identification of the cells of origin in dorsal root ganglia. *Brain Res.,* 126:149–153.
8. Elfvin, L-G., Hökfelt, T., and Goldstein, M. (1975): Fluorescence microscopical, immunohistochemical and ultrastructural studies on sympathetic ganglia of the guinea pig, with special reference to the SIF cells and their catecholamine content. *J. Ultrastruct. Res.,* 51:377–396.
9. Eränkö, O., editor (1976): *SIF Cells. Structure and Function of the Small Intensely Fluorescent Sympathetic Cells.* Fogarty International Center Proceedings No. 30. DHEW Publication No. (NIH) 76–942. Government Printing Office, Washington, D.C.
10. Henry, J.L. (1975): Substance P excitation of spinal nociceptive neurons. *Neurosci. Abstr.,* 1:390.
11. Hökfelt, T., Elde, R., Johansson, O., Terenius, L., and Stein, L. (1977): The distribution of enkephalin-immunoreactive cell bodies in the rat central nervous system. *Neurosci. Lett.,* 5:25–31.
12. Hökfelt, T., Elfvin, L-G., Elde, R., Schultzberg, M., Goldstein, M., and Luft, R. (1977): Occurrence of somatostatin-like immunoreactivity in some peripheral sympathetic noradrenergic neurons. *Proc. Natl. Acad. Sci. USA,* 74:3587–3591.
13. Hökfelt, T., Elfvin, L-G., Schultzberg, M., Fuxe, K., Said, S.I., Mutt, V., and Goldstein, M. (1977): Immunohistochemical evidence of vasoactive intestinal polypeptide-containing neurons and nerve fibers in sympathetic ganglia. *Neuroscience,* 2:885–896.
14. Hökfelt, T., Elfvin, L-G., Schultzberg, M., Goldstein, M., and Nilsson, G. (1977): On the occurrence of substance P-containing fibers in sympathetic ganglia: Immunohistochemical evidence. *Brain Res.,* 132:29–41.
15. Hökfelt, T., Kellerth, J-O., Nilsson, G., and Pernow, B. (1975): Experimental immunohistochemical studies on the localization and distribution of substance P in cat primary sensory neurons. *Brain Res.,* 100:235–252.
16. Job, C., and Lundberg, A. (1952): Reflex excitation of cells in the inferior mesenteric ganglion on stimulation of the hypogastric nerve. *Acta Physiol. Scand.,* 26:366–382.
17. Kristensson, K., Olsson, Y., and Sjöstrand, J. (1971): Axonal uptake and retrograde transport of exogenous proteins in the hypoglossal nerve. *Brain Res.,* 32:399–406.
18. Kuntz, A. (1953): *The Autonomic Nervous System.* Lea & Febiger, Philadelphia.
19. McLennan, H., and Pascoe, J.E. (1954): The origin of certain nonmedullated nerve fibers which form synapses in the inferior mesenteric ganglion of the rabbit. *J. Physiol. (Lond),* 124:145–156.
20. Njå, A., and Purves, D. (1977): Specific innervation of guinea pig superior cervical ganglion cells by preganglionic fibers arising from different levels of the spinal cord. *J. Physiol. (Lond),* 264:565–583.
21. Petras, J.M., and Faden, A.J. (1978): The origin of sympathetic preganglionic neurons in the dog. *Brain Res.,* 144:253–257.
22. Schultzberg, M., Hökfelt, T., Terenius, L., Elfvin, L-G., Lundberg, J.M., Brandt, J., Elde, R.P., and Goldstein, M. (1979): Enkephalin immunoreactive nerve fibers and cell bodies in sympathetic ganglia of the guinea pig and rat. *Neuroscience,* 4:249–270.
23. Skok, V.I. (1973): *Physiology of Autonomic Ganglia.* Igaku Shoin, Tokyo.
24. Szurszewski, J.H. (1977): Toward a new view of prevertebral ganglion. In: *Nerves and the Gut,* edited by F.P. Brooks and P.W. Evers, pp. 244–260. Charles B. Slack, Thorofare.

Histochemistry and Cell Biology of Autonomic Neurons, SIF Cells, and Paraneurons,
edited by O. Eränkö et al.
Raven Press, New York © 1980.

Peptide Neurons in the Autonomic Nervous System

Marianne Schultzberg, Tomas Hökfelt, and Jan M. Lundberg

Department of Histology, Karolinska Institute, S-10401 Stockholm 60, Sweden

Acetylcholine (ACh) and norepinephrine (NE) have long been regarded as the sole neurotransmitters of the autonomic nervous system. Recently a number of substances with different biologic activities have been isolated from mammalian tissue extracts. The peptidergic nature of these substances was revealed, and the development of radioimmunoassay made it possible to measure the levels of these peptides in tissues. Immunohistochemical studies have shown the occurrence of some of these peptides—i.e., substance P (SP), vasoactive intestinal polypeptide (VIP), enkephalin (ENK), somatostatin (SOM), and gastrin/cholecystokinin (GAS/CCK)—in neurons in the central and peripheral nervous system (for references see ref. 2). In addition, endocrine cells in the gastrointestinal mucosa were shown to contain some of the peptides (for references see ref. 2). The present study deals with the distribution of neurons containing SP, VIP, ENK, SOM, GAS/CCK, and neurotensin (NT) in some parts of the peripheral autonomic nervous system, i.e., the gastrointestinal tract, some sympathetic ganglia, and the adrenal glands.

ASPECTS ON METHODOLOGY

The distribution of peripheral peptide neurons was studied in rats and guinea pigs. The animals were perfused with formalin, and after dissection and rinsing the tissues were processed according to the indirect immunofluorescence technique of Coons and collaborators (7). Cryostat sections were incubated with specific antisera against SP, VIP, ENK, SOM, GAS/CCK, and NT followed by rinsing and incubation with a sheep antirabbit antiserum conjugated to fluorescein isothiocyanate (FITC). The sections were mounted and examined in a Zeiss Junior fluorescence microscope. The antibodies used in this study were raised in rabbits as described earlier [SP (31), VIP (33), ENK (37), SOM (35), GAS/CCK (26-28), NT (24)]. These antibodies have been characterized, and cross reactivity with a number of known peptides has been excluded. The possibility of cross reaction with unknown endogenous peptides and precursor molecules must be considered, however. We therefore use terms such as SP-like immunoreactivity, SP immunoreactive, etc. to indicate a positive immunoreaction with the antiserum in question.

Nerve fibers are generally easily detectable with the immunofluorescence technique used, whereas there is sometimes difficulty in visualizing peptide stores in neuronal cell bodies. Therefore a mitotic inhibitor (colchicine or vinblastine) has been used to increase the levels of peptides in cell bodies. Colchicine and vinblastine are known to arrest axonal transport (9,10,19,25).

Previous findings indicate the coexistence of a peptide and a classical neurotransmitter (5,20,23,30,36,37). Thus the question arises whether two or more peptides may occur in the same neurons. The "elution" technique of Tramu et al. (38) was used to elucidate this matter. We chose to look at the guinea pig proximal colon, where there are many peptide-immunoreactive cell bodies. Briefly, sections "stained" for one of the peptides were, after photography, treated with a solution containing potassium permanganate ($KMnO_4$) and sulfuric acid (H_2SO_4) to break the antigen–antibody bonds. The same sections were then incubated with antiserum against another peptide according to the scheme described above and subsequently photographed.

RESULTS AND DISCUSSION

Gastrointestinal Tract

The innervation of the gastrointestinal tract has long been considered cholinergic and noradrenergic. On the basis of biochemical and morphological studies, Burnstock (3) suggested that ATP is a neurotransmitter in a population of neurons innervating the gut. The ultrastructural studies by Baumgarten et al. (1) revealed the existence of a further type of nerve terminal containing large-diameter, opaque vesicles resembling those of neurosecretory cells. Gershon and collaborators (12,13,16,17) provided evidence for enteric serotonin neurons. Recent immunohistochemical studies showed the occurrence of peptide-containing neurons in the gut. This correlates well with the findings of Gabella (18) and Cooks and Burnstock (6), who described up to nine types of neurons with different ultrastructural characteristics.

Immunoreactivities to SP, VIP, ENK, SOM, and GAS/CCK were observed in nerve fibers in all regions of the tract studied, whereas NT immunoreactive fibers had a more limited distribution. Each peptide had a characteristic distribution pattern through the various layers of the gut wall. VIP immunoreactive fibers were numerous in all layers including smooth muscle layers, connective tissue layers, and ganglion plexuses in practically all regions. A very dense network of VIP fibers was observed especially in the lamina propria of the intestinal mucosa. SP immunoreactive fibers seem to be second most numerous. The SP fibers were mostly evenly distributed among the different layers but were forming an especially dense network around the ganglion cells of the submucous and myenteric plexuses. Whereas SP and VIP immunoreactive fibers were observed in high numbers in muscle (Fig. 1A,B) and connective tissue layers, ENK immunoreactive ones (Fig. 1C) were confined mainly to the smooth muscle layers and the myenteric plexus. The submucous layer and its ganglionic plexus received a very sparse innervation by ENK immunoreactive fibers as did the lamina propria. In comparison with the amount of SP, VIP, and ENK immunoreactive fibers, SOM immunoreactive ones were more sparse; e.g., the stomach was almost devoid of SOM fibers. In other parts of the tract, SOM immunoreactive fibers were mainly observed around ganglion cells of the submucous and myenteric plexuses (Fig. 1D). GAS/CCK immunoreactive fibers were even more sparse than the SOM immunoreactive ones and were most frequently seen in the lamina propria and the ganglionic plexuses.

With regard to cell bodies, only a few peptide immunoreactive cell bodies were observed in untreated animals, with the exception of VIP cells, which were seen in high numbers, especially in the rat ileum. Treatment with a mitotic inhibitor (colchicine or vinblastine) increased the number of detectable peptide cell bodies. SOM and VIP immunoreactive cells were observed in highest numbers in the submucous plexus, whereas SP and ENK immunoreactive ones were most numerous in the myenteric plexus.

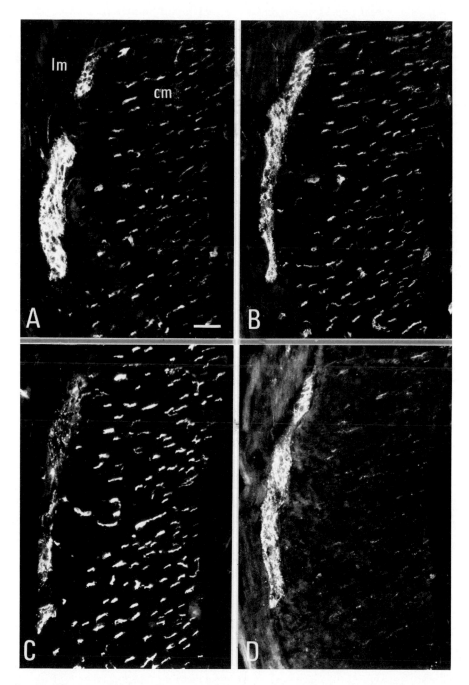

FIG. 1. Immunofluorescence micrographs of four consecutive transverse sections of the guinea pig rectum after incubation with SP antiserum **(A)**, VIP antiserum **(B)**, ENK antiserum **(C)**, and SOM antiserum **(D)**. Numerous SP **(A)**, VIP **(B)**, and ENK **(C)** immunoreactive fibers are observed in the circular smooth muscle layer (cm) and around the ganglion cells of the myenteric plexus. Only a small number of SOM immunoreactive fibers can be seen in the circular muscle layer, but a large number are observed in the myenteric plexus. The longitudinal muscle layer (lm) is more sparsely innervated by peptide immunoreactive fibers. Bar = 50 μm.

These findings show that peptide neurons to a large extent are intrinsic to the gastrointestinal tract, as previously indicated by studies on intestinal tissue cultures (34; see also 8). No NT immunoreactive cell bodies were observed. A possible coexistence of two or more peptides was studied with the "elution" and "restaining" technique of Tramu et al. (38). Although SOM and VIP immunoreactive cells on one hand, and SP and ENK immunoreactive ones on the other hand, were often found in the same ganglia, no evidence for a coexistence has yet been obtained (Fig. 2A,B). However, a GAS/CCK-like peptide was present in SOM but not in VIP immunoreactive neurons in the submucous plexus of the guinea pig.

Sympathetic Ganglia

Sympathetic ganglia were analyzed in several species with regard to possible occurrence of peptide neurons. Particular interest was devoted to prevertebral sympathetic ganglia connected to the gut. SOM-like immunoreactivity was found in a large proportion of the principal ganglion cells in the inferior mesenteric ganglion and the coeliac-superior mesenteric ganglion complex (Fig. 2D) (20). This suggests that SOM immunoreactive nerve fibers in the gut arise not only in intrinsic gut neurons but also in these prevertebral sympathetic ganglia. Preliminary studies by Lundberg et al. (*to be published*) showed an accumulation of SOM-like immunoreactivity on the ganglion side of the mesenteric nerve after ligation. A dense network

of VIP immunoreactive fibers observed in sympathetic ganglia (Fig. 2E) (21) may, on the other hand, originate in VIP cell bodies in the gastrointestinal wall. The VIP fibers were surrounding the majority of the principal ganglion cells (Fig. 2E). In some parts of the ganglia as well as among small intensely fluorescent (SIF) cells (15), a very sparse innervation by VIP immunoreactive fibers was seen. Incubation with SP antiserum revealed the presence of SP immunoreactive fibers in the prevertebral sympathetic ganglia (22), some of which had a varicose appearance whereas others were smooth. These may represent nerve terminals and axons, respectively, and may be of sensory nature (14). They may be involved in reflex arches. In none of the ganglia studied could SP immunoreactive fibers be observed around SIF cells. Intensely fluorescent ENK immunoreactive fibers were seen in the inferior mesenteric ganglion forming a dense network surrounding the principal ganglion cells (36). In the coeliac-superior mesenteric ganglion complex there was a similarly dense network, but the fibers exhibited a weaker fluorescence intensity. However, strongly fluorescent ENK immunoreactive fibers were seen in this ganglion in a patchy fashion, i.e., ENK fibers surrounding small groups of principal ganglion cells. Similar patches of ENK immunoreactive fibers were seen in the superior cervical ganglion. In the prevertebral sympathetic ganglia, areas composed of SIF cells were sparsely innervated by ENK immunoreactive fibers. The ENK immunoreactive fibers in all probability represent preganglionic fibers

▶

FIG. 2. Immunofluorescence micrographs of sections of the proximal colon (**A, B**), adrenal gland (**C**), coeliac ganglion (**D**), and inferior mesenteric ganglion (**E**) of the guinea pig after incubation with ENK antiserum (**A, C**), SP antiserum (**B**), SOM antiserum (**D**), and VIP antiserum (**E**). The proximal colon was treated with vinblastine. After photography and removal of the ENK antiserum (**A**) according to Tramu et al. (38), the section of the proximal colon shown in **A** was reincubated with SP antiserum (**B**). One or possibly two ENK immunoreactive cell bodies can be seen in the myenteric plexus (*arrows,* **A**). A "newly stained" SP immunoreactive cell is observed after incubation with SP antiserum (*arrowhead,* **B**). A large number of ENK immunoreactive cells are observed in the adrenal medulla (**C**). Note the occurrence of ENK immunoreactive nerve terminals (*arrows,* **C**). A large proportion of the principal ganglion cells of the coeliac ganglion show SOM-like immunoreactivity (**D**). A dense network of VIP immunoreactive nerve terminals (**E**) is observed around ganglion cells in the inferior mesenteric ganglion. Bars = 25 μm (**A, B**) and 50 μm (**C-E**).

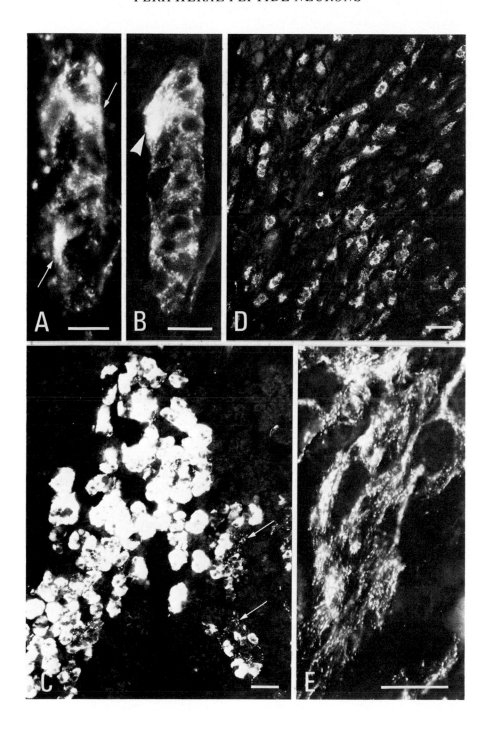

and thus have a central origin (Lundberg et al., *to be published*). ENK-like immunoreactivity was also observed in some cells. In the superior cervical ganglion of the rat, a few ENK immunoreactive cell bodies were seen, and their number increased after local treatment with colchicine. Furthermore, after colchicine or vinblastine treatment ENK-like immunoreactivity was also observed in a number of SIF cells.

Coexistence of Peptides and Classical Transmitters

Recently the question was raised whether a neuron can produce and release more than one transmitter substance (4). The concomitant occurrence of a peptide hormone and a biogenic amine in endocrine cells, particularly in the gut, is well known, and Pearse (32) proposed that these cells belong to a separate system. It was termed the APUD (amine content and/or precursor uptake and decarboxylation) system. As mentioned above, there are some cases in the peripheral nervous system where a peptide has been shown to occur in classical noradrenergic neurons and SIF cells (20,36). These results suggest that neurons can be included in the APUD system, and that coexistence may represent a more general phenomenon than hitherto assumed. A further case of coexistence is represented by ENK-like immunoreactivity in the majority of cells in the adrenal medulla of several species (Fig. 2C) (37). In the untreated rat, however, only a few adrenal medullary cells were ENK-immunoreactive, but sectioning of the splanchnic nerve dramatically increased the number of immunoreactive cells. Interestingly, nerve terminals in adrenal medulla were also ENK-like immunoreactive (Fig. 2C). Our findings are in good agreement with the results obtained by Di Giulio et al. (11), Yang et al. (39), and Lewis et al. (29). They demonstrated methionine-ENK and leucine-ENK with biochemical techniques in the adrenal medulla and

sympathetic ganglia. The function of these peptides in adrenal medullary cells is not yet understood. The occurrence of a peptide in cells containing a classical neurotransmitter is not exclusive of noradrenergic cells. In serotonergic cells the occurrence of a peptide (SP) has been shown (5,23). More recently VIP-like immunoreactivity was observed in some acetylcholinesterase (AChE)-rich cells of certain sympathetic ganglia, and similar patterns of VIP immunoreactive and AChE-rich nerve terminals were observed around sweat glands (30). The putative VIP-ACh cells may represent a case where a physiological meaning for coexisting transmitters can be advanced. It has been suggested (30) that VIP and ACh cooperate in sweat production, VIP mainly by increasing blood flow through a relaxation of smooth muscles around blood vessels and ACh mainly by stimulating sweat cells, either directly and/or via stimulation of myoepithelial cells. In view of the present findings, it may be interesting to look for further examples of coexistence of classical transmitters and peptides in the periphery. Perhaps some peptides may be present in the cholinergic neurons in the gut? Furthermore, two peptides may also coexist as suggested by our findings of a SOM-like and a GAS/CCK-like peptide in some intestinal neurons.

ACKNOWLEDGMENTS

The results reviewed in this chapter were published in original articles and reviews, where the origin and characteristics of the antisera were described. We are grateful for the generous supply of antisera from Drs. G. Nilsson (Stockholm), L. Terenius (Uppsala), J. Fahrenkrug (Copenhagen), J.F. Rehfeld (Århus), M. Brown (La Jolla), R.P. Elde (Minneapolis), and S.I. Said (Dallas). The present studies were supported by the Swedish Medical Research Council (04X-2887, 04X-4495), Magnus Bergvalls Stiftelse, Knut och Alice Wallenbergs Stiftelse, and Stiftelsen Clas Groschinskys Minnesfond.

REFERENCES

1. Baumgarten, H.G., Holstein, A.F., and Owman, Ch. (1970): Auerbach's plexus of mammals and man: Electron microscopic identification of three different types of neuronal processes in myenteric ganglia of the large intestine from rhesus monkey, guinea-pigs and man. *Z. Zellforsch.,* 106:376-397.
2. Bloom, S.R., editor (1978): *Gut Hormones.* Churchill Livingstone, Edinburgh.
3. Burnstock, G. (1972): The purinergic nervous system. *Pharmacol. Rev.,* 24:506-581.
4. Burnstock, G. (1976): Do some nerve cells release more than one transmitter? *Neuroscience,* 1:239-248.
5. Chan-Palay, V., Jonsson, G., and Palay, S.L. (1978): Serotonin and substance P coexist in neurons of the rats' central nervous system. *Proc. Natl. Acad. Sci. USA,* 75:1582-1586.
6. Cooks, R.D., and Burnstock, G. (1976): The ultrastructure of Auerbach's plexus in the guinea-pig. I. Neuronal elements. *J. Neurocytol.,* 5:171-194.
7. Coons, A.H. (1958): Fluorescent antibody methods. In: *General Cytochemical Methods,* edited by J.F. Danielli, pp. 399-422. Academic Press, New York.
8. Costa, M., Patel, Y., Furness, J.B., and Arimura, A. (1977): Evidence that some intrinsic neurons of the intestine contain somatostatin. *Neurosci. Lett.,* 6:215-222.
9. Dahlström, A. (1968): Effect of colchicine on transport of amine storage granules in sympathetic nerves of rat. *Eur. J. Pharmacol.,* 5:111-112.
10. Dahlström, A. (1971): Effects of vinblastine and colchicine on monoamine containing neurons of the rat, with special regard to the axoplasmic transport of amine granules. *Acta Neurophathol. (Berl) (Suppl.),* 5:226-237.
11. Di Giulio, A.M., Yang, H-Y.T., Lutold, B., Fratta, W., Hong, J., and Costa, E. (1978): Characterization of enkephalin-like material extracted from sympathetic ganglia. *Neuropharmacology,* 17:989-992.
12. Dreyfus, C.F., Bornstein, M.B., and Gershon, M.D. (1977): Synthesis of serotonin by neurons of the myenteric plexus in situ and in organotypic tissue culture. *Brain Res.,* 128:125-139.
13. Dreyfus, C.F., Sherman, D.L., and Gershon, M.D. (1977): Uptake of serotonin by intrinsic neurons of the myenteric plexus grown in organotypic tissue cultures. *Brain Res.,* 128:109-123.
14. Elfvin, L-G., and Dahlsgaard, C.J. (1977): Retrograde transport of horseradish peroxidase in afferent fibers of the inferior mesenteric ganglion of the guinea pig: Identification of the cells of origin in dorsal root ganglia. *Brain Res.,* 126:149-153.
15. Eränkö, O., editor (1976): *SIF Cells. Structure and Function of the Small Intensely Fluorescent Sympathetic Cells.* Fogarty International Center Proceedings No. 30. DHEW Publication No. (NIH) 76-942. Government Printing Office, Washington, D.C.
16. Gershon, M.D., and Altman, R.F. (1971): An analysis of the uptake of 5-hydroxytryptamine by the myenteric plexus of the small intestine of the guinea-pig. *J. Pharmacol. Exp. Ther.,* 179:29-41.
17. Gershon, M.D., Dreyfus, C.F., Pickel, V.M., Joh, T.H., and Reis, D.J. (1977): Serotonergic neurons in the peripheral nervous system: Identification in gut by immunohistochemical localization of tryptophan hydroxylase. *Proc. Natl. Acad. Sci. USA,* 74:3086-3089.
18. Gabella, G. (1972): Fine structure of the myenteric plexus in the guinea-pig ileum. *J. Anat.,* 111:69-97.
19. Hökfelt, T., and Dahlström, A. (1971): Effects of two mitosis inhibitors (colchicine and vinblastine) on the distribution and axonal transport of noradrenaline storage particles, studied by fluorescence and electron microscopy. *Z. Zellforsch.,* 119:460-482.
20. Hökfelt, T., Elfvin, L-G., Elde, R., Schultzberg, M., Goldstein, M., and Luft, R. (1977): Occurrence of somatostatin-like immunoreactivity in some peripheral sympathetic noradrenergic neurons. *Proc. Natl. Acad. Sci. USA,* 74:3587-3591.
21. Hökfelt, T., Elfvin, L-G., Schultzberg, M., Fuxe, K., Said, S.I., Mutt, V., and Goldstein, M. (1977): Immunohistochemical evidence of vasoactive intestinal polypeptide-containing neurons and nerve fibers in sympathetic ganglia. *Neuroscience,* 2:885-896.
22. Hökfelt, T., Elfvin, L-G., Schultzberg, M., Goldstein, M., and Nilsson, G. (1977): On the occurrence of substance P-containing fibers in sympathetic ganglia: Immunohistochemical evidence. *Brain Res.,* 132:29-41.
23. Hökfelt, T., Ljungdahl, Å., Steinbusch, H., Verhofstad, A., Nilsson, G., Brodin, E., Pernow, B., and Goldstein, M. (1978): Immunohistochemical evidence of substance P-like immunoreactivity in some 5-hydroxytryptamine-containing neurons in the rat central nervous system. *Neuroscience,* 3:517-538.
24. Kobayashi, R.M., Brown, M., and Vale, W. (1977): Regional distribution of neurotensin and somatostatin in rat brain. *Brain Res.,* 126:584-588.
25. Kreutzberg, G. (1969): Neuronal dynamics and axonal flow. IV. Blockage of intra-axonal enzyme transport. *Proc. Natl. Acad. Sci. USA,* 62:722-728.
26. Larsson, L-I. (1977): Ontogeny of peptide-producing nerves and endocrine cells of the gastro-duodeno-pancreatic region. *Histochemistry,* 54:133-142.
27. Larsson, L-I., and Rehfeld, J.F. (1977): Evidence for a common evolutionary origin of gastrin and cholecystokinin. *Nature,* 269:335-338.
28. Larsson, L-I., and Rehfeld, J.F. (1977): Characterization of antral gastrin cells with region-specific antibodies. *J. Histochem. Cytochem.,* 25:1317-1321.
29. Lewis, R.V., Stern, A.S., Rossier, J., Stein, S., and Udenfriend, S. (1979): Putative enkephalin precursors in bovine adrenal medulla. *Biochem. Biophys. Res. Commun. (in press).*
30. Lundberg, J.M., Hökfelt, T., Schultzberg, M., Uvnäs-Wallensten, K., Köhler, C., and Said, S. (1979): Occurrence of vasoactive intestinal poly-

peptide (VIP)-like immunoreactivity in certain cholinergic neurons of the cat: Evidence from combined immunohistochemistry and acetylcholine esterase staining. *Neuroscience,* 4:1539–1559.

31. Nilsson, G., Pernow, B., Fisher, G., and Folkers, K. (1975): Presence of substance P-like immunoreactivity in plasma from man and dog. *Acta Physiol. Scand.,* 94:542–544.

32. Pearse, A.G.E. (1969): The cytochemistry and ultrastructure of polypeptide hormone-producing cells of the APUD series and the embryologic, physiologic and pathologic implications of the concept. *J. Histochem. Cytochem.,* 17:303–313.

33. Said, S.I., and Faloona, G.R. (1975): Elevated plasma tissue levels of VIP in the watery diarrhea syndrome due to pancreatic, bronchogenic and other tumors. *N. Engl. J. Med.,* 293:155–160.

34. Schultzberg, M., Dreyfus, C.F., Gershon, M.D., Hökfelt, T., Elde, R., Nilsson, G., Said, S., and Goldstein, M. (1978): VIP-, enkephalin-, substance P-, and somatostatin-like immunoreactivity in neurons intrinsic to the intestine: Immunohistochemical evidence from organotypic tissue cultures. *Brain Res.,* 155:239–248.

35. Schultzberg, M., Hökfelt, T., Nilsson, G., Terenius, L., Rehfeld, J.F., Brown, M., Elde, R., Goldstein, M., and Said, S. (1980): Distribution of peptide and catecholamine neurons in the gastro-intestinal tract of rat and guinea-pig: Immunohistochemical studies with antisera to substance P, VIP, enkephalins, somatostatin, gastrin cholecystokinin, neurotensin and dopamine β-hydroxylase. *Neuroscience (in press).*

36. Schultzberg, M., Hökfelt, T., Terenius, L., Elfvin, L.G., Lundberg, J.M., Brandt, J., Elde, R., and Goldstein, M. (1979): Enkephalin immunoreactive nerve terminals and cell bodies in sympathetic ganglia of the guinea-pig and rat. *Neuroscience,* 4:249–270.

37. Schultzberg, M., Lundberg, J.M., Hökfelt, T., Terenius, L., Brandt, J., Elde, R., and Goldstein, M. (1978): Enkephalin-like immunoreactivity in gland cells and nerve terminals of the adrenal medulla. *Neuroscience,* 3:1169–1186.

38. Tramu, G., Pillez, A., and Leonardelli, J. (1978): An efficient method of antibody elution for the successive or simultaneous location of two antigens by immunocytochemistry. *J. Histochem. Cytochem.,* 26:322–324.

39. Yang, H-Y.T., Di Giulio, A.M., Fratta, W., Hong, J.S., Majane, E.A., and Costa, E. (1979): Enkephalin in bovine adrenal gland: Multiple molecular forms of met⁵-enkephalin immunoreactive peptides. *Neuropharmacology (in press).*

Histochemistry and Cell Biology of Autonomic Neurons, SIF Cells, and Paraneurons,
edited by O. Eränkö et al.
Raven Press, New York © 1980.

Immunohistochemical Distribution of Enkephalins: Interactions with Catecholamine-Containing Systems

Richard J. Miller and *Virginia M. Pickel

*Department of Pharmacological and Physiological Sciences, University of Chicago, Chicago, Illinois 60637; and Cornell University Medical College, *Department of Neurology and Laboratory of Neurobiology, New York, New York 10021*

Since the discovery of the opioid penta-peptides known as the enkephalins and related larger opioid peptides of the endorphin class (8,14), intensive investigation into the localization and neurobiological properties of these substances has taken place. In particular we are rapidly developing a knowledge of the distribution at the light and electron microscopic levels of the opioid peptides. The knowledge of such a distribution not only gives us some neuroanatomical rationale for the previously known pharmacology of narcotics but has also lead to further speculations about the possible functions of these novel neuromodulatory substances. Indeed, there is good reason to believe at this time that the enkephalins do act as some type of neurotransmitter or neuromodulator. Electron microscopic localization has revealed that the enkephalins are localized in granules in nerve endings (19) and gut endocrine cells (2). Moreover, they may be released by depolarizing stimuli from neural tissue in a calcium-dependent fashion (3,9), and microiontophoretic application of these substances onto opiate-sensitive neurons has revealed that they produce modulatory effects (18).

Of course, no set of neurons in the nervous system functions *in vacuo*. That is to say, all neurons interact with other systems and may be modulated by, or may modulate, further neurons or tissues. In the case of the enkephalins and endorphins, their profound behavioral effects suggested to some authors that some of their actions may be mediated by interaction with dopaminergic systems. In particular, their cataleptic (4) and neuroendocrine effects (17) may fit into this category. In the present communication we examine certain anatomical localizations of enkephalinergic systems in relation to catecholamine-containing systems. We also present some evidence that interactions between enkephalinergic and dopaminergic systems are critical for certain narcotic-induced functions seen *in vivo* following drug administration. We argue that similar functions might be carried out normally by opioid substances operating *in situ*.

GENERAL ANATOMICAL LOCALIZATION OF THE ENKEPHALINS

Our expanding knowledge about the localization of enkephalinergic systems has revealed that in several cases these exist in close proximity to previously known catecholamine-containing systems. Moreover, in certain cases it is possible that the enkephalins may even be found within the same cells as certain biogenic amines. Such a situation appears to be the case, for example, in the adrenal medulla and in certain small intensely fluorescent (SIF) cells in sympathetic ganglia, e.g., the superior cervical ganglion (22). In addition, endocrine cells in the gut mucosa stain for enkephalin and serotonin (2). Several authors observed that many cells of the adrenal medulla exhibit an intense enkephalin immunohistochemical reaction. Moreover, such enkephalin-like immunoreactivity appears to be localized in the granular fraction of the adrenal, as does epinephrine (23). Thus the enkephalins and epinephrine appear to be localized in granules in many cases within the same cells of the adrenal medulla. It was recently shown that cholinergic stimuli release enkephalin and epinephrine from the adrenal medulla of dogs (23). It should be pointed out that on biochemical analysis not all of the enkephalin-like immunoreactivity from the adrenal actually chromatographs at the same position as leucine- or methionine-enkephalin and appears to be of considerably higher molecular weight. This of course illustrates one of the problems associated with immunohistochemistry: that one may often localize precursors or other molecules which possess cross reactivity to the antisera employed. Thus at this stage, in most cases it is safe only to speak of the localization of "enkephalin-like immunoreactivity." Although in many cases enkephalin itself is responsible for this immunoreactivity, in some cases (e.g., the adrenal medulla), some part of the immunoreactivity may be due to cross-reacting molecules of greater or lesser relevance. Enkephalin-like immunoreactivity in sympathetic ganglia and adrenal cells is dealt with elsewhere (10). With respect to the central nervous system, it is clear that in certain areas catecholamine-containing and enkephalin-containing neurons exist in close proximity. This is particularly true in the case of the median eminence of the hypothalamus, the basal ganglia, nucleus locus ceruleus, area postrema, and in certain areas of the limbic system such as the central nucleus of the amygdala. The existence of high concentrations of enkephalins in most of these areas has been confirmed by radioimmunoassay (11,16).

NUCLEUS LOCUS CERULEUS

Examples of areas where enkephalinergic fibers interact with noradrenergic systems are the nucleus locus ceruleus and the A2 region of the midbrain (20). The nucleus locus ceruleus contains noradrenergic cell bodies which project rostrally to various areas, including the Purkinje cells of the cerebellum. Immunocytochemical localization of enkephalin at the light microscopic level indicates that the locus ceruleus and A2 nuclei contain neuronal varicosities which are positive for enkephalin. In addition, these nuclei contain nerve endings staining positively for another neuropeptide known as substance P (Fig. 1). Immunocytochemical staining with antibodies against tyrosine hydroxylase (TH) shows that in the locus ceruleus and the A2 region

►

FIG. 1. Ultrastructural localization of met[5]-enkephalin-like immunoreactivity in A2 region of the medulla. **A:** Axon terminal labeled specifically for enkephalin forms an asymmetrical contact (*open arrow*) with a dendrite (d). An intense accumulation of reaction product is associated with LDVs (*arrows*). SCVs are also labeled. An unlabeled axon terminal forming a synapse with the same dendrite contains LDVs (*arrow*) and SCVs. Lead precipitate (l) is also seen. Bar = 0.5 μm. **B:** Enkephalin-positive reaction in an axon terminal containing LDVs (*black and white arrow*) and SCVs or small dense vesicles forms an asymmetrical contact (*open arrow*) with the small dendrite or dendritic spine (d). **Inset:** A small axon labeled for met[5]-enkephalin. Labeled vesicles (*arrow*) are not as evident as in labeled axons showing substance P-like immunoreactivity. Bar = 0.5 μm. (From ref. 19.)

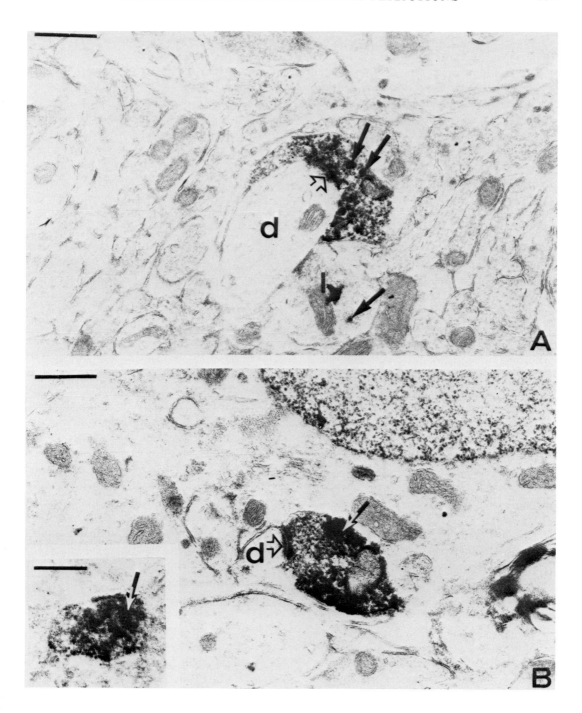

TH is primarily localized to perikarya and dendrites of intrinsic neurons. Axon terminals showing positive reactions for substance P and enkephalin are morphologically similar to each other and to one type of axon terminal which forms synapses with dendrites labeled for TH. This type of axon terminal always forms asymmetrical synaptic junctions and contains large dense vesicles and many small clear vesicles. In addition to the finding of enkephalinergic terminals in the locus ceruleus, it has also been shown that this area is rich in opiate receptors, as demonstrated by autoradiography. Moreover, microiontophoresis of opiates onto locus ceruleus neurons causes a depression of neuronal firing which can be blocked by naloxone (1). This is presumably a manifestation of the action of opiates via opiate receptors in this area. Thus it may be that incoming enkephalinergic fibers regulate the activity of noradrenergic neurons in this nucleus. It is not completely clear, however, as to what the actual *in vivo* consequences of these particular interactions are. However, in the case of the enkephalin/dopamine interactions described below, the functional consequences are much clearer.

BASAL GANGLIA

The profound cataleptic effects produced by many opiates and opioid peptides are somewhat reminiscent of those produced by neuroleptics (4). The finding that the basal ganglia and in particular the globus pallidus are very rich in enkephalin-like immunoreactivity (11,16,21) suggests that these behavioral effects of narcotics may be produced by interaction with dopaminergic systems, particularly those of the nigrostriatal pathway. Moreover, a good deal of enkephalin is found in the nucleus accumbens and the central amygdaloid nucleus, where dopaminergic systems are also located (21). Thus these may also represent potential points of interaction. Immunohistochemical localization of enkephalin at the light microscopic level

shows many labeled neuronal perikarya in the caudate nucleus of the rat. The greatest accumulation of neuronal structures is in the ventral and caudolateral portions of this nucleus. Labeled perikarya measure 10 to 15 μm in diameter. It was previously shown that a pathway running from the caudate nucleus to the globus pallidus in the rat contains enkephalin. Thus Cuello and Paxinos (5) showed that by making a cut between the caudate and the globus pallidus most of the enkephalin-like immunoreactivity in the pallidum is eventually lost. At the electron microscopic level, neuronal perikarya, dendrites, and axons of caudate neurons are labeled (Fig. 2). The distribution of reaction product in dendrites and perikarya is diffuse and of relatively low intensity. The dendrites have many spiny processes which form asymmetrical synapses with unlabeled axon terminals containing small clear vesicles. In contrast to the perikarya and dendrites, a dense accumulation of reaction product is present in a few myelinated and numerous unmyelinated axons and axonal varicosities. Approximately 25% of the labeled varicosities form asymmetrical junctions, primarily with unlabeled dendrites and rarely with unlabeled perikarya and axons. The myelinated axons showing enkephalin-like immunoactivity are sometimes localized in discrete fiber bundles containing other unlabeled myelinated axons. This is especially evident in the zone where the internal capsule fibers join the neostriatum and globus pallidus.

One might now ask if any functional interaction between dopaminergic and enkephalinergic systems occurs with respect to the nigrostriatal pathway. Preliminary data do in fact suggest that dopamine agonists and antagonist drugs may affect the concentration of enkephalin found within the basal ganglia. This is reflected in measurements of the turnover rate of enkephalins within this structure. Thus Costa and his colleagues observed that long-term neuroleptic treatment selectively increases the enkephalin content of the rat striatum and nucleus accumbens (12). More-

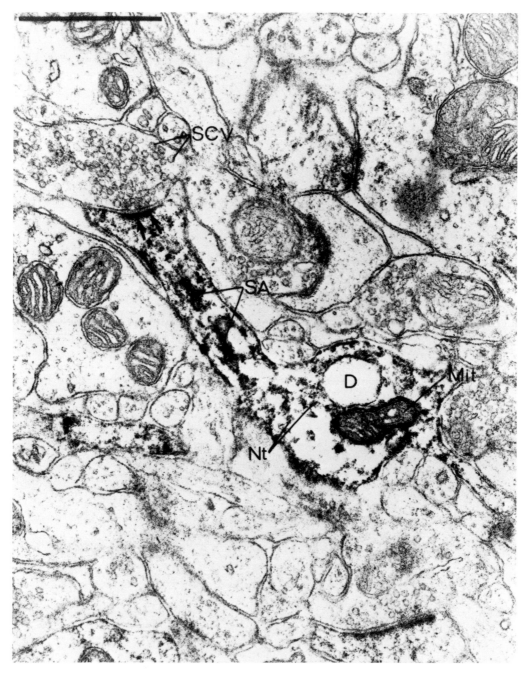

FIG. 2. A: Ultrastructural localization of ELI in a peripheral dendrite (d). The dendrite contains many neurotubules (Nt) and a mitochondrion (Mit), and forms a spiney apparatus (SA). The dendritic spine has a well-defined postjunctional plaque (*arrows*) with an unlabeled terminal bouton containing many small, clear vesicles (SCV). Bar = 1,000 nm.

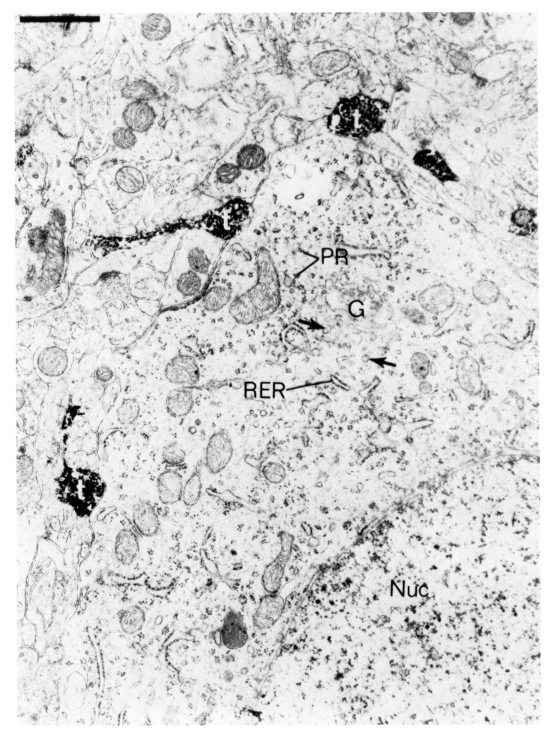

FIG. 2. B: Ultrastructural localization of ELI in terminal boutons (t) associated with unlabeled neuronal perikarya. (Nuc) Nucleus. (PR) Polyribosomes. (G) Golgi apparatus. (RER) Rough endoplasmic reticulum. (*Arrows*) Coated vesicles. Bar = 1,000 nm.

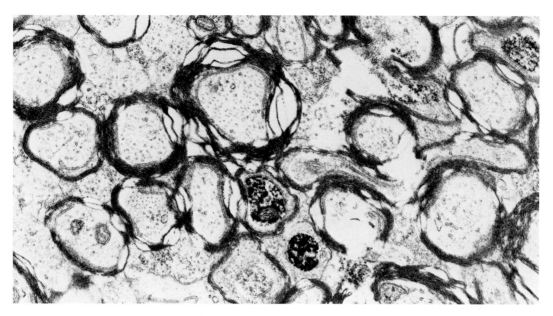

FIG. 2. C: Electron micrograph showing ELI in the axoplasm of two myelinated axons A×1 and A×2, which are located among numerous other unlabeled myelinated axons. Bar = 1,500 nm.

over, such chronic treatment also accelerates the biosynthesis of enkephalin in the striatum (13). With respect to dopaminergic neurons, several authors showed that opiates and opioid peptides accelerate the turnover rate of dopamine in the nigrostriatal pathway (6) (Fig. 3), although the mechanism by which this occurs is not entirely clear. Moreover, tolerance to this effect is also seen in animals made tolerant-dependent on the action of opiates. Thus it can be seen from the above discussion that, with respect to the basal ganglia, enkephalinergic and dopaminergic systems exist in close proximity to one another. Moreover, in the functional sense the two sets of neurons may alter each other's activity (20). This supports the hypothesis that certain effects of opiates and opioid peptides may be mediated by the dopaminergic systems within the striatum.

HYPOTHALAMUS

Many enkephalinergic cell groups and nerve terminals are found within the hypo-thalamus. For example, a considerable number of enkephalinergic nerve terminals are localized within the external layer of the median eminence of the rat (21). This, of course, is where the nerve terminals of many hypothalamic peptidergic neurons are found. Many peptides found in this area play a role in controlling the secretion of hormones from the pituitary gland. In particular, substances such as somatostatin, luteinizing hormone-releasing hormone (LHRH), and thyrotropin-releasing hormone (TRH) are secreted into the portal hypophyseal system and carried to the anterior lobe of the pituitary gland, where they act on hormone-secreting cells. In the case of prolactin, however, it is dopamine which is released from nerve terminals in the external layer of the median eminence and which travels to the anterior lobe of the pituitary. At this point, the dopamine inhibits the secretion of prolactin from pituitary lactotrophs. It has been observed for many years that narcotic drugs such as morphine have a variety of neuroendocrine effects, including the ability to increase the secretion of vaso-

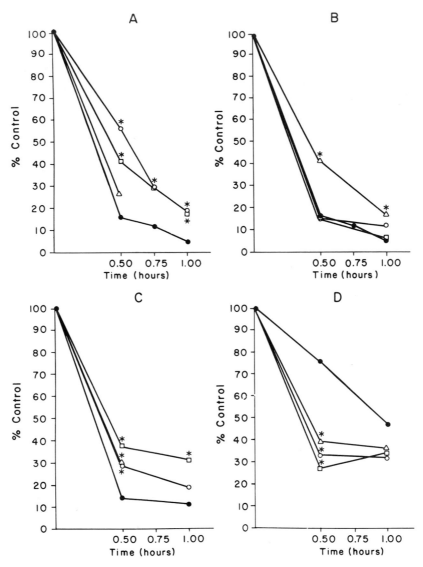

FIG. 3. A: Effects of morphine on the AMT-induced decline in dopamine concentration in the median eminence. Values are expressed as a percentage of the median eminence dopamine concentration of untreated animals, 124.1 ± 8.8 (SEM) ng/mg of protein ($N = 10$). Animals were injected with saline, AMT, and then saline or morphine as described in the text. (●) Saline ($N = 7$). (△) Morphine at 1.67 mg/kg ($N = 3$). (○) Morphine at 5 mg/kg ($N = 7$). (□) Morphine at 15 mg/kg ($N = 6$). **B**: Effects of naloxone + morphine on the AMT-induced decline in median eminence dopamine concentration. (●) Saline ($N = 7$). (△) Morphine at 15 mg/kg ($N = 7$). (□) Naloxone at 20 mg/kg ($N = 4$). (○) Morphine at 15 mg/kg + naloxone at 20 mg/kg ($N = 5$). **C**: Effects of morphine on the AMT-induced decline in frontal cortex dopamine concentration. Values are expressed as a percentage of the dopamine concentration in the frontal cortex of untreated animals: 0.412 ± 0.012 ng/mg protein. (●) Saline ($N = 6$). (△) Morphine at 1.67 mg/kg ($N = 4$). (○) Morphine at 5 mg/kg ($N = 7$). (□) Morphine at 15 mg/kg ($N = 6$). **D**: Effects of morphine on the AMT-induced decline in neostriatal dopamine concentration. Values are expressed as a percentage of the dopamine concentration in the neostriatum of untreated animals, 91.7 ± 7.6 ng/mg protein. (●) Saline ($N = 6$). (△) Morphine at 1.67 mg/kg ($N = 3$). (○) Morphine at 5 mg/kg ($N = 6$). (□) Morphine at 15 mg/kg ($N = 6$). In all cases, the SEM was 4% or less of the control value. (*) Data are significantly different ($p < 0.05$) from the corresponding AMT-saline-treated control.

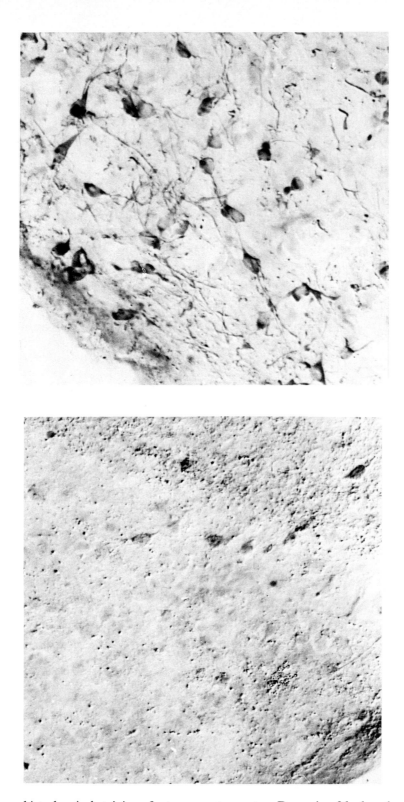

FIG. 4. Immunohistochemical staining of rat area postrema. **top**: Dopamine-β-hydroxylase immunoreactivity. Note the many cell bodies and some terminal varicosities. **bottom**: Enkephalin immunoreactivity. Note the lightly stained cell bodies and many terminal varicosities. (Micrographs were prepared by Mr. David Armstrong and used with his permission.)

pressin, growth hormone, and prolactin from the pituitary gland (17). The localization of enkephalinergic terminals in the median eminence provides a neuroanatomical basis for these effects and suggests the hypothesis that one role for enkephalinergic systems *in situ* may be a neuroendocrine modulatory one. It is also apparent that enkephalins do not have a direct action on the pituitary gland in releasing pituitary prolactin (7). Consequently, it is hypothesized that enkephalins act via some mechanism in the hypothalamus. The finding of a close apposition at the light microscope level between enkephalinergic and dopaminergic nerve terminals in the median eminence makes it possible that the release of dopamine into the hypophyseal portal system may be modulated by enkephalinergic neurons. We investigated the effect of endogenous and exogenous opiates on the turnover rate of dopamine in the median eminence of the rat brain and observed that morphine and the opioid peptides are able to slow the turnover of dopamine in this tissue, implying that they decrease the release of dopamine from the nerve terminals of the tuberoinfundibular dopaminergic neurons (Fig. 3). This would result in a release of tonic inhibitory dopaminergic control of pituitary prolactin release, resulting in an increased output of prolactin. Further indications that this hypothesis is true come from experiments in animals made tolerant to the actions of morphine. In such animals, it can be seen that there is a decreased ability of a dose of morphine to increase pituitary prolactin release. Moreover, in these same animals, morphine is also unable to slow the turnover of dopamine in the tuberoinfundibular system. Thus tolerance to the prolactin-releasing effects and the dopamine-turnover-modulating effects of opiates develops at the same time, indicating a connection between the two.

AREA POSTREMA

One further effect of narcotic agents is that they are powerful emetics (15). It is therefore interesting to note that enkephalin-containing neurons were recently localized within the area postrema of the rat (Fig. 4). In addition, neuronal perikarya and terminals staining with antisera against dopamine-β-hydroxylase, and therefore probably noradrenergic, are also found in the area postrema. This area, known to be concerned in the production of emetic effects, is thus another point of potential interaction between enkephalinergic and catecholaminergic systems.

CONCLUSIONS

The above discussion illustrates certain instances where enkephalin- and catecholamine-containing systems are known to exist in close proximity, and in some of these cases it is thought that the two systems may actually interact directly. This implies that some of the pharmacological effects of exogenous opiates may be mediated by interaction with catecholamine-containing systems. Thus some enkephalinergic systems may modulate the activity of catecholamine-containing systems *in situ.* We do not wish to suggest that enkephalinergic systems and dopamine-containing systems always exist together. However, because of our well-advanced understanding of the biochemistry and neuroanatomy of catecholamine-containing neurons, they provide excellent models for examining the way in which the newly discovered enkephalin-containing systems interact with other neuronal elements.

ACKNOWLEDGMENTS

This research was supported by NIH grants, MH-24285, NSO-G911, HL-18974, and LDA-02121-01. V.M.P. holds a Research Cancer Development Award (MH 00078).

REFERENCES

1. Aghajanian, G.K. (1978): Tolerance of locus coeruleus neurons to morphine supression withdrawal response by clonidine. *Nature,* 276:186–188.

2. Alumets, J., Hakanson, R., Lundler, F., and Chang, K.J. (1978): Leu-enkephalin-like material in neurons and enterochromaffin cells in gut. *Histochemistry,* 56:187-196.

3. Bayon, A., Rossier, J., Mauss, A., Bloom, F.E., Iversen, L.L., Ling, N., and Guillemin, R. (1978): In vitro release of [5-methionine] enkephalin and [5-leucine] enkephalin from rat globus pallidus. *Proc. Natl. Acad. Sci. USA,* 75:3505-3506.

4. Bloom, F.E., Segal D., Ling, N., and Guillemin, R. (1977): Endorphins: Profound behavioural effects in rats suggests new etiological factors in mental illness. *Science,* 196:629-630.

5. Cuello, C., and Paxinos, G. (1978): Evidence for a long leu-enkephalin striatopallidal pathway in rat brain. *Nature,* 271:178-180.

6. Deyo, S., Swift, R., and Miller, R.J. (1979): Morphine and enkephalins modulate hypothalamic dopamine turnover. *Proc. Natl. Acad. Sci. USA (in press).*

7. Grandison, L., and Guidotti, A. (1977): Regulation of prolactin release by endogenous opiates. *Nature,* 270:357-359.

8. Guillemin, R. (1978): Peptides in brain: The new endocrinology neurone. *Science* 202:390-402.

9. Henderson, G., Hughes, J., and Kosterlitz, H.W. (1979): In vitro release of leu & met enkephalin from corpus striatum. *Nature (Lond.),* 271:677-679.

10. Pelto-Huikko, M., Helen, P., Hervonen, A., Linnoila, I., Pickel, V.M., and Miller, R.J. (1980): Localization of (Met⁵)- and (Leu⁵) enkephalin in nerve terminals and SIF cells in adult human sympathetic ganglia. *This volume.*

11. Hong, J.S., Yang, H.Y., Fratta, W., and Costa, E. (1977): Determination of methionine enkephalin in discrete regions of rat brain. *Brain Res.,* 134:383-386.

12. Hong, J.S., Yang, H.Y., Fratta, W., and Costa, E. (1978): Rat striatal methionine enkephalin content after chronic treatment with cataleptogenic and noncataleptogenic antischizophrenic drugs. *J. Pharmacol. Exp. Ther.,* 205:161-167.

13. Hong, J.S., Yang, H.Y., Gillian, J.C., DiGuilio, A.N., Fratta, W., and Costa, E. (1979): Chronic treatment with haloperidol accelerates biosynthesis of enkephalins in rat striatum. *Brain Res.,* 160:192-195.

14. Hughes, J., Smith, T.W., Kosterlitz, H.W., Fothergill, A., Morgan, B.A., and Morris, H. (1975): Identification of two related pentapeptides from brain with potent opiate agonist activity. *Nature,* 258:577-581.

15. Jaffe, J.H. (1976): Narcotic analgesics. In: *The Pharmacological Basis of Therapeutics,* edited by L.S. Goodman, and A. Gilman, p. 237. Macmillan, New York.

16. Kobayashi, R., Palkovits, M., Miller, R.J., Chang, K.J., and Cuatrecases, P. (1978): Hypophysectomy does not alter rat brain enkephalin distribution. *Life Sci.,* 22:379-389.

17. Meites, J., Bruni, J.F., Van Vugt, D.A., and Smith, A.E. (1979): Relations of endogenous opioid peptides and morphine to neuroendocrine functions. *Life Sci.,* 24:325-336.

18. Nicoll, A., Siggins, G.R., Ling, N., Bloom, F.E., and Guillemin, R. (1977): Neuronal actions endorphins and enkephalins among brain regions: A comparative microiontophoretic study. *Proc. Natl. Acad. Sci. USA,* 196:629-630.

19. Pickel, V., Joh, T.H., Reis, D.J., Leeman, S.E., and Miller, R.J. (1979): Localization of substance P and enkephalin in axon terminals related to dendrites of catecholaminergic neurons. *Brain Res.,* 160:387-400.

20. Pollard, H., Llorens-Cortes, C., and Schwartz, J.C. (1977): Enkephalin receptors on dopaminergic neurone in rat striatum. *Nature,* 268:165-167.

21. Sar, N., Stumpf, W.E., Miller, R.J., Chang, K.J., and Cuatrecases, P. (1978): Immunohistochemical localization enkephalin in rat brain and spinal cord. *J. Comp. Neurol.,* 182:17-38.

22. Schultzberg, N., Hökfelt, T., Lundberg, J.M., Terenius, L., Elfin, C., and Elde, R. (1978): Enkephalin-like immunoreactivity in nerve terminals, in sympathetic ganglia and adrenal medullary gland cells. *Acta Physiol. Scand.,* 103:455-472.

23. Viveros, O.H., Dilberto, E.J., Hazum, E., and Chang, K.J. (1979): Enkephalin-like peptides in the adrenal medulla: Co-storage and co-secretion with catecholamines. *Mol. Pharmacol. (in press).*

Histochemistry and Cell Biology of Autonomic Neurons, SIF Cells, and Paraneurons,
edited by O. Eränkö et al.
Raven Press, New York © 1980.

Opioid Mechanisms in Regulation of Cerebral Monoamines *In Vivo*

Liisa Ahtee and L. M. J. Attila

Department of Pharmacy, Division of Pharmacology, University of Helsinki, SF-00170 Helsinki 17, Finland

Narcotic analgesics increase the turnover of dopamine and 5-hydroxytryptamine (5-HT) in the brain *in vivo*, and naloxone antagonizes these effects (for references see refs. 3,6). The effects of β-endorphin on brain monoamine synthesis are remarkably similar to those of morphine. Furthermore, the pure narcotic antagonists naloxone and naltrexone decrease the synthesis of cerebral catecholamines and 5-HT (5). Thus it is probable that opiate receptors and their endogenous ligands are involved in the regulation of cerebral monoamine synthesis. Chronic morphine treatment produces tolerance toward several effects of exogenous β-endorphin (9). Therefore chronic administration of narcotic analgesics could make the opiate receptors, which are involved in the control of cerebral monoamine neurons, tolerant to endogenous opioid ligands. When these receptors would no longer respond to opioid ligands, it is likely that cerebral monoamine synthesis and release would decrease. Recently Przewlocki et al. (7) reported that chronic morphine administration decreases the cerebral concentration of opioid peptides. Such a decrease could also lead to decreased cerebral monoamine synthesis and release.

To test these ideas we treated rats chronically with narcotic analgesics for 3 to 8 weeks. Their cerebral concentrations of do-pamine and its main metabolite homovanillic acid (HVA), as well as those of 5-HT and its metabolite 5-hydroxyindoleacetic acid (5-HIAA), were determined. The rate of α-methyl-*p*-tyrosine (αMT)-induced depletion of dopamine was measured in the limbic forebrain and striatum of these rats. Some of the rats were challenged with a single dose ("test dose") of morphine or methadone to study the development of tolerance.

Short communications of these experiments were given at the XVI Scandinavian Congress of Physiology and Pharmacology (2,4). Complete details will be published at a future date. The methadone experiments were described in detail earlier (1).

MATERIAL AND METHODS

Male Wistar rats receiving a standard diet, with water freely available, were housed 8 to 10 rats per cage at 22° to 24°C with a 12-hr light-dark cycle. The daily morphine dose was divided into two doses which were injected subcutaneously at 800 to 900 hr (8 to 9 a.m.) and 1700 to 1800 hr (5 to 6 p.m.). The daily morphine dose was increased gradually so that the final daily dose was 40 mg/kg (3-week experiment), 95 mg/kg (5-week experiment), or 100 mg/kg (8-week experiment). The control rats received a corresponding volume of 0.9% NaCl solution (saline). The last morphine (20 or 50 mg/kg) or saline injection (except the test dose) was given 12 hr to 6 days before decapitation. The test dose of morphine (10 or 30 mg/kg s.c.) was given 2 or 2.5 hr before decapitation. Control rats were injected with similar volume of saline at corresponding times.

For estimation of the release rate of cerebral dopamine, the rats were injected intraperitoneally with αMT (200 mg/kg) 2 hr before decapitation. For 5-HIAA estimations the rats were treated with probenecid, 200 mg/kg i.p., 2 hr before decapitation. Dopamine, HVA, 5-HT, and 5-HIAA were estimated spectrophotofluorimetrically.

Complete details of the experimental pro-

cedures in the morphine experiments will be published at a future date. The methadone experiments were described earlier (1).

RESULTS AND DISCUSSION

Dopamine

Chronic or acute morphine or methadone administration did not alter the cerebral dopamine concentration. In the limbic forebrain of rats treated chronically with morphine for 20 days, the αMT-induced release of dopamine was significantly retarded at 1 to 2 days, but not any more at 3 to 6 days, after morphine withdrawal (Fig. 1). Three

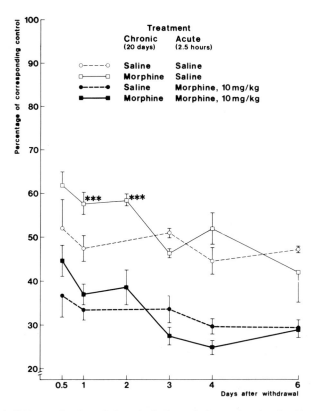

FIG. 1. αMT-induced (200 mg/kg i.p., 2 hr) depletion of dopamine in the limbic forebrain of rats treated chronically with saline or morphine for 20 days and withdrawn from the chronic treatment for 0.5 to 6 days. One-half of the rats were then injected with saline (*thin lines*) and the other half with morphine, (10 mg/kg s.c.; *thick lines*) 2.5 hr before decapitation. The dopamine concentrations are given as the percentage of the corresponding control value (2.19 ± 0.09 to 2.73 ± 0.10 µg/g). The *points* represent the mean value of two to seven determinations; the *vertical lines* indicate the SEM. The significance of the difference from the corresponding chronic saline group was determined by Student's *t*-test: (***) $p < 0.001$.

weeks' chronic morphine administration did not alter the rate of αMT-induced dopamine depletion in the striatum (Fig. 2). The test dose of morphine (10 mg/kg, 2.5 hr), however, decreased the dopamine concentration in about equal proportion in the brain of control rats (treated chronically with saline) and in the brain of chronically morphine-treated rats (Figs. 1 and 2).

In an earlier experiment (1), it was found that after 8 weeks' chronic methadone treatment the striatal HVA concentration was reduced by 50% [from 0.66 ± 0.05 μg/g (N = 5) to 0.36 ± 0.04 μg/g (N = 7)] at 19 hr after methadone withdrawal. However, acutely administered methadone (10 mg/kg, 2 hr) increased the striatal HVA concentration to about the same concentration in control rats treated chronically with saline (to 1.68 ± 0.11 μg/g, N = 3) and in chronically metha-

done-treated rats (to 1.51 ± 0.09 μg/g, N = 8).

These results further support the hypothesis that opiate receptors and their endogenous ligands are involved in the regulation of cerebral dopamine neurons. It seems that the dopamine neurons in the limbic forebrain are more easily affected by chronic administration of narcotic analgesics than those of the striatum. However, long enough treatment seems to affect these neurons too (1). Interestingly, Rosenman and Smith (8) found that the rate of synthesis of catecholamines from labeled tyrosine in the whole brain of mice treated chronically with morphine was decreased at 18 to 54 hr after withdrawal of morphine. They also found that acute administration of morphine to chronically treated mice still increased the synthesis of cerebral catecholamines (8).

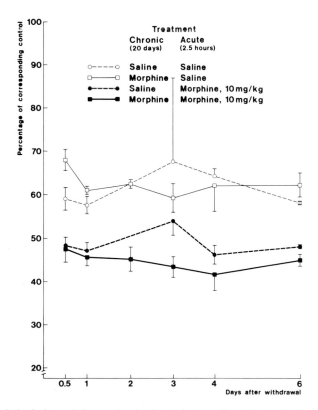

FIG. 2. αMT-induced depletion of dopamine in the striatum of rats treated chronically with saline or morphine for 20 days. For explanations see legend to Fig. 1. The dopamine concentrations in the control striata were 4.57 ± 0.24 to 4.95 ± 0.23 μg/g.

5-Hydroxytryptamine

In rats treated chronically with morphine for 8 weeks, the cerebral 5-HT concentration was decreased by 12% at 27 to 28 hr after morphine withdrawal, and the 5-HIAA concentration by 16% (Table 1). The test dose of morphine (30 mg/kg, 2 hr) increased the cerebral 5-HIAA concentration by 30 to 40% in the control rats and the chronically morphine-treated rats. The increase was slightly less in the morphine-treated rats (Table 1). One group of rats was treated chronically with morphine for 5 weeks. After morphine withdrawal the cerebral 5-HT and 5-HIAA concentrations of these rats were similar to those of the chronically saline-treated control rats (results not shown).

Thus long enough morphine administration significantly decreases the cerebral 5-HT and 5-HIAA concentrations. These findings suggest that endogenous opioid mechanisms which are weakened by chronic morphine administration are involved in the regulation of synthesis and release of cerebral 5-HT.

SUMMARY AND CONCLUSIONS

To study the role of the opiate receptors and their endogenous ligands in the regulation of cerebral monoamine synthesis and release, male Wistar rats were treated chronically with morphine or methadone for 3 or 8 weeks. The rate of αMT-induced depletion of limbic and striatal dopamine, the striatal HVA, and the whole brain 5-HT and 5-HIAA concentrations were determined. Some of the rats were challenged with a single dose ("test dose") of morphine or methadone to study the development of tolerance.

In the limbic forebrain but not in the striatum of rats treated chronically with morphine for 3 weeks, the αMT-induced dopamine depletion rate was retarded 1 to 2 days after morphine withdrawal. After 8 weeks of chronic methadone treatment, the striatal HVA concentration was reduced by 50% at 19 hr after methadone withdrawal. After 8 weeks of chronic morphine treatment, the cerebral 5-HT and 5-HIAA concentrations were decreased by 15 to 20% at 27 to 28 hr after morphine withdrawal. Thus these experiments suggest that the endogenous opioid mechanisms which are weakened by chronic administration of narcotic analgesics are involved in the regulation of synthesis and release of dopamine and 5-HT.

However, acute administration of narcotic analgesics increased the rate of dopamine depletion and the concentrations of HVA and 5-HIAA in the brains of chronically treated rats to about the same extent as in the brains of rats which received the narcotic analgesics for the first time in their lives. Therefore it is probable that tolerance of opiate receptors is not a major reason for decreased release of cerebral dopamine and 5-HT after chronic administration of narcotic analgesics. Reduced cerebral opioid peptide concentrations could be responsible for at least part of this decrease.

These results also indicate that chronic treatment of rats with narcotic analgesics more readily alters the function of their cerebral dopamine neurons than that of 5-HT neurons. This agrees well with the fact that the doses of narcotic analgesics needed acutely to alter cerebral 5-HT turnover are larger than those required to change dopamine turnover (3,5).

TABLE 1. *Cerebral 5-HT and 5-HIAA concentrations in rats treated chronically with saline or morphine*

Chronic treatment: 4_1 weeks ($8\ 4_2$)	Levels after acute treatment (test dose)	
	Saline	Morphine: 30 mg/kg, 2 hr
5-HT (ng/g)		
Saline	559 ± 5	571 ± 23
Morphine	492 ± 20*	454 ± 8**
5-HIAA (ng/g)		
Saline	774 ± 24	1,081 ± 5
Morphine	646 ± 52*	831 ± 22***

For 5-HIAA estimations the rats were treated with probenecid (200 mg/kg i.p.) 2 hr before decapitation. The values are the mean ± SEM; $N = 3$ or 4. Compared to the corresponding saline group: *$p<0.05$, **$p<0.01$, ***$p<0.001$.

ACKNOWLEDGMENTS

These investigations were supported in part by grants from the Medical Research Council, the Academy of Finland, and the Sigrid Jusélius Foundation. The skilled assistance of Anita Havas is highly appreciated.

REFERENCES

1. Ahtee, L. (1974): Catalepsy and stereotypies in rats treated with methadone; relation to striatal dopamine. *Eur. J. Pharmacol.*, 27:221–230.
2. Ahtee, L. (1979): Chronic morphine administration decreases cerebral 5-HT and 5-HIAA concentration in rats. *Acta Physiol. Scand. [Suppl.]*, 473:65.
3. Ahtee, L., and Carlsson, A. (1979): Dual action of methadone on 5-HT synthesis and metabolism. *Naunyn Schmiedebergs Arch. Pharmacol.*, 307:51–56.
4. Attila, L.M.J., and Ahtee, L. (1979): Effects of chronic morphine administration on catecholamines in different parts of the rat brain. *Acta Physiol. Scand. [Suppl.]*, 473:67.
5. Garcia-Sevilla, J.A., Ahtee, L., Magnusson, T., and Carlsson, A. (1978): Opiate-receptor mediated changes in monoamine synthesis in rat brain. *J. Pharm. Pharmacol.*, 30:613–621.
6. Kuschinsky, K. (1977): Opiate dependence. *Prog. Pharmacol.*, 1:1–39.
7. Przewlocki, R., Duka, T., and Herz, A. (1979): Changes in rat endorphin levels after chronic morphine treatment. *Naunyn Schmiedebergs Arch. Pharmacol. (Suppl.)*, 307:R67.
8. Rosenman, S.J., and Smith, C.B. (1972): ^{14}C-Catecholamine synthesis in mouse brain during morphine withdrawal. *Nature*, 240:153–155.
9. Tseng, L., Loh, H.H., and Li, C.H. (1977): Human-β-endorphin: Development of tolerance and behavioral activity in rats. *Biochem. Biophys. Res. Commun.*, 74:390–396.

Histochemistry and Cell Biology of
Autonomic Neurons, SIF Cells, and
Paraneurons,
edited by O. Eränkö et al.
Raven Press, New York © 1980.

Intrinsic Amine-Handling Neurons in the Intestine

J. B. Furness, M. Costa, and P. R. C. Howe

*Center for Neuroscience and Departments of Human Morphology, Human Physiology, and Medicine,
School of Medicine, Flinders University, Bedford Park, 5042 Australia*

We examined the properties and distributions of amine-handling neurons in the small intestine of the guinea pig (10,22). These neurons, which are intrinsic to the intestine, are also found in the colon and in the rabbit and rat. Experiments designed to examine these neurons were performed on segments of ileum after the terminals of noradrenergic axons, which are of extrinsic origin (20), had degenerated following extrinsic denervation (Fig. 1) or treatment with 6-hydroxydopamine. The nerve cell bodies and their processes were revealed by fluorescence histochemical methods after exposure of the nerves (*in vitro* or *in vivo*) to the precursor amino acid L-DOPA or the aromatic amines dopamine or 6-hydroxytryptamine (Fig. 1). To reveal these neurons, it was necessary to inhibit degradation of the amines by monoamine oxidase (MAO). If aromatic L-amino acid decarboxylase (AADC) is inhibited using benserazide (RO 4-4602), the neurons no longer become fluorescent after treatment with L-DOPA. The intrinsic amine-handling neurons thus contain the enzymes AADC and MAO and have uptake mechanisms for catecholamines and indoleamines. However, these neurons do not normally contain histochemically detectable levels of the fluorophore-forming amines 5-hydroxytryptamine, norepinephrine, dopamine, or epinephrine (1,5,14,18,22,34) (Fig. 1). The amine-handling neurons represent about 11% of the nerve cell bodies in the submucous plexus and 0.4% of those in the myenteric plexus of the guinea pig small intestine. Varicose axons from the amine-handling neurons ramify among the fluorescent and nonfluorescent nerve cell bodies of both ganglionated plexuses. The axons are also found in the nerve strands which run between the ganglia, in the tertiary component of the myenteric plexus, and in the deep muscular plexus at the base of the circular muscle coat. A sparse supply of axons is associated with submucous arterioles and the lamina propria of the mucosa.

Our present knowledge suggests three possibilities for the transmitter present in the intrinsic amine-handling neurons: a catecholamine, an indoleamine, or a peptide. The suggestion of a peptide is justified because it has the amine-handling properties (particularly uptake and decarboxylation of aromatic amino acids) of peptide-hormone-secreting endocrine cells (32; Fujita et al., *this volume*).

Various lines of evidence suggest that the neurons do not utilize catecholamines as neurotransmitters. They have quite different properties from known catecholamine neurons. Firstly, they are resistant to the degenerative actions of 6-hydroxydopamine, even when this toxic chemical is administered in

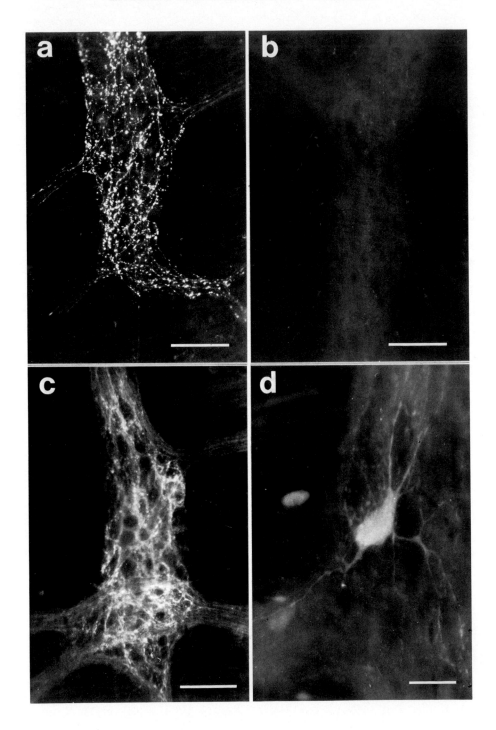

very large amounts (10,22). Their capacity to take up catecholamines is much poorer than catecholamine neurons. Whereas noradrenergic nerve endings in the intestine can be revealed after depletion by reserpine and loading with norepinephrine (1 to 2 mg/kg i.p.) (5), such doses of norepinephrine do not reveal the amine-handling neurons. They were demonstrated only by injecting dopamine in doses greater than 20 mg/kg. Such high doses of norepinephrine are lethal in guinea pigs, and so parallel experiments with this agent could not be performed. If any of the amine-handling neurons were to synthesize a catecholamine, they would be expected to contain the enzyme tyrosine hydroxylase (TH) and possibly other synthesizing enzymes such as dopamine-β-hydroxylase (DBH) and phenylethylamine N-methyltransferase (PNMT), in addition to AADC and MAO. They might also be expected to contain detectable stores of these amines. TH levels were determined by a radiometric assay in normal and extrinsically denervated segments of intestine (24). TH activity was concentrated in the myenteric and submucous plexuses in the normal ileum but could not be detected after extrinsic denervation. In the normal ileum, the immunohistochemical localization of DBH corresponded to the localization of noradrenergic nerves determined by the glyoxylic acid fluorescence histochemical technique. The DBH disappeared after extrinsic denervation.

Levels of norepinephrine (NE), dopamine (DA), and epinephrine (E) were determined using a COMT radioenzymatic method (29).

In the normal ileum, the concentrations of NE in the submucosa and myenteric plexus were 12.5 ± 0.5 and 4.3 ± 0.2 ng/mg protein, respectively. After extrinsic denervation these levels fell to about 3% of normal. The levels of DA and E in the normal tissue were less than 10% of the NE levels. Extrinsic denervation had a smaller effect on these amines, reducing their concentrations to 30 to 50% of normal. The residual levels of all three amines after extrinsic denervation were similar. These remaining stores, unlike the stores of NE in extrinsic nerves, were not depleted by reserpine (1 mg/kg i.p., 18 hr). This shows that the residual amines are not stored in a vesicular compartment with similar properties to catecholamine storage granules. Studies on subcellular fractions prepared from the intestine showed that the residual amines, unlike the NE in extrinsic nerves, were recovered in the soluble fraction rather than in the microsomal (vesicular) fraction. Thus it seems likely that the small amounts of NE, E, and DA which are found in the extrinsically denervated intestine are in a cytoplasmic pool; and since TH and DBH are not found in the denervated intestine, it is postulated that NE and E are not synthesized locally but may be taken up from the circulation, perhaps into the intrinsic amine-handling neurons. Some DA could also reach the intestine through the circulation, or it could be produced by AADC from circulating L-DOPA.

There are at least six peptides contained in neurons in the intestine: somatostatin, substance P, enkephalins, vasoactive intestinal

◄

FIG. 1. Fluorescence histochemical localization of aromatic amines in whole mount preparations of the guinea pig small intestine, prepared by the glyoxylic acid method (21). **a:** Ganglion of the myenteric plexus from a normal segment of ileum showing noradrenergic axons. **b:** Myenteric ganglion from a segment of ileum of a guinea pig that received two injections of 6-hydroxydopamine (200 μg/kg) 8 and 4 days before sacrifice. At 30 min before killing, the animal was injected with pargyline (100 μg/kg i.p.). All noradrenergic nerves have degenerated and, despite the inhibition of monoamine oxidase, no specific amine fluorescence could be detected. **c:** Myenteric ganglion from a segment of ileum of a guinea pig treated with 6-hydroxydopamine as in **b.** At 30 min before killing the animal was injected with pargyline (100 μg/kg) and L-DOPA (170 mg/kg). A large number of axons of intrinsic origin show fluorescence of the dopamine formed by the enzymatic decarboxylation of L-DOPA. **d:** An amine-handling neuron in the myenteric plexus demonstrated by the fluorescence of the dopamine synthesized from the L-DOPA precursor. Bars in **a–c** = 50 μm. Bar in **d** = 20 μm.

polypeptide, gastrin/cholecystokinin tetra-peptide, and bombesin (23). The distributions of the first four of these peptides have been examined in detail (9,11-13,23). These studies show that each of the peptides is contained in a separate group of enteric neurons and that none of these groups has a distribution similar to the intrinsic amine-handling neurons. The results also indicate that peptide neurons do not necessarily have amine-handling properties.

The available evidence suggests that the intrinsic amine-handling neurons are not peptide-producing cells and that they do not utilize catecholamines as neurotransmitters. Therefore the possibility that they synthesize an indoleamine, perhaps 5-hydroxytryptamine (5-HT), should be considered carefully. 5-HT was first suggested to be an intestinal neurotransmitter more than 10 years ago (4,25). Since that time Gershon and his colleagues continued to produce evidence that there are intrinsic neurons in the intestine that can take up and store 5-HT and which contain the enzymes tryptophan hydroxylase (TPH), as well as AADC and MAO (16,17, 26,27). The possibility of an indoleamine being an intestinal transmitter was reviewed recently (7). Perhaps the strongest evidence for this hypothesis is the finding that certain intestinal neurons contain TPH. This was shown immunohistochemically using antibodies raised in rabbits against purified rat TPH (27). Immunoreactive cell bodies were found in the submucous and myenteric plexuses of guinea pigs, mice, and rats. Furthermore, it was shown that radiolabeled 5-HT is produced from labeled tryptophan in isolated strips of the guinea pig myenteric plexus attached to the longitudinal muscle (16). These strips are free of the 5-HT-producing mucosal enterochromaffin cells. The synthesis was inhibited by parachlorophenylanine. A yellow fluorescence, consistent with the presence of 5-HT, was found in the myenteric plexus after incubation with tryptophan. It is puzzling that histochemical studies, combined in some cases with spectrophotometric measurements, failed to reveal endogenous stores of 5-HT, even when MAO inhibitors were used to enhance the stores (1,14,18,22). This suggests that any 5-HT which is normally formed is either stored in very low concentrations or converted to another compound.

There are a number of pharmacological studies suggestive of 5-HT being an intestinal neurotransmitter (4,6,15,19,28,33,35,37). In the first of these studies, it was found that activation of gastric enteric inhibitory nerves by vagal stimulation was partly antagonized by nicotinic blocking agents and partly by desensitizing receptors for 5-HT. Complete block was achieved only when acetylcholine receptors were blocked and 5-HT receptors desensitized. However, these results were not confirmed in subsequent experiments using a slightly different experimental regimen (3). An explanation of this discrepancy, based on more recent observations, has been suggested (7). It is now known that noradrenergic nerves from the superior cervical ganglion run with the vagus, and when stimulated these nerves cause the release of 5-HT from enterochromaffin cells (2,36). The mucosa was present in experiments in which a 5-HT involvement was implicated in the response but was not present in the other experiments.

When the intestine is distended, the circular muscle contracts on the oral side and relaxes on the anal. The pharmacology of the ascending excitatory component of this response was examined recently (6). It was found that the contraction was due to a nerve-mediated reflex, probably involving at least three neurons in series: a neuron sensitive to stretch, an interneuron, and a final neuron acting on the muscle. The reflex response was only partly blocked by the muscarinic antagonist hyoscine, and the remaining component of the response was blocked by exposure of the intestine to 5-HT or by methysergide. It was suggested that the action of 5-HT was due to desensitization of its own receptors, thus preventing the contribution of a 5-HT-like neurotransmitter to the reflex. However, when 5-HT was applied to the circular muscle in concentrations up to 10^{-4} g/ml, it caused a contraction in fewer than 10% of experiments (8). Therefore it seems likely that the substance released was similar to, but not actually, 5-HT. On the other hand, 5-HT could cause an antagonism of the response by some other mechanism, e.g., the prejunctional inhibition of transmitter release.

When intracellular electrical recordings are made from neurons in the myenteric plexus, slow depolarizing responses can be recorded if interganglionic nerve strands are stimulated (30,37). These slow potentials are caused by a decrease in potassium conductance and can be mimicked by administration of 5-HT or substance P (28,30,31,37). The slow potentials and the effects of 5-HT were blocked by exposure to 5-HT and by methysergide, but the actions of substance P were not changed (28,37). There is thus circumstantial evidence for the release of a 5-HT-like compound at neuroneuronal synapses in the gut wall. The effects of methysergide and 5-HT on the ascending excitatory reflex could also be interpreted in terms of actions at neuroneuronal synapses if it is assumed that the residual response after hyoscine *in vitro* is due to overflow of the neurotransmitter from the enteric ganglia to the circular muscle (7,8).

CONCLUSIONS

The enteric nervous system contains a set of neurons with amine-handling properties. Present evidence suggests that these neurons do not synthesize catecholamines and that their amine-handling properties are not incidental to their using peptides as neurotransmitters. It is possible that the nerves utilize an indoleamine other than 5-HT.

ACKNOWLEDGMENTS

This work was supported by grants from the Australian Research Grants Committee and the National Health and Medical Research Council of Australia. We thank Venetta Esson, Inta Strazdins, and Pat Vilimas for their excellent technical assistance.

REFERENCES

1. Ahlman, H., and Enerbäck, L. (1974): A cytofluorometric study of the myenteric plexus in the guinea-pig. *Cell Tissue Res.,* 153:419-434.
2. Ahlman, H., Lundberg, J., Dahlström, A., and Kewenter, J. (1976): A possible vagal adrenergic release of serotonin from enterochromaffin cells in the cat. *Acta Physiol. Scand.,* 98:366-375.
3. Beani, L., Bianchi, C., and Crema, A. (1971): Vagal non-adrenergic inhibition of guinea-pig stomach. *J. Physiol. (Lond),* 217:259-279.
4. Bülbring, E., and Gershon, M.D. (1967): 5-Hydroxytryptamine participation in the vagal inhibitory innervation of the stomach. *J. Physiol. (Lond),* 192:823-846.
5. Costa, M., and Furness, J.B. (1971): Storage, uptake and synthesis of catecholamines in the intrinsic adrenergic neurones in the proximal colon of the guinea-pig. *Z. Zellforsch. Mikrosk. Anat.,* 120:364-385.
6. Costa, M., and Furness, J.B. (1976): The peristaltic reflex: An analysis of the nerve pathways and their pharmacology. *Naunyn Schmiedebergs Arch. Pharmacol.,* 294:47-60.
7. Costa, M., and Furness, J.B. (1979): On the possibility that an indoleamine is a neurotransmitter in the gastrointestinal tract. *Biochem. Pharmacol.,* 28:565-571.
8. Costa, M., and Furness, J.B. (1979): The sites of action of 5-hydroxytryptamine in nerve-muscle preparations from the guinea-pig small intestine and colon. *Br. J. Pharmacol.,* 65:237-248.
9. Costa, M., Cuello, C., Furness, J.B., and Franco, R. (1979): Distribution of enteric neurons showing immunoreactivity for substance P in the guinea-pig ileum. *Neuroscience,* 5:323-331.
10. Costa, M., Furness, J.B., and McLean, J.R. (1976): The presence of aromatic L-amino acid decarboxylase in certain intestinal nerve cells. *Histochemistry,* 48:129-143.
11. Costa, M., Furness, J.B., Buffa, R., and Polak, J.M. (1980): Distribution of enteric neurons showing immunoreactivity for enkephalins in the guinea-pig intestine. *In preparation.*
12. Costa, M., Furness, J.B., Buffa, R., and Said, I. (1980): Distribution of enteric neurons showing immunoreactivity for vasoactive intestinal polypeptide (VIP) in the guinea-pig intestine. *Neuroscience (in press).*
13. Costa, M., Furness, J.B., Llewellyn-Smith, I.J., Davies, B., and Oliver, J. (1980): An immunohistochemical study of the projections of somatostatin containing neurons in the guinea-pig intestine. *Neuroscience (in press).*

14. Diab, I.M., Dinerstein, R.J., Watanabe, M., and Roth, L.J. (1976): (^{3}H)Morphine localization in myenteric plexus. *Science,* 193:689-691.

15. Dingledine, R., and Goldstein, A. (1976): Effects of synaptic transmission blockade on morphine action in the guinea-pig myenteric plexus. *J. Pharmacol. Exp. Ther.,* 196:97-106.

16. Dreyfus, C.F., Bornstein, M.B., and Gershon, M.D. (1977): Synthesis of serotonin by neurons of the myenteric plexus in situ and in organotypic tissue culture. *Brain Res.,* 128:125-139.

17. Dreyfus, C.F., Sherman, D.L., and Gershon, M.D. (1977): Uptake of serotonin by intrinsic neurons of the myenteric plexus grown in organotypic tissue culture. *Brain. Res.,* 128:109-123.

18. Dubois, A., and Jacobowitz, D.M. (1974): Failure to demonstrate serotonergic neurons in the myenteric plexus of the rat. *Cell Tissue Res.,* 150:493-496.

19. Furness, J.B., and Costa, M. (1973): The nervous release and the action of substances which affect intestinal muscle through neither adrenoreceptors nor cholinoreceptors. *Philos. Trans. R. Soc. Lond. [Biol],* 265:123-133.

20. Furness, J.B., and Costa, M. (1974): The adrenergic innervation of the gastrointestinal tract. *Ergeb. Physiol.,* 69:1-51.

21. Furness, J.B., and Costa, M. (1975): The use of glyoxylic acid for the fluorescence histochemical demonstration of peripheral stores of noradrenaline and 5-hydroxytryptamine in whole mounts. *Histochemistry,* 41:335-352.

22. Furness, J.B., and Costa, M. (1978): Distribution of intrinsic nerve cell bodies and axons which take up aromatic amines and their precursors in the small intestine of the guinea-pig. *Cell. Tissue Res.,* 188:527-543.

23. Furness, J.B., and Costa, M. (1979): Types of nerves in the enteric nervous system. *Neuroscience,* 5:1-21.

24. Furness, J.B., Costa, M., and Freeman, C.G. (1979): Absence of tyrosine hydroxylase activity and dopamine β-hydroxylase immunoreactivity in intrinsic nerves of the guinea-pig ileum. *Neuroscience,* 4:305-310.

25. Gershon, M.D., Drakontides, A.B., and Ross, L.L. (1965): Serotonin: Synthesis and release from the myenteric plexus of the mouse intestine. *Science,* 149:197-199.

26. Gershon, M.D., Dreyfus, C.F., Pickel, V.M., Joh, T.H., and Reis, D.J. (1977): Serotonergic neurons in the peripheral nervous system: Identification in gut by histochemical localization of tryptophan hydroxylase. *Proc. Natl. Acad. Sci. USA,* 74:3086-3089.

27. Gershon, M.D., Robinson, R., and Ross, L.L. (1976): Serotonin accumulation in the guinea-pig myenteric plexus: Ion dependence, structure activity relationship and the effects of drugs. *J. Pharmacol. Exp. Ther.,* 198:548-561.

28. Grafe, P., Mayer, C.J., and Wood, J.D. (1979): Evidence that substance P does not mediate slow synaptic excitation within the myenteric plexus. *Nature,* 297:720-721.

29. Howe, P.R.C., Provis, J.C., Costa, M., Furness, J.B., and Chalmers, J.P. (1979): The properties of catecholamine stores which remain in the intestine after extrinsic denervation. *(submitted).*

30. Katayama, Y., and North, R.A. (1978): Does substance P mediate slow synaptic excitation within the myenteric plexus? *Nature,* 274:387-388.

31. Katayama, Y., North, R.A., and Williams, J.T. (1979): The action of substance P on neurones of the myenteric plexus of the guinea-pig small intestine. *Proc. R. Soc. Lond. [Biol] (in press).*

32. Pearse, A.G.E. (1977): The diffuse neuroendocrine system and the APUD concept: Related endocrine peptides in brain, intestine, pituitary, placenta and anuran cutaneous glands. *Med. Biol.,* 55:115-125.

33. Rattan, S., and Goyal, R.K. (1978): Evidence of 5-HT participation in vagal inhibitory pathway to opossum LES. *Am. J. Physiol.,* 234:E273-E276.

34. Robinson, R.G., and Gershon, M.D. (1971): Synthesis and uptake of 5-hydroxytryptamine by the myenteric plexus of the guinea-pig ileum: A histochemical study. *J. Pharmacol. Exp. Ther.,* 178:311-324.

35. Singh, I., and Singh, A. (1970): Neurones histaminergiques, kininergiques et hydroxytraminergiques dans les ganglions intramuraux du muscle gastrique et du muscle vesical de la Grenouille. *J. Physiol. (Paris),* 62:431-439.

36. Tansy, M.F., Rothman, G., Bartlett, J., Farber, P., and Hohenleitner, F.J. (1971): Vagal adrenergic degranulation of enterochromaffin cell system in guinea-pig duodenum. *J. Pharm. Sci.,* 60:81-85.

37. Wood, J.B., and Mayer, C.J. (1979): Serotonergic activation of tonic-type enteric neurons in guinea-pig small bowel. *J. Neurophysiol.,* 42:582-593.

Histochemistry and Cell Biology of Autonomic Neurons, SIF Cells, and Paraneurons,
edited by O. Eränkö et al.
Raven Press, New York © 1980.

Immunocytochemical Demonstration of the Catecholamine-Synthesizing Enzymes and Neuropeptides in the Catecholamine-Storing Cells of Human Fetal Sympathetic Nervous System

*Antti Hervonen, †Virginia M. Pickel, †Tong H. Joh, †Donald J. Reis, ‡Ilona Linnoila, §L. Kanerva, and **Richard J. Miller

*Department of Biomedical Sciences, University of Tampere, 33100 Tampere 52, Finland; †Laboratory of Neurobiology, Department of Neurology, Cornell Medical College, New York, New York 10021; ‡Endocrinology Group, Laboratory of Pulmonary Function and Toxicology, National Institute of Environmental Health Sciences, Research Triangle Park, North Carolina; §Department of Anatomy, University of Helsinki, Finland; and **Department of Pharmacological and Physiological Sciences, University of Chicago, Chicago, Illinois*

The functional role and physiological characteristics of the catecholamine-storing cells, small intensely fluorescent (SIF) cells and paraganglia, in the autonomic nervous system is still a matter of debate. The multiplicity of the terms used in this volume is one sign of these different views. In the present study the term SIF cell refers to small intensely fluorescent cells (2,3) within the sympathetic ganglia, and the term paraganglia refers to separate clusters of catecholamine-storing cells outside the sympathetic ganglia (1,5).

The possibility of peptide (hormone) production by the human SIF cells and paraganglia was pointed out in our earlier studies (6–11). Recently Schultzberg et al. (16,17) reported enkephalin-like immunoreactivity in the adrenal medulla and SIF cells of rat and guinea pig. Human fetal tissues offer the only possible means for systematic mapping of the existence of opioid peptides in man, although developmental aspects should be considered.

At least two major problems remain to be solved: (a) Which catecholamine is stored in human SIF cells and paraganglia? (b) Which of the known neuropeptides, if any, is present in these tissues?

One indirect method of elucidating the catecholamine (neurotransmitter) is by immunocytochemical localization of the respective synthesizing enzymes: tyrosine hydroxylase (TH) in dopaminergic and noradrenergic cells, dopamine-β-hydroxylase (DBH) in noradrenergic cells, and phenylethanolamine-N-methyltransferase (PNMT) in epinephrine containing cells. In the present study the immunocytochemical localization of these enzymes and leu[5]-enkephalin, met[5]-enkephalin, and substance P (SP) was combined with catecholamine fluorescence to further evaluate the nature of the human SIF cells and paraganglia.

MATERIALS AND METHODS

Small pieces of superior cervical ganglia, other ganglia of the sympathetic trunk, as well as solitary paraganglia were prepared from six human fetuses (estimated ages 14 to 22 weeks). The pieces were either frozen in liquid nitrogen and processed for formaldehyde-induced fluorescence (FIF) according to the principles given by Eränkö (2) or were fixed 2 to 12 hr in 4% formalin freshly made from paraformaldehyde powder. The frozen dried tissues were embedded in paraffin, and after fluorescence microscopy the deparaffinized sections were processed according to the immunoglobulin-enzyme bridge method (13); the formalin-fixed tissues were sectioned on a cryostat or on a vibrating microtome (Vibratome) and processed further for the peroxidase–antiperoxidase (PAP) immunocytochemical procedure as described in detail elsewhere (14,15). The two immunocytochemical methods gave identical results. Details of the methods as well as the description of the antibodies are given elsewhere (11).

RESULTS AND DISCUSSION

A summary of the main results obtained with human fetal material is presented in this chapter. Some of the findings were also reported in a symposium on Regulation and Function of Neural Peptides, Gardone Riviera, Brescia, Italy, 1979.

Using the FIF method, SIF cells can be clearly distinguished from the weakly fluorescent neuroblasts (Fig. 1). The neuroblasts were clearly labeled for TH; the greater ac-cumulation of the reaction product was found in the most mature neurons (Fig. 2), although the strongest FIF was seen in the smaller, evidently more immature neuroblasts. The labeling of the neuroblasts for DBH was light, whereas the small cells along the blood vessels were heavily labeled (Figs. 3 and 4). These cells were evidently SIF cells as shown by the deparaffinized FIF sections. The main catecholamine of the human fetal SIF cells is thus evidently norepinephrine rather than dopamine. PNMT could not be demonstrated in any structures within the ganglia or in the paraganglia. A considerable portion of the SIF cells also showed enkephalin-like immunoreactivity (ELI) (Figs. 5 and 6), but no SP-positive structures were seen.

The paraganglia are numerous in the human fetus and in adult man (8,9). The clusters of catecholamine-storing cells are well vascularized and emit bright FIF (Figs. 7 and 8). Surprisingly, not all of these cells were labeled for TH in the larger paraganglia (Fig. 9), although all cells showed clear labeling for DBH (Fig. 10). ELI was found only in some cells within the paraganglia (Figs. 11 and 12), but no positive reaction was obtained with SP or PNMT antibodies.

The results suggest that the main catecholamine in human SIF cells and paraganglia is norepinephrine rather than dopamine. This does not support the hypothesis of the dopaminergic regulation of the postsynaptic potentials in man. Mainly norepinephrine was found in the human fetal paraganglia in earlier studies (5).

ELI was localized in the same cells that show strong FIF and are labeled with DBH antibody. This suggests the coexistence of

▶

FIG. 1. FIF of catecholamines in the human fetal superior cervical ganglion. The neuroblasts emit very weak FIF, whereas the SIF cells (*arrows*) show intense FIF. ×102.

FIG. 2. TH labeling in neuroblasts. The largest of the developing neurons showed the strongest labeling. ×255.

FIGS. 3 and 4. DBH labeling of the ganglia. The neuroblasts were very weakly stained, whereas the SIF cells were clearly labeled (*arrows*). Occasionally some larger perikarya (*large arrow*) showed labeling. ×340.

FIGS. 5 and 6. Enkephalin-like immunoreactivity in the SIF cells (*arrows*). Blood vessels are marked with small triple arrows. ×340.

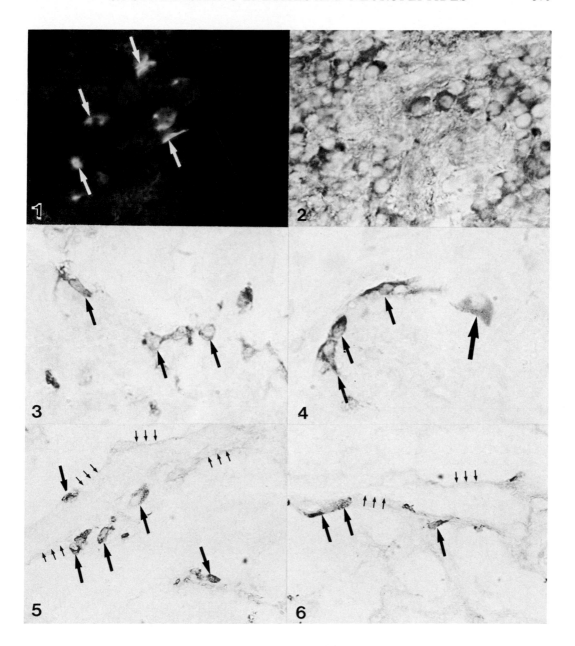

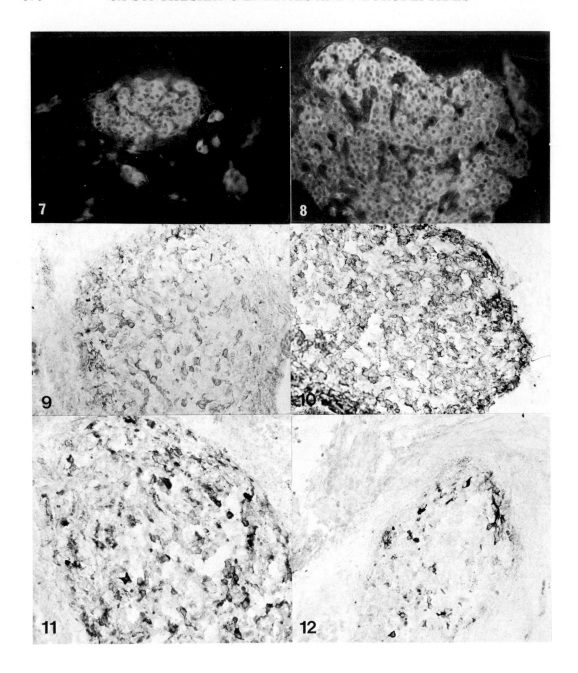

biogenic amine and enkephalin in human tissues. A similar situation prevails in the sympathetic ganglia of the rat and guinea pig (16) and in the human adult sympathetic ganglia (6) and adrenal medulla (11,12). Recent results from adult ganglia also showed that met[5]- and leu[5]-enkephalin can be demonstrated in the same cells (11). Furthermore, ACTH[(1-24)] antibody gives clear labeling of all cells in the human fetal paraganglia (Linnoila and Hervonen, *to be published*). These findings indicate that possibly more than one of the known peptides (hormones) might be present in the human fetal catecholamine-storing cells, at least in the paraganglia.

The human catecholamine-storing cells, whether localized in adrenal medulla (11), within the sympathetic ganglia as SIF cells, or as separate paraganglionic clusters, should be re-evaluated as paraneurons in light of the present findings (4). The release of norepinephrine might be accompanied by release of the opioid peptide enkephalin. The physiological importance of human SIF cells and paraganglia might be more concerned with endocrine functions other than just releasing norepinephrine.

REFERENCES

1. Coupland, R.E. (1965): *The Natural History of the Chromaffin Cell.* Longmans, London.
2. Eränkö, O., editor (1976): *SIF-cells. Structure and Function of the Small, Intensely Fluorescent Cells.* DHEW Publ. No (NIH) 76-942. Government Printing Office, Washington, D.C.
3. Eränkö, O., and Eränkö, L. (1971): Small intensely fluorescent granule-containing cells in the sympathetic ganglion of the rat. *Prog. Brain Res.,* 34:39-51.
4. Fujita, T. (1977): Concept of paraneurons. *Arch. Histol. Jpn.,* 40:1-12.

5. Hervonen, A. (1971): Development of catecholamine storing cells in human fetal paraganglia and adrenal medulla. *Acta Physiol. Scand. [Suppl.],* 386:1-94.
6. Hervonen, A., Linnoila, R.I., Pickel, V.M., Helen, P., Peltohuikko, M., and Miller, R.J. (1979): Localization of (met 5) and leu (5) enkephalin in nerve terminals in human sympathetic ganglia. *Neuroscience. (in press).*
7. Hervonen, A., Partanen, S., Vaalasti, A., Partanen, M., Kanerva, L., and Alho, H. (1978): The distribution and endocrine nature of the abdominal paraganglia of adult man. *Am. J. Anat.,* 153:563-572.
8. Hervonen, A., Vaalasti, A., Partanen, S., and Kanerva, L. (1978): Endocrine nature of the paraganglia of man. *Experientia,* 34:111-112.
9. Hervonen, A., Vaalasti, A., Partanen, M., Kanerva, L., and Hervonen, H. (1978): Effects of ageing on the histochemically demonstrable catecholamines and acetycholinesterase of human sympathetic ganglia. *J. Neurocytol.,* 7:11-23.
10. Hervonen, A., Vaalasti, A., Partanen, M., Kanerva, L., and Vaalasti, T. (1976): The paraganglia, a persisting endocrine system in man. *Am. J. Anat.,* 146:207-210.
11. Linnoila, R.I., DiAugustine, R.P., Hervonen, A., and Miller, R.J. (1979): Distribution of (leu[5]) and (met[5]) enkephalin-, VIP-, and substance P-like immunoreactivity in human adrenal glands. *Neuroscience (in press).*
12. Lundberg, J.M., Hamberger, B., Schultzberg, M., Hökfelt, T., Grandberg, P-O., Efendic, S., Terenius, L., Goldstein, M., and Luft, R. (1979): Enkephalin and somatostatin like immunoreactivities in human adrenal medulla and pheochromocytoma. *Proc. Natl. Acad Sci. (USA),* 76:4079-4083.
13. Petrusz, P., DiMeo, P., Ordronneau, P., Weaver, C., and Keefer, D.A. (1975): Improved immunoglobulin-enzyme bridge method for light microscopic demonstration of hormone containing cells of the rat adenohypophysis. *Histochemistry,* 46:9-26.
14. Pickel, V.M., Joh, T.H., and Reis, D.J. (1975): Ultrastructural localization of tyrosine hydroxylase in noradrenergic neurons of brain. *Proc. Natl. Acad. Sci. (USA),* 72:649-663.
15. Pickel, V.M., Joh, T.H., Reis, D.J., Leeman, S.E., and Miller, R.J. (1979): Electron microscopic localization of substance P and enkephalin in axon ter-

◄
FIGS. 7 and 8. FIF of the paraganglia within the sympathetic ganglion (Fig. 7) and outside the sympathetic ganglion (Fig. 8). Note the numerous sinusoidal capillaries between the intensely fluorescent cells. ×255.
FIG. 9. TH labeling varies from cell to cell, some of the paraganglionic cells being negative, although all show intense FIF. ×255.
FIG. 10. DBH labeling shows patterns similar to those seen with FIF of the consecutive sections. All cells are labeled, although the degree of staining varies. ×255.
FIGS. 11 and 12. ELI in the paraganglia: Some cells, approximately 10% of the total amount in the cross section, are strongly labeled, whereas the majority are not stained. ×255.

minals related to dendrites of catecholaminergic neurones. *Brain Res.,* 160:387–400.

16. Schultzberg, M., Hökfelt, T., Terenius, L., Elfvin, L-G., Lundberg, J.M., Brandt, J., Elde, R.P., and Goldstein, M. (1979): Enkephalin immunoreactive nerve fibres and cell bodies in sympathetic ganglia of the guinea-pig and rat. *Neuroscience,* 4:249–270.

17. Schultzberg, M., Lundberg, J.M., Hökfelt, T., Terenius, L., Brandt, J., Elde, R.P., and Goldstein, M. (1979): Enkephalin immunoreactivity in gland cells and nerve terminals of the adrenal medulla. *Neuroscience,* 3:1169–1186.

*Histochemistry and Cell Biology of
Autonomic Neurons, SIF Cells, and
Paraneurons,*
edited by O. Eränkö et al.
Raven Press, New York © 1980.

Localization of (Met⁵)- and (Leu⁵)-Enkephalin in Nerve Terminals and SIF Cells in Adult Human Sympathetic Ganglia

*Markku Pelto-Huikko, *Antti Hervonen, *Pauli Helen, †Ilona Linnoila, ‡Virginia M. Pickel, and §Richard J. Miller

Department of Biomedical Sciences, University of Tampere, 33100 Tampere 52, Finland; †Endocrinology Group, Laboratory of Pulmonary Function and Toxicology, National Institutes of Environmental Health Sciences, Research Triangle Park, North Carolina; ‡Laboratory of Neurobiology, Department of Neurology, Cornell University Medical College, New York, New York 10021; and §Department of Physiological and Pharmacological Sciences, University of Chicago, Chicago, Illinois

Hughes et al. (5) were the first to isolate from brain extracts two opiate-like pentapeptides: [met⁵]-enkephalin and [leu⁵]-enkephalin. Since then specific radioimmunoassays for [Met⁵]- and [leu⁵]-enkephalin have been developed and used for the measurement of these two peptides in the brain (9). Nerve terminals with enkephalin-like immunoreactivity (ELI) have also been found in the peripheral autonomic nervous system. Enkephalins have been immunohistochemically demonstrated in the nerves of the gastrointestinal tract in various mammals including man (1,3,7,13,15). Schultzberg et al. (15) found ELI in the adrenal medulla of the rat, guinea pig, and cat, in the medullary gland cells and nerve terminals. Recently the same authors reported the presence of ELI in nerve fibers and cell bodies in the sympathetic ganglia of the guinea pig and rat (14).

We studied human sympathetic ganglia obtained from surgery to elucidate the distribution of ELI in the human autonomic nervous system.

MATERIAL AND METHODS

Pieces of the lower cervical (stellate) ganglion as well as upper thoracic ganglia (Th 1–4) were received from six patients suffering of peripheral vascular disturbances of the upper extremity. The ganglia were immediately processed further for different morphological methods: formaldehyde-induced fluorescence (FIF) of catecholamines, electron microscopy, and immunohistochemistry.

Histochemical Demonstration of Catecholamines

Slices of ganglia were frozen in liquid nitrogen and freeze-dried at $-40°C$ for 6 to 7 days at a final vacuum of 10^{-4} torr. Excess amounts of phosphorus pentoxide were used as a water trap. After drying, the desiccators containing the tissues were warmed to above room temperature and rapidly moved to the vapor chamber. They were exposed to vapor generated from paraformaldehyde powder previously equilibrated to 70% relative humidity for 1 hr at a temperature of 80°C. The vapor-treated pieces were embedded in paraffin under vacuum and sectioned on a serial microtome at 10 μm. The sections were examined and photographed using a Leitz (Orthoplan) microscope equipped with an epi-illuminator and an appropriate filter set.

Immunohistochemistry

Antisera

Antisera were raised in rabbits against synthetic [met⁵]- and [leu⁵]-enkephalin (Wellcome Research Laboratories, Beckenham, Kent, U.K.) which had been coupled to bovine serum albumin by glutaraldehyde. These antisera demonstrated by radioimmunoassay less than 1% cross reactivity against [leu⁵]- and [met⁵]-enkephalin, respectively. No cross reactivity for the antisera could be detected against β-lipotropin (β-LPH) and adrenocorticotropic hormone (ACTH). A more detailed description of the specificity of these antisera is given elsewhere (9).

Immunohistochemistry with Freeze-Dried Sections

After the demonstration of catecholamines, immunohistochemical studies (IHC) were carried out on the same (or serial) deparaffinized sections with the improved immunoglobulin-enzyme bridge method for the light microscope (10) as described previously (7). Satisfactory staining was obtained with any of the primary antisera at a dilution of 1:2,000. Controls included incubations with normal rabbit serum and antigen-inactivated serum (50 μg of the corresponding peptide in 1 ml of diluted antiserum).

Immunohistochemistry with Formalin-Fixed Tissue

Fresh slices of ganglia were immersed for 12 hr in 4% formalin (prepared from para-

formaldehyde powder with 0.1 M phosphate buffer pH 7.3) for 12 hr and washed in several changes of buffer for 24 hr. The tissues were then sectioned on a cryostat or on a vibrating microtome (Vibratome) at 20 to 40 μm. IHC on Vibratome sections was performed with the modification (11) of Sternberg's peroxidase–antiperoxidase (PAP) technique (16). For details see Pickel et al. (12).

RESULTS

The neurons of human sympathetic ganglia show varying FIF. Some of the perikarya emitted FIF of moderate intensity, but a continuous series of decreasing fluorescence intensities to totally nonfluorescent neurons was found in every ganglion (4). Brightly fluorescent varicosities were found occasionally between the principal neurons (Fig. 1). SIF cells are rare in adult human sympathetic ganglia, but a few examples were found in each ganglion studied. Similar results were obtained with both immunohistochemical labeling procedures.

Small dots of ELI were found around the principal neurons (Figs. 2–4). These dots often formed a varicose fiber in close contact with the neuron perikaryon. The density of the ELI network varied greatly from ganglion to ganglion and from patient to patient. In some ganglia practically all of the principal neurons were in apparent contact with the ELI-stained structures, whereas in others little positive labeling was seen. Varicosities showed considerably more ELI than the intervaricose segments of the same fiber (Figs. 3 and 4). The labeling of the varicose terminals and fibers was not seen in sections incubated with normal rabbit serum or blocked antiserum controls. In contrast to this specific labeling, the controls as well as the sections incubated with enkephalin antiserum show a reaction for endogenous peroxidase located in pigment granules in the perikarya of the neurons and in satellite cells. The specific and nonspecific labeling are

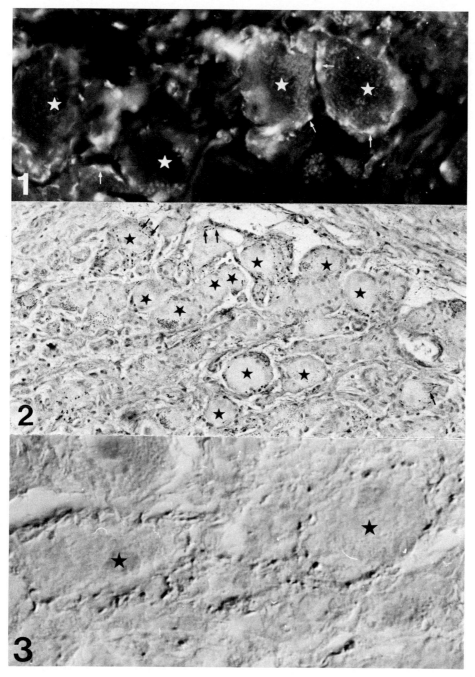

FIG. 1. FIF of the human stellate ganglion. The perikarya (*stars*) are surrounded by fluorescent varicosities (*arrows*). ×400.

FIG. 2. ELI in the same ganglion. The neurons marked with *stars* are in contact with positive, often varicose fibers (*arrows*). ×260.

FIG. 3. A tight network of ELI varicosities surrounding two principal neurons. Nomarski objective. ×500.

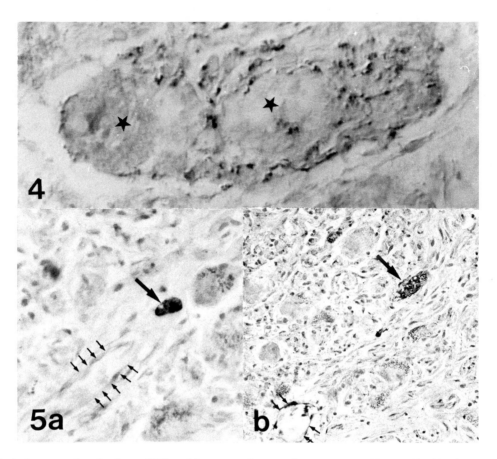

FIG. 4. An example of a dense ELI-positive network around two neurons. Nomarski objective. ×500.
FIG. 5. A small group of SIF cells close to a blood vessel is stained heavily in consecutive sections with leu⁵-enkephalin (**a**) and met⁵-enkephalin (**b**) antibodies. ×260.

readily distinguishable by comparison of the enkephalin-labeled and control sections.

The few SIF cells found in adult ganglia showed labeling with both enkephalin antibodies used (Fig. 5). No other stained perikarya were found. Schultzberg et al. (14) recently showed ELI in the ganglia of rat and guinea pig sympathetic trunk. The variation between the ganglia in the present material might also be due to the autonomic dysfunction leading to the patients' disease (Raynaud's syndrome). The origin of the fibers showing ELI is not clear: perikarya showing ELI were demonstrated in the spinal cord by Hökfelt et al. On the other hand, the origin

of fibers showing ELI in the prevertebral ganglia of rat and guinea pig is probably peripheral (8,14). Although the peripheral tissues supplied by these ganglia have not been studied in man or other species, the problem of the origin of the fibers requires further histochemical data for elucidation. The lack of ELI-positive perikarya other than SIF cells within the ganglia suggests an extraganglionic origin. The few SIF cells found did not give rise to processes that could explain the network of intraganglionic varicosities.

The patterns of the varicose fibers showing ELI were similar after incubations with leu⁵- and met⁵-enkephalin antibodies. ELI was found in the same SIF cell after both incubations. The cross reactivity between the antibodies is less than 1% but could still be the reason for the labeling in either of the two incubations. However, the staining with both antibodies was intense. Furthermore, there may also be some cross reactivity with a larger precursor compound recently shown in other peripheral tissues (2,6). We demonstrated similar dual labeling in the human adrenal medulla (7), where groups of the cells were stained for only leu⁵-enkephalin or only met⁵-enkephalin. These findings with the same antibodies suggest that the presence of both types of ELI in SIF cells might really represent the presence of the two peptides in the same cells. Moreover, biogenic amines and ELI have been shown to coexist in the SIF cells in adult human sympathetic ganglia. The patterns of adrenergic varicosities surrounding the neurons are similar to the varicosities showing ELI.

ACKNOWLEDGMENTS

This research was supported by NIH grants MH 24585, NS06911, HL18974, and R.J. Miller LDA02121-01. V.M.P. holds a Research Career Development Award (MH00078).

REFERENCES

1. Alumets, J., Håkanson, F., Sundler, F., and Chang, K-J. (1979): Leu-enkephalin-like material in nerves and enterochromaffin cells in the gut. *Histochemistry*, 56:187–196.
2. DiGiulio, A.N., Yang, H.Y., Lutold, B., Fratta, W., Hang, J., and Costa, E. (1978): Characterization of enkephalin-like material extracted from sympathetic ganglia. *Neuropharmacology*, 17:989–992.
3. Elde, R., Hökfelt, T., Johansson, O., and Terenius, L. (1976): Immunohistochemical studies using antibodies to leucine-enkephalin: Initial observations on the nervous system of the rat. *Neuroscience*, 1:349–351.
4. Hervonen, A., Vaalasti, A., Partanen, M., Kanerva, L., and Hervonen, H. (1978): Effects of ageing on the histochemically demonstrable catecholamines and acetylcholinesterase of human sympathetic ganglia. *J. Neurocytol.*, 7:11–23.
5. Hughes, J., Smith, T.W., Kosterlitz, H.W., Fothergill, L.A., Morgan, B.A., and Morris, H.R. (1975): Identification of two related pentapeptides from the brain with potent opiate agonist activity. *Nature*, 258:577–579.
6. Lewis, R.V., Stein, S., Gerber, L.D., Rubinstein, M., and Udenfriend, S. (1978): High molecular weight opioid containing proteins in striatum. *Proc. Natl. Acad. Sci. USA*, 75:4021–4023.
7. Linnoila, R.I., DiAugustine, R.P., Miller, R.J., Chang, K.J., and Cuatercasas, P. (1978): An immunohistochemical and radioimmunological study of the distribution of (met⁵)- and (leu⁵)-enkephalin in the gastrointestinal tract. *Neuroscience*, 3:1187–1196.
8. Lundberg, J.M., Hökfelt, T., Nilsson, G., Terenius, L., Elde, R., Goldstein, M., Reheeld, J., and Said, S. (1979): Peripheral peptide and catecholamine-containing neurons in the vagus, splanchnic, hypogastric and pelvic nerves: Occurrence and distribution of substance P-, enkephalin-, VIP-, gastrin-, CCK- and somatostatin-like immunoreactivity and catecholamine synthesizing enzymes in nerve trunks and ganglia. *Neuroscience (in press)*.
9. Miller, R.J., Chang, K., Cooper, B., and Cuatercasas, P. (1978): Radioimmunoassay and characterization of enkephalins in rat tissues. *J. Biol. Chem.*, 253, No. 2: 531–538.
10. Petrusz, P., Dimeo, P., Ordronneau, P., Weaver, C., and Keefer, D.A. (1975): Improved immunoglobulin-enzyme bridge method for light microscopic demonstration of hormone-containing cells rat adenohypophysis. *Histochemistry*, 46:9–26.
11. Pickel, V.M., Joh, T.H., and Reis, D.J. (1975): Ultrastructural localization of tyrosine hydroxylase in noradrenergic neurons of brain. *Proc. Natl. Acad. Sci. USA*, 72:649–663.
12. Pickel, V.M., Joh, T.H., Reis, D.J., Leeman, S.E., and Miller, R.J. (1979): Electron microscopic localization of substance P and enkephalin in axon terminals related to dendrites of catecholaminergic neurones. *Brain Res.*, 160:387–400.
13. Polak, J.M., Sullivan, S.N., Bloom, S.R., Facer, P., and Pearse, A.G.E. (1977): Enkephalin-like immunoreactivity in the human gastrointestinal tract. *Lancet*, 1:972.
14. Schultzberg, M., Hökfelt, T., Terenius, L., Elfvin, L-G., Lundberg, J.M., Brandt, J., Elde, R.P., and Goldstein, M. (1979): Enkephalin immunoreactive nerve fibres and cell bodies in sympathetic ganglia of the guinea pig and rat. *Neuroscience*, 4:249–270.
15. Schultzberg, M., Lundberg, J.M., Hökfelt, T., Terenius, L., Brandt, J., Elde, R.P., and Goldstein, M. (1978): Enkephalin-like immunoreactivity in gland cells and nerve terminals of the adrenal medulla. *Neuroscience*, 3:1169–1186.
16. Sternberger, L.A. (1974): *Immunocytochemistry*. Prentice-Hall, In., Englewood Cliffs, N.J.

Subject Index